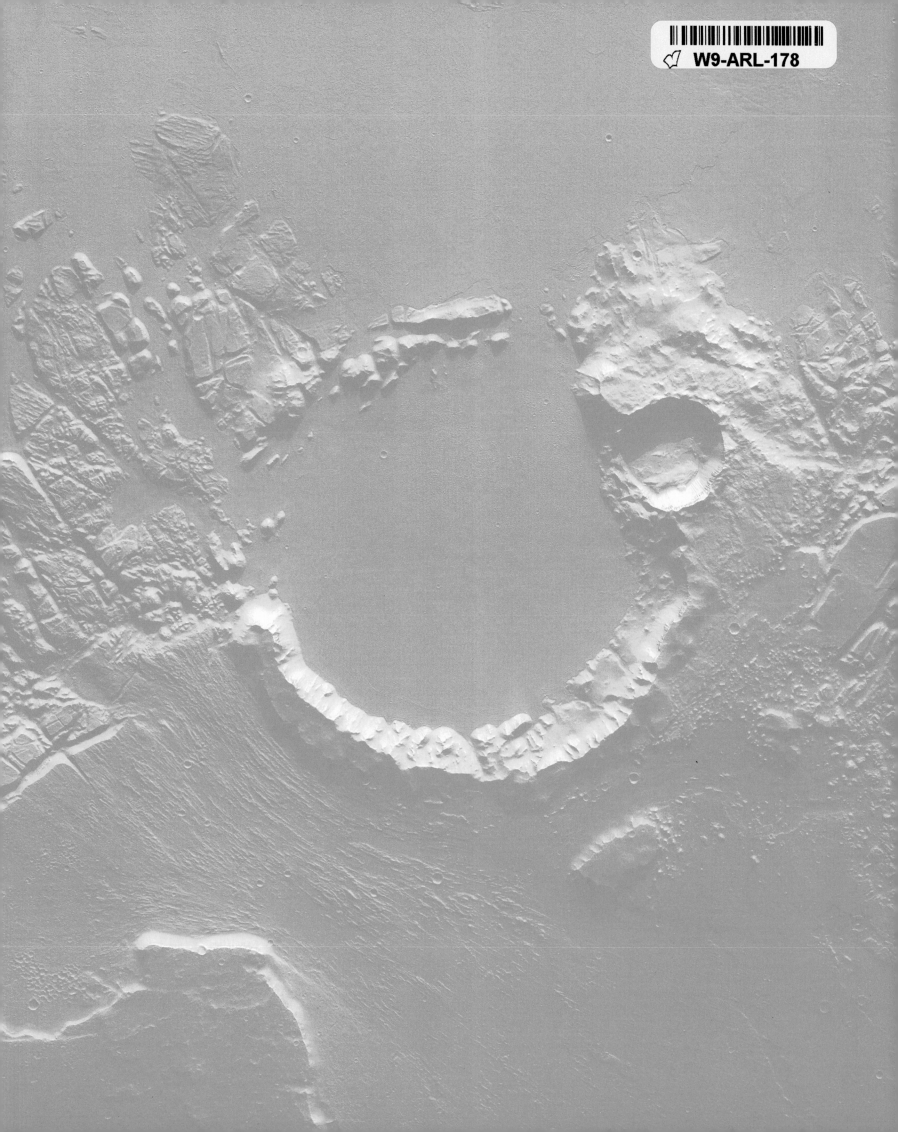

THE
PLANETS

SMITHSONIAN

THE PLANETS

CONTENTS

LONDON, NEW YORK, MELBOURNE,
MUNICH, AND DELHI

Senior Editor Ben Morgan
Senior Designer Smiljka Surla
Project Editor Lizzie Davey
Editors Ann Baggaley, Ruth O'Rourke-Jones, Steve Setford
US Editor Christine Heilman
Designers Kathy Gammon, Spencer Holbrook, Fiona Macdonald,
Simon Murrell, Steve Woosnam-Savage
Editorial Assistant Olivia Stanford
Illustrators Peter Bull, Infomen, Maltings Partnership,
Kees Veenenbos
Managing Editor Paula Regan
Managing Art Editor Owen Peyton Jones
Producer, Pre-Production Nikoleta Parasaki
Senior Producer Mary Slater
DK Picture Library Rob Nunn
Jacket Editor Maud Whatley
Jacket Designer Mark Cavanagh
Jacket Design Development Manager Sophia MTT
Art Director Phil Ormerod
Associate Publishing Director Liz Wheeler
Publishing Director Jonathan Metcalf

First American Edition, 2014
Published in the United States by DK Publishing
4th floor, 345 Hudson Street
New York, New York 10014

14 15 16 17 18   10 9 8 7 6 5 4 3 2 1
001–192970–Sept/2014

Published in Great Britain by Dorling Kindersley Limited.

A catalog record for this book is available from
the Library of Congress.

ISBN: 978-1-4654-2464-8

DK books are available at special discounts when purchased in bulk
for sales promotions, premiums, fund-raising, or educational use.
For details, contact: DK Publishing Special Markets, 345 Hudson
Street, New York, New York 10014 or SpecialSales@dk.com.

Printed and bound in China by Leo Paper Products

Discover more at
www.dk.com

6   Foreword

8   FAMILY OF THE SUN
10   Our place in space
12   Around the Sun
14   Birth of the solar system
16   Formation of the planets
18   Size and scale
20   Our solar system

22   OUR STAR
24   The Sun
26   Sun structure
28   Storms on the Sun
30   Sun rays
32   The solar cycle
34   Solar eclipses
36   Story of the Sun
38   Missions to the Sun

40   ROCKY WORLDS
42   Neighboring worlds
44   Mercury
46   Mercury structure
48   Mercury up close
50   Mercury mapped
52   Destination Carnegie Rupes
54   The winged messenger
56   Missions to Mercury
58   Venus
60   Venus structure
62   Venus up close

64   Venus mapped
66   Destination Maxwell Montes
68   The planet of love
70   Missions to Venus
72   Earth
74   Earth structure
76   Tectonic Earth
78   Earth's changing surface
80   Water and ice
82   Life on Earth
84   Earth from above
86   Our planet
88   The Moon
90   Moon structure
92   Earth's companion
94   Moon mapped
96   Destination Hadley Rille
98   Earthrise
100   Lunar craters
102   Highlands and plains
104   Story of the Moon
106   Missions to the Moon
108   Apollo project
110   Mars
112   Mars structure
114   Mars mapped
116   Water on Mars
118   Destination Valles Marineris
120   Martian volcanoes
122   Destination Olympus Mons
124   Dunes of Mars
126   Polar caps

128 The moons of Mars

130 The Red Planet

132 Missions to Mars

134 Roving on Mars

136 Exploring Mars

138 **Asteroids**

140 The asteroid belt

142 Near-Earth asteroids

144 Missions to asteroids

146 GAS **GIANTS**

148 Realm of giants

150 **Jupiter**

152 Jupiter structure

154 Jupiter up close

156 The Jupiter system

158 Io

159 Europa

160 Galilean moons

162 Ganymede

163 Callisto

164 Destination Enki Catena

166 King of the planets

168 Missions to Jupiter

170 **Saturn**

172 Saturn structure

174 Saturn's rings

176 Destination Saturn's rings

178 Saturn up close

180 Saturn in the spotlight

182 The Saturn system

184 Saturn's major moons

186 Destination Ligeia Mare

188 Cassini's view

190 Destination Enceladus

192 Lord of the rings

194 Missions to Saturn

196 **Uranus**

198 Uranus structure

200 The Uranus system

202 Destination Verona Rupes

204 **Neptune**

206 Neptune structure

208 The Neptune system

210 Destination Triton

212 The blue planets

214 Voyagers' grand tour

216 OUTER **LIMITS**

218 The Kuiper belt

220 Dwarf planets

222 Comets

224 Comet orbits

226 Missions to comets

228 Cosmic snowballs

230 Prophets of doom

232 Worlds beyond

234 REFERENCE

236 Solar system data

244 Glossary

248 Index

255 Acknowledgments

This trademark is owned by the Smithsonian Institution and is registered in the United States Patent and Trademark Office.

Smithsonian
Established in 1846, the Smithsonian—the world's largest museum and research complex—includes 19 museums and galleries and the National Zoological Park. The total number of artifacts, works of art, and specimens in the Smithsonian's collection is estimated at 137 million. The Smithsonian is a renowned research center, dedicated to public education, national service, and scholarship in the arts, sciences, and history.

Consultants
Maggie Aderin-Pocock, MBE, is a space scientist, an honorary research associate at University College London, and co-host of the BBC TV series *The Sky at Night*.

Ben Bussey is a planetary scientist and physicist at Johns Hopkins University in Baltimore, Maryland. A specialist in remote sensing, he participated in the Near-Earth Asteroid Rendezvous–Shoemaker (NEAR) mission and is co-author of *The Clementine Atlas of the Moon*.

Andrew K. Johnston is a geographer at the Center for Earth and Planetary Studies at the Smithsonian National Air and Space Museum in Washington, DC. He is author of *Earth from Space* and co-author of the *Smithsonian Atlas of Space Exploration*.

Authors
Heather Couper, CBE, is a former head of the Greenwich Planetarium in London, and past president of the British Astronomical Association. Asteroid 3922 Heather is named after her.

Robert Dinwiddie specializes in writing educational and illustrated reference books on scientific topics.

John Farndon is the author of many books on science, nature, and ideas. He has been shortlisted four times for the children's Science Book Prize.

Nigel Henbest is an astronomer, former editor of the *Journal of the British Astronomical Association*, and author. He has written more than 38 books and more than 1,000 articles on space and astronomy.

David W. Hughes is Emeritus Professor of Astronomy at the University of Sheffield, UK. He has published over 200 research papers on asteroids, comets, meteorites, and meteors, and has worked for the European, British, and Swedish space agencies.

Giles Sparrow is an author and editor specializing in astronomy and space science. He is a Fellow of the Royal Astronomical Society.

Carole Stott is an astronomer and author who has written more than 30 books about astronomy and space. She is a former head of astronomy at the Royal Observatory at Greenwich, London.

Colin Stuart is a writer specializing in physics and space. He is a Fellow of the Royal Astronomical Society.

Martian crater
Spacecraft such as NASA's Mars Reconnaissance Orbiter give us an intimate view of worlds we can only dream of visiting in person. This image of a meteorite crater in the Arabia Terra region of Mars reveals incredible details, including "painted" stripes formed where dust has cascaded down the slope toward the center.

FOREWORD

The amazing diversity of worlds in our solar system has inspired people for generations. Our immediate neighborhood in space includes a star powered by nuclear fusion, large worlds of swirling gases, smaller planets made of rock and metal, and countless tiny bodies.

In the farther reaches of the solar system, four large gas planets orbit the Sun: Jupiter, Saturn, Uranus, and Neptune. Four smaller terrestrial planets orbit closer to home: Earth, Venus, Mars, and Mercury. Also nearby is the main belt of asteroids. Other tiny, ice-covered bodies, mostly found in the realms beyond the planets, orbit in a few distinct groups at the edge of the Sun's gravitational influence.

Although it is small compared to the Sun and the gas planets, the most important place to us is, of course, Earth. It is the only world so far found to support life, and in our exploratory missions across the solar system we have yet to find anywhere quite like home.

People have stood on only one other world besides Earth. Astronauts reached the surface of the Moon in the 1960s in one of the greatest stories of human enterprise. We have also sent spacecraft to other planets, acquiring a vast amount of data. Our robotic machines crawl over the surface of Mars and return images of a dusty, dry world, but one that reminds us of the desert landscapes on Earth. Venus, cloaked in thick, hot clouds, seems a very alien place in comparison. Other intriguing places that continue to fascinate us include Europa and Enceladus, ice-covered moons of outer planets that both contain layers of liquid water under the surface.

The exotic beauty of our solar system has captured the imagination of people everywhere. This book shows in detail what each world has in common, what sets each apart from the others, and how they all fit together within our small region of the universe. I hope that *The Planets* fulfills part of your dreams of discovery.

Andrew K. Johnston
Smithsonian National Air and Space Museum

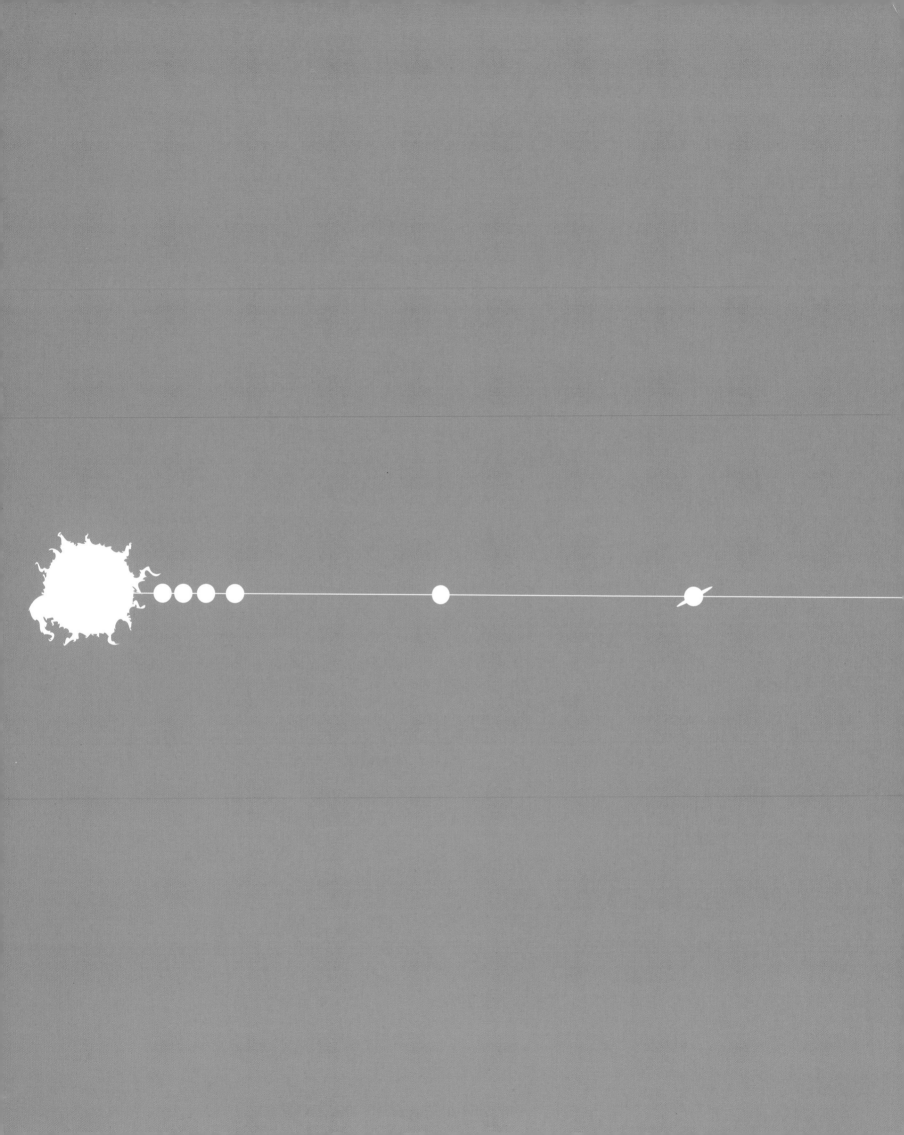

FAMILY OF THE **SUN**

Our Sun is just one of around 200 billion stars that make up the Milky Way—the vast, spiral galaxy we call home. The Sun lies about halfway out from the galactic heart in a minor spiral arm, orbiting the center once every 200 million years at the brisk pace of 120 miles (200 km) per second. Like thousands of other stars, it is surrounded by a family of smaller objects trapped in its vicinity by

OUR PLACE IN **SPACE** ——————————◯

gravity, just as the Sun is caught by the pull of the Milky Way. The largest of these objects are known to us as planets, and their wandering journeys through the night sky have earned them ancient names. Most of the planets detected near other stars are vast, boiling worlds with wayward orbits—habitats impossible for life. Not so in our solar system. Its eight planets follow stable, almost circular paths around the Sun. The innermost planets—Mercury, Venus, Earth, and Mars—are small, solid globes of rock and iron. In contrast, the outer worlds— Jupiter, Saturn, Uranus, and Neptune—are bloated giants formed of gas and liquid, each accompanied by a large retinue of moons, like a solar system in miniature. Less easily observed, but far more numerous, are the many smaller objects that populate the dark recesses of the solar system, from dwarf planets like Pluto to comets and asteroids— leftover rubble from the primordial cloud of debris from which the planets formed.

◁ **Milky Way**
Our galaxy is believed to be spiral in shape, but because we view it from within, we see it edge-on. Best seen on the darkest, clearest nights—far from cities and other forms of light pollution—it appears as a milky band across the sky. The bright patches are huge, luminous nebulae—glowing clouds of gas and dust in which new stars and planets are taking shape. The rift that appears to divide the Milky Way in two is a darker cloud, about 300 light-years from Earth, that blocks the light from more distant stars behind it.

AROUND **THE SUN**

THE SUN'S GRAVITY HOLDS IN THRALL A DIVERSE ASSORTMENT OF CELESTIAL OBJECTS. AS WELL AS THE EIGHT PLANETS, WITH THEIR OWN FAMILIES OF RINGS AND MOONS, THE SOLAR SYSTEM COMPRISES BILLIONS OF PIECES OF ROCKY AND ICY DEBRIS.

The planets all orbit the Sun in the same direction, and in almost the same flat plane. Closest to the Sun's heat are four small, rocky worlds: Mercury, Venus, Earth, and Mars. In the chilly farther reaches of the solar system lie the giant planets: Jupiter, Saturn, Uranus, and Neptune. They are composed mostly of substances more volatile than rock, such as hydrogen, helium, methane, and water.

The asteroids, most of which reside between Mars and Jupiter, are lumps of rocky debris left over from the birth of the planets. The edge of the planetary system is marked by icy chunks—comets and the Kuiper belt objects—that have survived from the earliest days of the solar system.

▽ **Orbits**
The planets travel along paths around the Sun that are not perfectly circular but slightly elliptical (oval). Smaller bodies typically follow much more elliptical orbits, tipped up from the plane in which the planets move. Most extreme are the comets, which trace very long, thin elliptical orbits from the outer limits of the solar system, some of them tipped up at a right angle. Certain comets, including Halley, travel around the Sun in the opposite direction to the planets.

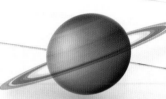

Saturn

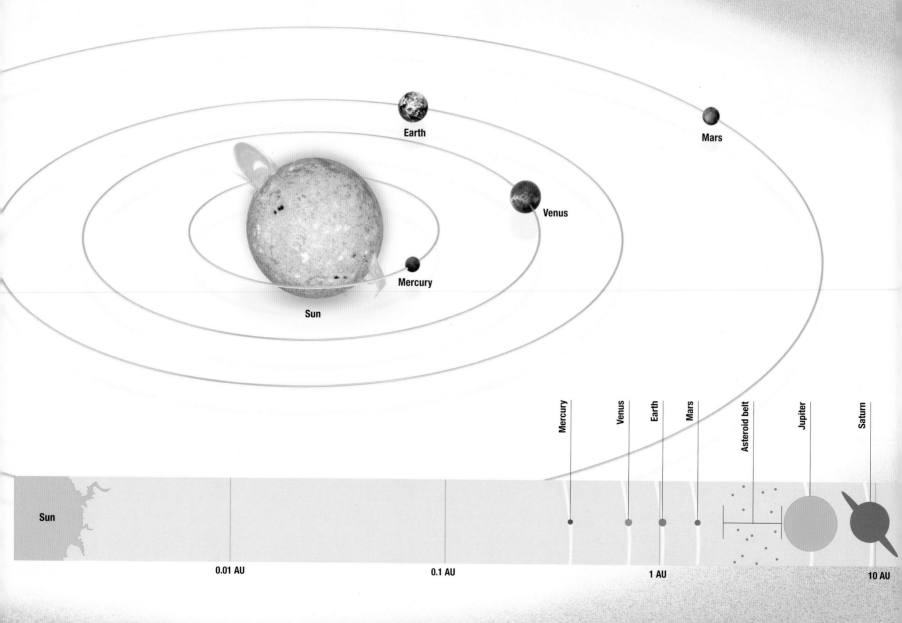

Earth

Mars

Venus

Mercury

Sun

Mercury

Venus

Earth

Mars

Asteroid belt

Jupiter

Saturn

Sun

0.01 AU

0.1 AU

1 AU

10 AU

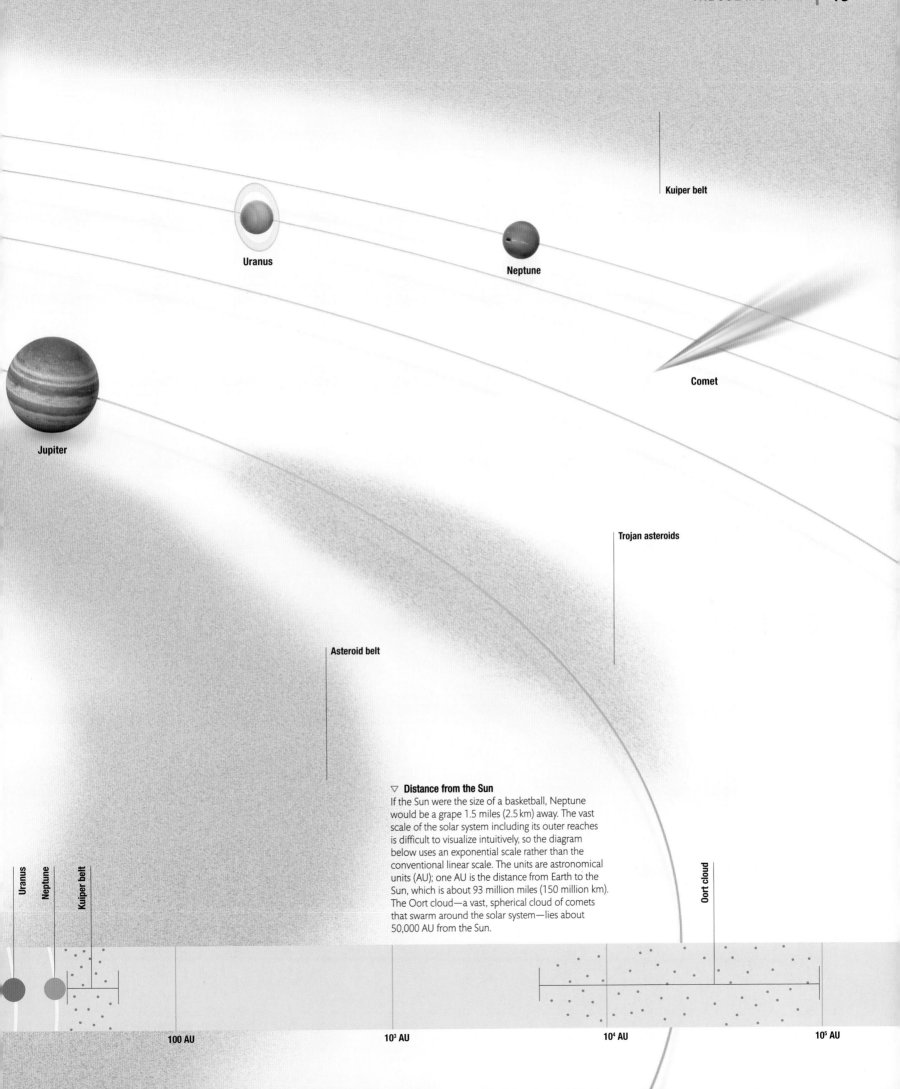

Kuiper belt

Uranus

Neptune

Comet

Jupiter

Trojan asteroids

Asteroid belt

▽ **Distance from the Sun**
If the Sun were the size of a basketball, Neptune
would be a grape 1.5 miles (2.5 km) away. The vast
scale of the solar system including its outer reaches
is difficult to visualize intuitively, so the diagram
below uses an exponential scale rather than the
conventional linear scale. The units are astronomical
units (AU); one AU is the distance from Earth to the
Sun, which is about 93 million miles (150 million km).
The Oort cloud—a vast, spherical cloud of comets
that swarm around the solar system—lies about
50,000 AU from the Sun.

Uranus

Neptune

Kuiper belt

Oort cloud

100 AU

10^3 AU

10^4 AU

10^5 AU

BIRTH OF THE
SOLAR SYSTEM

CREATED OUT OF GAS AND DUST, THE SUN FIRST SHONE AS A STAR WITHIN A RING OF DEBRIS—THE LEFTOVERS FROM ITS FORMATION. THESE MATERIALS SLOWLY GREW FROM TINY PARTICLES INTO ASTEROIDS, MOONS, AND PLANETS.

Five billion years ago, the solar system did not exist. Our galaxy, the Milky Way, was already 8 billion years old, and within it generations of stars had lived and died, seeding space with gas and dust that assembled into huge, dark clouds. Then, on the outskirts of the galaxy, something started to stir. An exploding star—a supernova—squeezed a neighboring dark cloud, which then began to collapse under its own gravity. Deep within, denser clumps of gas started to coagulate into thousands of protostars. As each one of these shrank, they heated up until nuclear reactions began in their cores and stars were born.

Many of these newly hatched stars were surrounded by whirling disks of gas and icy dust. In one case in particular—the newborn Sun—we know that this material, over millions of years, created the planets of our solar system.

Solar system nursery

Sheltered from the dangerous radiation of space, the new solar system developed in the depths of a giant bank of interstellar smog. This cloud was composed mainly of hydrogen and helium gas left over from the Big Bang and polluted with specks of soot and cosmic dust ejected from dying stars. It was so cold that gases such as methane, ammonia, and water vapor froze onto the tiny dust particles. These microscopic hailstones, whirling around the young Sun, were the seeds from which the planets would eventually grow.

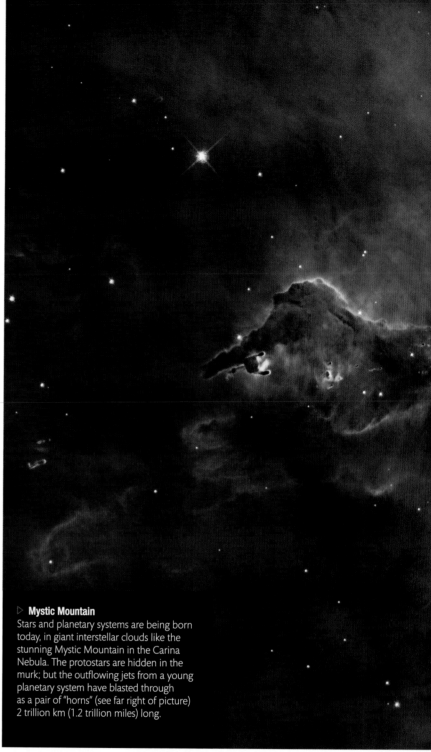

▷ **Mystic Mountain**
Stars and planetary systems are being born today, in giant interstellar clouds like the stunning Mystic Mountain in the Carina Nebula. The protostars are hidden in the murk; but the outflowing jets from a young planetary system have blasted through as a pair of "horns" (see far right of picture) 2 trillion km (1.2 trillion miles) long.

99.8 percent of **the Solar System's mass** is found in the Sun.

△ **Sun's secret birth**
Hidden in a nebula rich with chemical compounds, known as a molecular cloud, the embryonic Sun was no more than a collapsing clump of gas. As it contracted, this clump heated up to become a protostar.

△ **Bipolar outflow**
The protostar began to rotate, generating a strong magnetic field that forced streamers of gas away in opposite directions. The gas collapsing around the protostar turned ever faster and flattened out.

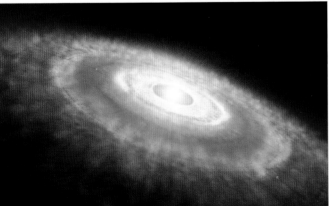

△ **Lighting up**
The protostar grew hot enough to ignite nuclear reactions, and the Sun began to shine. Its heat boiled away the ice nearby, leaving only rocky dust in the inner disk. But icy grains still survived on the outer edges.

▷ **Space rubble**
The rubble left over from the building of the solar system still falls to Earth as meteorites. The rare stony meteorites known as carbonaceous chondrites have remained unchanged since the birth of the planets. By analyzing the radioactive atoms in them, scientists can pinpoint the exact age of the solar system: 4.5682 billion years old. The oldest meteorites contain chondrules, glassy drops of melted rock formed in the heat generated by the development of the solar system.

FORMATION OF THE PLANETS

THE EIGHT PLANETS OF OUR SOLAR SYSTEM, NOW ORBITING SERENELY, WERE BORN IN A MAELSTROM OF COLLIDING DEBRIS LEFT OVER FROM THE SUN'S FORMATION.

The interstellar cloud that gave birth to the Sun was not used up entirely when our star formed. A disk of residual debris was left in orbit around the Sun like rings around Saturn, forming a "solar nebula." This material would eventually form the planets.

In the cold outer regions of the solar nebula, the debris consisted largely of tiny grains of frozen water, methane, and ammonia—hydrogen compounds too volatile to condense into ice in the inner solar system. Closer in, however, the Sun's heat boiled away volatile compounds, leaving only particles of rock and metal. As a result, the planets that formed in different parts of the solar nebula grew from very different materials. Inside the "frost line"—the point beyond which volatile compounds can survive the Sun's heat—the rocky debris gave rise to four small terrestrial planets with cores of metal. Beyond the frost line, icy debris coalesced into hot globes of spinning fluid, swollen to gigantic proportions by hydrogen and helium gas from the solar nebula.

Debris from the era of planet formation still litters the solar system in the form of asteroids, comets, and Kuiper belt objects (icy bodies beyond Neptune). Disturbed by the wanderings of Jupiter and Saturn, some of this icy rubble may even have delivered water to the once-dry Earth, kick-starting the chemical process that gave rise to life.

The **gas giant planets** account for nearly **99 percent** of the mass orbiting the Sun.

When worlds collide
In the first 100 million years after the Sun formed, protoplanets frequently collided as they whirled around the Sun. Mercury may owe its huge core to a catastrophic impact that stripped the nascent planet of its rocky mantle. Venus's anomalous clockwise spin—the opposite of most planets—may be the result of another collision. A protoplanet also seems to have hit Earth, almost splitting our world apart; the incandescent spray from this impact formed the Moon.

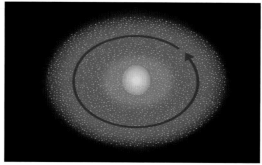

△ **Solar nebula**
The solar nebula started out as a homogeneous disk of gas and dust. As the dust particles jostled together in space, they became electrostatically charged and began to stick to one another. Closer to the Sun, they built up from grains of rock and metal to form rocky boulders similar in composition to asteroids. Beyond the frost line, they gradually enlarged into masses of ice.

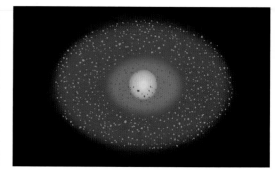

△ **Planetesimals form**
When two solid lumps orbiting the Sun collided at high speed, they smashed into each other. However, if the encounter was slow, gravity pulled them together. Overall, the process of construction was more frequent than destruction, so these chunks slowly grew by an inch or two per year. Eventually, they developed into bodies a few miles in diameter, called planetesimals.

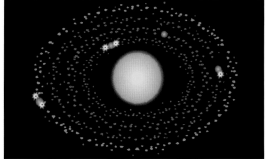

△ Rocky planets evolve
A million years after the birth of the solar system, the region near the Sun swarmed with 50–100 rocky bodies similar in size to Earth's Moon. As these protoplanets hurtled around the Sun, crashing into one another like bumper cars, collisions became ever more violent. The bigger protoplanets came out best, scooping up their smaller competitors. Only four would eventually survive, forming today's rocky planets.

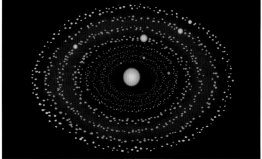

△ Gas giants expand
Beyond the frost line, the abundance of icy material created larger bodies. Fast-growing Jupiter developed sufficient gravity to pull in gas from the solar nebula and build up into a massive hydrogen-helium world. Saturn followed suit. However, in the outer reaches of the solar system, where material was sparse, Uranus and Neptune grew more slowly. Residual debris around the gas giants condensed, creating moons.

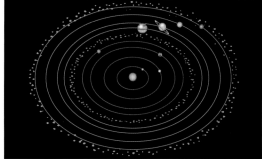

△ Planets migrate to modern positions
Originally, Uranus may have been the outermost planet, but the orbits of Jupiter and Saturn gradually changed, and when Saturn's "year" became exactly twice that of Jupiter, the resulting gravitational resonance threw Neptune farther out, followed by Uranus. These outer planets, in turn, threw icy planetesimals all over the solar system, bombarding the inner planets and forming today's Kuiper belt.

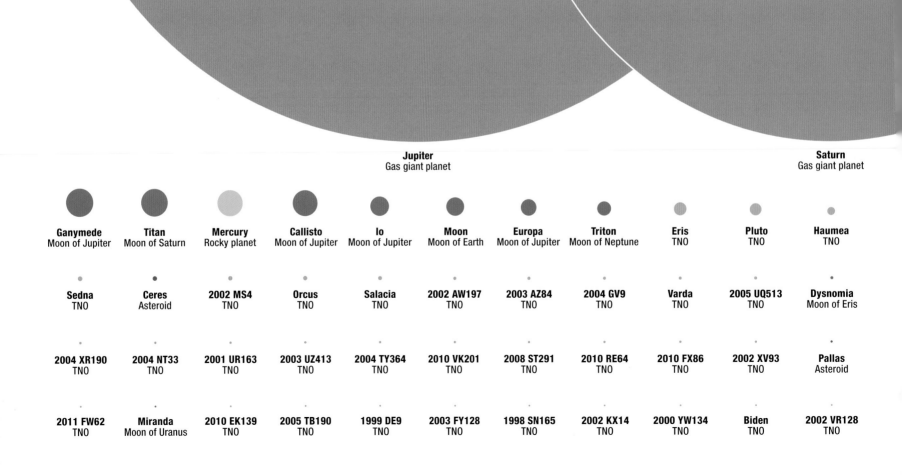

Jupiter
Gas giant planet

Saturn
Gas giant planet

Ganymede Moon of Jupiter	**Titan** Moon of Saturn	**Mercury** Rocky planet	**Callisto** Moon of Jupiter	**Io** Moon of Jupiter	**Moon** Moon of Earth	**Europa** Moon of Jupiter	**Triton** Moon of Neptune	**Eris** TNO	**Pluto** TNO	**Haumea** TNO
Sedna TNO	**Ceres** Asteroid	**2002 MS4** TNO	**Orcus** TNO	**Salacia** TNO	**2002 AW197** TNO	**2003 AZ84** TNO	**2004 GV9** TNO	**Varda** TNO	**2005 UQ513** TNO	**Dysnomia** Moon of Eris
2004 XR190 TNO	**2004 NT33** TNO	**2001 UR163** TNO	**2003 UZ413** TNO	**2004 TY364** TNO	**2010 VK201** TNO	**2008 ST291** TNO	**2010 RE64** TNO	**2010 FX86** TNO	**2002 XV93** TNO	**Pallas** Asteroid
2011 FW62 TNO	**Miranda** Moon of Uranus	**2010 EK139** TNO	**2005 TB190** TNO	**1999 DE9** TNO	**2003 FY128** TNO	**1998 SN165** TNO	**2002 KX14** TNO	**2000 YW134** TNO	**Biden** TNO	**2002 VR128** TNO

The Sun
Star

SIZE AND **SCALE**

THIS GRAPHIC SHOWS THE RELATIVE SIZES OF THE 100 LARGEST BODIES IN THE SOLAR SYSTEM, FROM THE SUN AND PLANETS TO THE NUMEROUS OTHER OBJECTS THAT ARE PART OF OUR STAR'S FAMILY.

On a cosmic scale, the Sun is the only substantial body in the solar system, so much larger than anything else that our own planet is a mere dot beside it. The largest of the planets by far are the gas giants, the biggest of which, Jupiter, could swallow Earth 1,300 times over. Farther down the scale come the rocky inner planets and then a miscellany of other bodies: moons, asteroids, and icy objects that populate the region beyond Neptune (trans-Neptunian objects). Diminution in size does not proceed neatly by class; Pluto, for example, is outsized by seven moons, and even Mercury is smaller than the two largest moons. Some of the largest asteroids and trans-Neptunian objects have sufficient mass to form a spherical shape and are therefore also classified as dwarf planets.

KEY
- Star
- Gas giant planet
- Rocky planet
- Moon
- Asteroid
- Trans-Neptunian object (TNO)

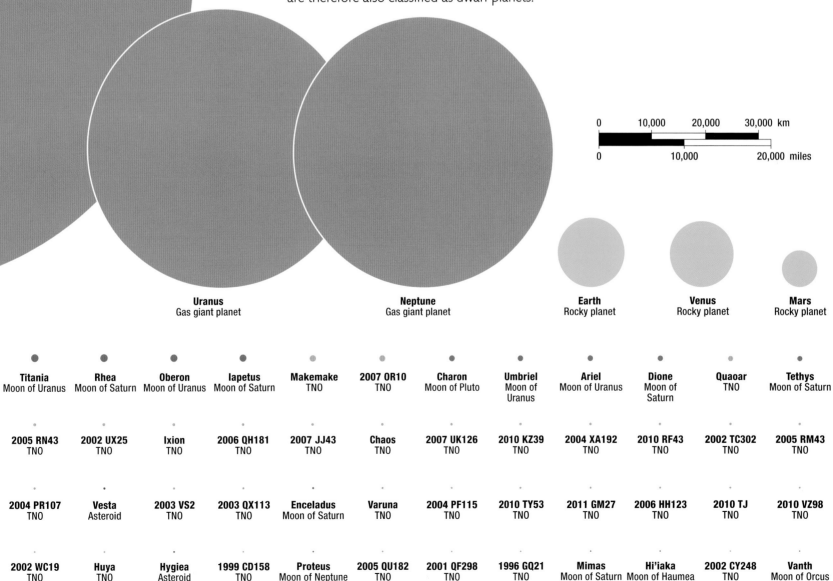

| 0 | 10,000 | 20,000 | 30,000 km |
| 0 | 10,000 | | 20,000 miles |

Uranus
Gas giant planet

Neptune
Gas giant planet

Earth
Rocky planet

Venus
Rocky planet

Mars
Rocky planet

| **Titania** Moon of Uranus | **Rhea** Moon of Saturn | **Oberon** Moon of Uranus | **Iapetus** Moon of Saturn | **Makemake** TNO | **2007 OR10** TNO | **Charon** Moon of Pluto | **Umbriel** Moon of Uranus | **Ariel** Moon of Uranus | **Dione** Moon of Saturn | **Quaoar** TNO | **Tethys** Moon of Saturn |

| **2005 RN43** TNO | **2002 UX25** TNO | **Ixion** TNO | **2006 QH181** TNO | **2007 JJ43** TNO | **Chaos** TNO | **2007 UK126** TNO | **2010 KZ39** TNO | **2004 XA192** TNO | **2010 RF43** TNO | **2002 TC302** TNO | **2005 RM43** TNO |

| **2004 PR107** TNO | **Vesta** Asteroid | **2003 VS2** TNO | **2003 QX113** TNO | **Enceladus** Moon of Saturn | **Varuna** TNO | **2004 PF115** TNO | **2010 TY53** TNO | **2011 GM27** TNO | **2006 HH123** TNO | **2010 TJ** TNO | **2010 VZ98** TNO |

| **2002 WC19** TNO | **Huya** TNO | **Hygiea** Asteroid | **1999 CD158** TNO | **Proteus** Moon of Neptune | **2005 QU182** TNO | **2001 QF298** TNO | **1996 GQ21** TNO | **Mimas** Moon of Saturn | **Hi'iaka** Moon of Haumea | **2002 CY248** TNO | **Vanth** Moon of Orcus |

OUR **SOLAR SYSTEM**

FOR CENTURIES, PEOPLE BELIEVED EARTH WAS AT THE CENTER OF THE COSMOS, WITH HEAVENLY BODIES IN ORBIT AROUND US. WHEN THIS MODEL WAS FINALLY OVERTURNED, IT LED TO A REVOLUTION IN SCIENCE.

The greatest conceptual breakthrough in our understanding of the solar system was the idea that Earth orbits the Sun, rather than vice versa. The heliocentric (sun-centered) model of the solar system was difficult to accept for several reasons. Common sense suggests the Sun moves across the sky; a stationary Sun implies that the apparently fixed and solid Earth must be moving and rotating. Moreover, the ancient Greek model of an Earth-centered solar system generated good predictions of planetary movements, supporting the faulty theory. And when the heliocentric model was shown to be more accurate, it faced resistance from the prevailing religious notion that Earth was the center of creation.

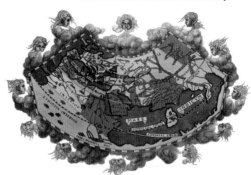
Medieval recreation of ancient Greek world map

c. 3000–500 BCE

Flat Earth
Early philosophers in Egypt and Mesopotamia believe Earth is flat and surrounded by sea, an idea later adopted by the Greeks. The Greek philosopher Thales claims that land floats on the ocean and that earthquakes are caused by giant waves.

c. 500 BCE

Spherical Earth
Pythagoras is the first of the Greek philosophers to suggest Earth is a sphere. Around 330 BCE, Aristotle offers further evidence: Earth's shadow during a lunar eclipse is round, and new stars appear as a person travels over Earth's curved surface.

Ceres, first known asteroid

Newton's *Principia*

Sputnik 1

1957

First satellite
The Space Age begins when the Soviet Union sends the first artificial satellite, Sputnik 1, into orbit around Earth. Two years later, the Soviet spacecraft Luna 3 sends back the first photographs of the far side of the Moon.

1801

Asteroids identified
While making routine observations, Italian astronomer Guiseppe Piazzi comes across a rocky body orbiting between Mars and Jupiter. Named Ceres, this is the first, and largest, asteroid to be discovered. In 2006, Ceres is also classified as a dwarf planet.

1781

Discoveries beyond Saturn
German-born British astronomer William Herschel discovers Uranus, a planet beyond Saturn, doubling the size of the known solar system. A variation in the new-found planet's orbit will eventually lead astronomers to discover Neptune, in 1846.

Apollo 11 Moon landing

Viking 1 image of Mars

1962

Voyage to Venus
NASA's Mariner 2 passes Venus, becoming the first spacecraft to fly past another planet. It records Venus's scorching temperature, which is too high to sustain life. In 1964, Mariner 4 flies past Mars and reveals a cold, barren, cratered world.

1969

First on the Moon
US astronaut Neil Armstrong becomes the first person to set foot on another world. Analysis of rocks brought back to Earth by Apollo astronauts suggests the Moon formed as a result of a massive impact between Earth and another planet.

1976

Landing on Mars
Viking 1 and Viking 2, the first spacecraft to land successfully on Mars, send back breathtaking images. They monitor the weather over two Martian years, analyze the composition of the atmosphere, and test the soil, inconclusively, for signs of life.

Early geocentric model of the cosmos

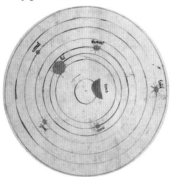

Copernicus's model of the solar system

▷ **C. 400 BCE**

Central fire
Greek philosopher Philolaus proposes that Earth and the Sun orbit a hidden "central fire." Aristarchus later claims the Sun is the center, and that the stars do not move relative to each other because they are so far away. His ideas are subsequently ignored.

▷ **C. 150 BCE**

The Ptolemaic system
Greek astronomer and geographer Claudius Ptolemy puts forward his geocentric theory, which places Earth at the center of the cosmos. Belief in the Ptolemaic system dominates astronomy for the next 1,400 years.

▷ **1543 CE**

Copernican revolution
Just before his death, the Polish astronomer and mathematician Nicolaus Copernicus publishes his revolutionary heliocentric model of the solar system, putting the stationary Sun at the center.

Galileo Galilei

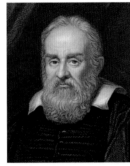

An elliptical orbit around the Sun

◁ **1687**

Planetary orbits explained
English scientist Isaac Newton publishes his supremely important *Principia*, laying the foundations of modern physics. He shows how gravity keeps planets in elliptical orbits around the Sun, and derives three laws of motion, explaining how forces work.

◁ **1633**

Astronomer on trial
The Catholic Church puts Italian astronomer Galileo Galilei on trial for teaching Copernicus's theory. His pioneering telescopic observations support the Sun-centered model. Galileo is forced to recant and is put under house arrest.

◁ **1609**

Kepler's laws
German mathematician Johannes Kepler calculates that the planets follow non-circular, elliptical orbits and alter speed according to their distance from the Sun. Kepler's laws resolve flaws in the Copernican model and later inspire Isaac Newton's discoveries.

Voyager 1 image of Jupiter

Nucleus of Halley's Comet

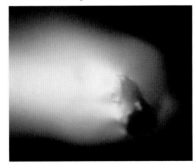

Saturn, as viewed by Cassini

▷ **1979**

Flyby of Jupiter
In a trail-blazing mission, Voyager 1 flies by Jupiter and its moons. The US craft reveals erupting volcanoes on the moon Io and an icy crust on Europa. Sister craft Voyager 2, launched two years earlier, will go on to pass Uranus (1986) and Neptune (1989).

▷ **1986**

Close encounter with a comet
Intercepting Halley's Comet at 150,000 mph (240,000 km/h), the European spacecraft Giotto takes the first close-up pictures of a comet's nucleus. They reveal a dark-coated lump of ice 9 miles (15 km) wide. Giotto then visits a second comet, Grigg-Skjellerup.

▷ **2004**

Orbit of Saturn
NASA's Cassini-Huygens spacecraft, launched in 1997, enters orbit around Saturn and later lands a probe onto the moon Titan. Cassini witnesses a huge storm in Saturn's clouds and discovers icy geysers erupting from the moon Enceladus.

OUR **STAR**

THE **SUN**

THE SUN IS THE HOTTEST, LARGEST, AND MOST MASSIVE OBJECT IN THE SOLAR SYSTEM. ITS INCANDESCENT SURFACE BATHES ITS FAMILY OF PLANETS IN LIGHT, AND ITS IMMENSE GRAVITATIONAL FORCE CHOREOGRAPHS THEIR ORBITS.

The Sun is a typical star, little different from billions of others in our galaxy, the Milky Way. It dominates everything around it, accounting for 99.8 percent of the solar system's mass. Compared with any of its planets, the Sun is immense. Earth would fit inside the Sun over one million times; even the biggest planet, Jupiter, is a thousandth of the Sun's volume. Yet the Sun is by no means the biggest star; VY Canis Majoris, known as a hypergiant, could hold almost 3 billion Suns.

Our star will not be around forever. Now approximately halfway through its life, in about 5 billion years it will turn into a red giant, swelling and surging out toward the planets. Mercury and Venus will be vaporized. The Earth may experience a similar fate, but even if our planet is not engulfed, it will become a sweltering furnace under the intense glare of a closer Sun. Eventually, the Sun will shake itself apart and puff its outer layers into space, leaving behind a ghostly cloud called a planetary nebula.

Energy traveling from the Sun's core takes **100,000 years** to reach the surface and appear as light.

THE SUN DATA

Diameter	865,374 miles (1,393,684 km)
Mass (Earth = 1)	333,000
Energy output	385 million billion gigawatts
Surface temperature	10,000°F (5,500°C)
Core temperature	27 million °F (15 million °C)
Distance from Earth	93 million miles (150 million km)
Polar rotation period	34 Earth days
Age	about 4.6 billion years
Life expectancy	about 10 billion years

▷ **Photosphere**
Photographed in wavelengths of light visible to the human eye, the Sun appears to have a smooth, spherical surface, speckled by cooler areas called sunspots. This apparent surface, called the photosphere, is illusory. It is merely the point in the Sun's vast atmosphere at which hot gas becomes transparent, letting light flood through.

▷ **Chromosphere**
The photosphere merges into an upper, hotter layer called the chromosphere. This ultraviolet image from NASA's Solar Dynamics Observatory reveals structures in both layers. The granular pattern is caused by convection cells—pockets of hot gas rising and sinking within the Sun.

▷ **Corona**
Extending far beyond the chromosphere is the Sun's tenuous outer atmosphere, the corona, revealed here by ultraviolet imaging. Invisible to the naked eye except during a solar eclipse, the corona is even hotter than the chromosphere and seethes with activity as eruptions of plasma burst through it.

Energy from the Sun's surface, or photosphere, escapes as visible light.

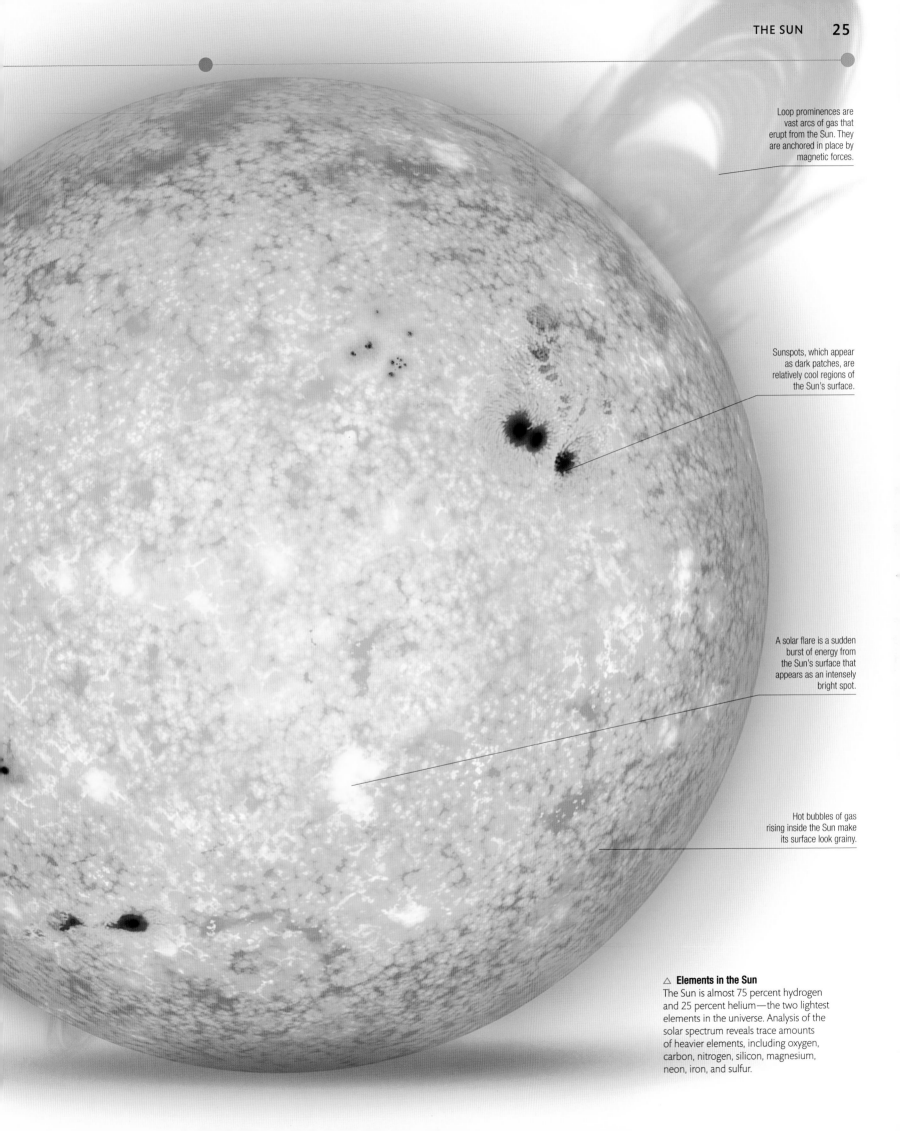

Loop prominences are vast arcs of gas that erupt from the Sun. They are anchored in place by magnetic forces.

Sunspots, which appear as dark patches, are relatively cool regions of the Sun's surface.

A solar flare is a sudden burst of energy from the Sun's surface that appears as an intensely bright spot.

Hot bubbles of gas rising inside the Sun make its surface look grainy.

△ **Elements in the Sun**
The Sun is almost 75 percent hydrogen and 25 percent helium—the two lightest elements in the universe. Analysis of the solar spectrum reveals trace amounts of heavier elements, including oxygen, carbon, nitrogen, silicon, magnesium, neon, iron, and sulfur.

SUN STRUCTURE

**IT MAY SEEM LIKE AN UNCHANGING YELLOW BALL
IN THE SKY, BUT THE SUN IS INCREDIBLY DYNAMIC.
A GIANT NUCLEAR FUSION REACTOR, IT FLOODS THE
SOLAR SYSTEM WITH ITS BRILLIANT ENERGY.**

The Sun has no solid surface—it is made of gas, mostly
hydrogen. Intense heat and pressure split the gas atoms
into charged particles, forming an electrified state of
matter known as plasma. Inside the Sun, density and
temperature rise steadily toward the core, where the
pressure is more than 100 billion times greater than
atmospheric pressure on Earth's surface. In this extreme
environment, unique in the solar system, nuclear fusion
occurs. Hydrogen nuclei are fused together to form
helium nuclei, and a fraction of their mass is lost as
energy, which percolates slowly to the Sun's outer layers
and then floods out into the blackness of space,
eventually reaching Earth as light and warmth.

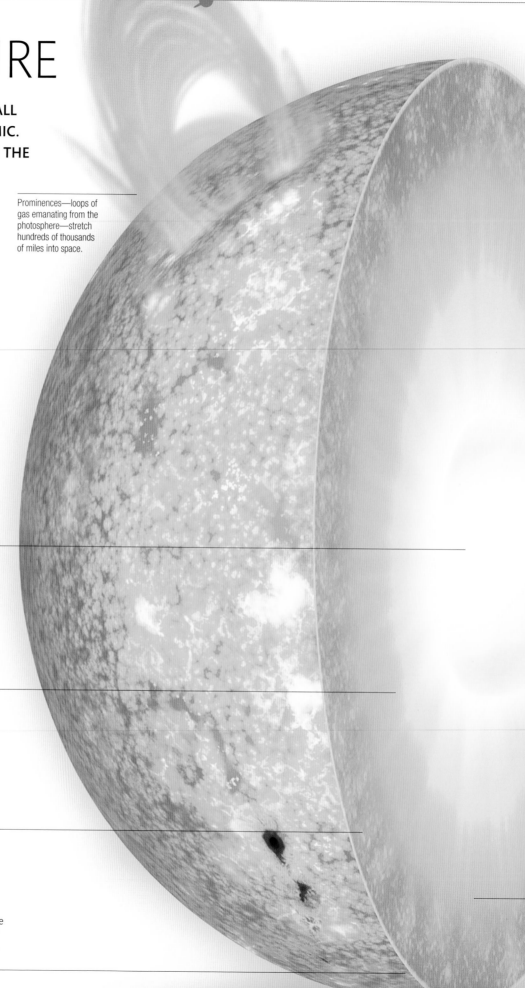

Prominences—loops of
gas emanating from the
photosphere—stretch
hundreds of thousands
of miles into space.

Core
Making up the inner fifth of the Sun, the core is
where nuclear fusion creates 99 percent of the Sun's
energy. The center of the core, where hydrogen has
been fused, is mostly helium. The temperature in the
core is 27 million °F (15 million °C).

Radiative zone
Light energy works its way slowly up through the
radiative zone, colliding with atomic nuclei and being
reradiated billions of times. The radiative zone is so
densely packed with matter that energy from the
core can take as long as 100,000 years to reach the
surface. The radiative zone accounts for 70 percent
of the Sun's radius, and temperatures range from
3.5 to 27 million °F (1.5 to 15 million °C).

Convective zone
In the convective zone, pockets of hot gas
expand and rise toward the solar surface. The
process, known as convection, carries the
energy upward much faster than in the radiative
zone. Temperatures here vary from 10,000 to
3.5 million °F (5,500 to 1.5 million °C).

Photosphere
The photosphere—a region only 60 miles
(100 km) thick—is the apparent surface of the
Sun. This is where energy reaches the top of
the convective zone and escapes into space.
The temperature here is 10,000°F (5,500°C).

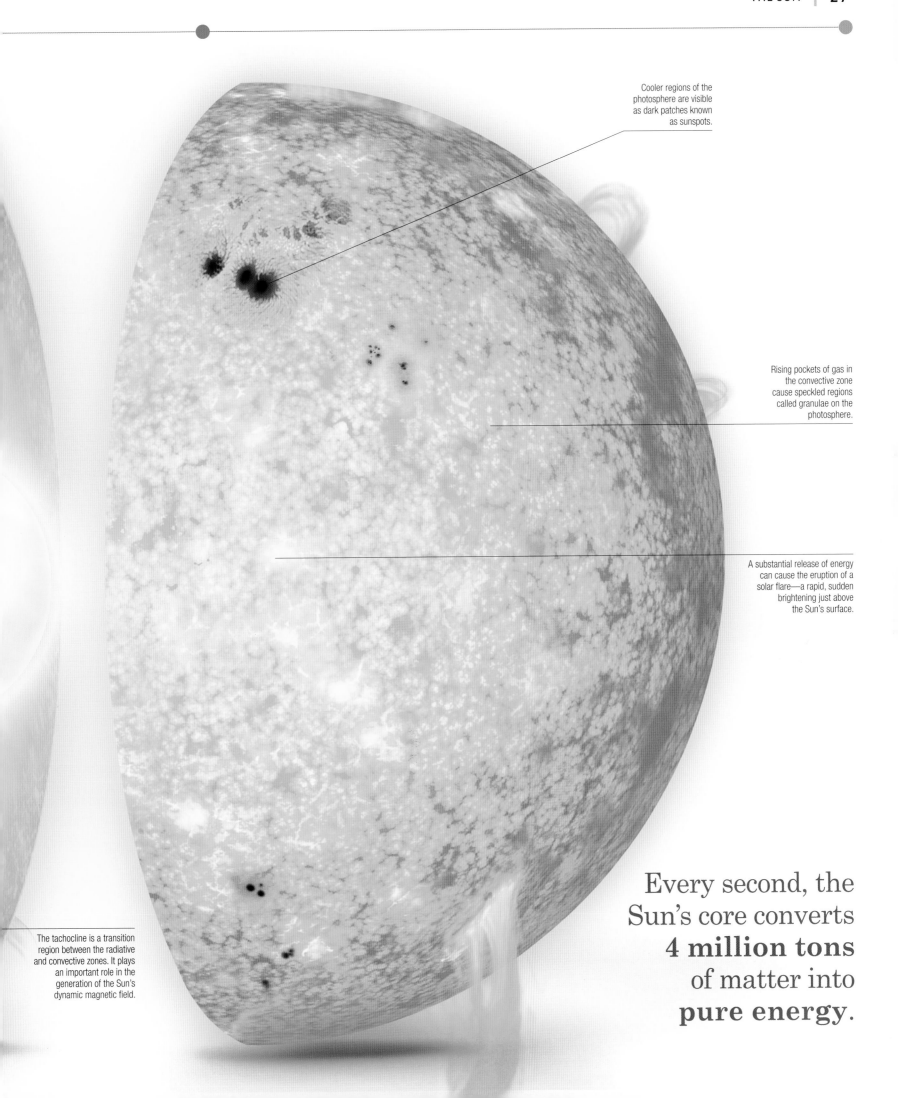

Cooler regions of the photosphere are visible as dark patches known as sunspots.

Rising pockets of gas in the convective zone cause speckled regions called granulae on the photosphere.

A substantial release of energy can cause the eruption of a solar flare—a rapid, sudden brightening just above the Sun's surface.

The tachocline is a transition region between the radiative and convective zones. It plays an important role in the generation of the Sun's dynamic magnetic field.

Every second, the Sun's core converts **4 million tons** of matter into **pure energy**.

STORMS ON **THE SUN**

A SEETHING BALL OF PLASMA, THE SUN IS NEVER THE SAME FROM ONE DAY TO THE NEXT. THE SOLAR SURFACE IS IN CONSTANT MAGNETIC TURMOIL, RESULTING IN THE BIGGEST EXPLOSIVE EVENTS IN THE SOLAR SYSTEM.

Heat and light are not all that the Sun gives to its family of orbiting worlds. Our star regularly hurls vast swarms of electrically charged particles out into the solar system in violent solar storms. For 150 years, astronomers have been able to observe these events from Earth, but it is only in the last 20 years that they have been Sun-watching at closer quarters, using a suite of telescopes launched into space. These instruments are capable of seeing the Sun even when our spinning planet turns ground-based instruments away from it. A thorough understanding of this space weather is crucial as our world becomes ever more reliant on technology—an intense burst of solar activity aimed directly at Earth can disable power grids and wreck satellite circuitry.

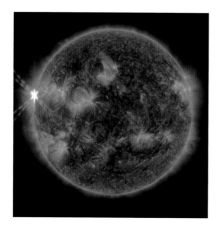

◁ **Solar flares**
Like light bouncing off a gleaming surface, areas of the Sun suddenly and rapidly brighten from time to time. Such events, known as solar flares, often signal the coming onslaught of a coronal mass ejection. The ultraviolet image shown on the left, taken by NASA's Solar Dynamics Observatory, captures a solar flare erupting from the left limb of the Sun.

▷ **Prominences**
The Sun's magnetic field lines sometimes tangle so much that they "snap," releasing their pent-up energy. When this happens, sprawling loops of hot plasma known as prominences erupt from the solar surface, following the magnetic field lines and tracing out vast and beautiful loops. These flamelike plumes can extend 300,000 miles (500,000 km) into space, and last from several days to months. Prominences often take a distinctive arch shape but can emerge in other forms, too, including pillars and pyramids. If they erupt Earthward, so that we see them in front of the Sun rather than against the darkness of space, they are referred to as filaments. This sequence of five photographs shows the eruption of a solar prominence as it gradually bulges out from the surface of the Sun before flaring into full splendor.

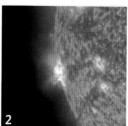

1

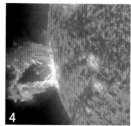

2

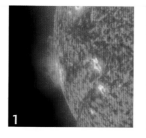

3

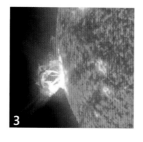

4

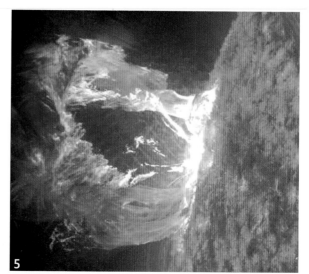

5

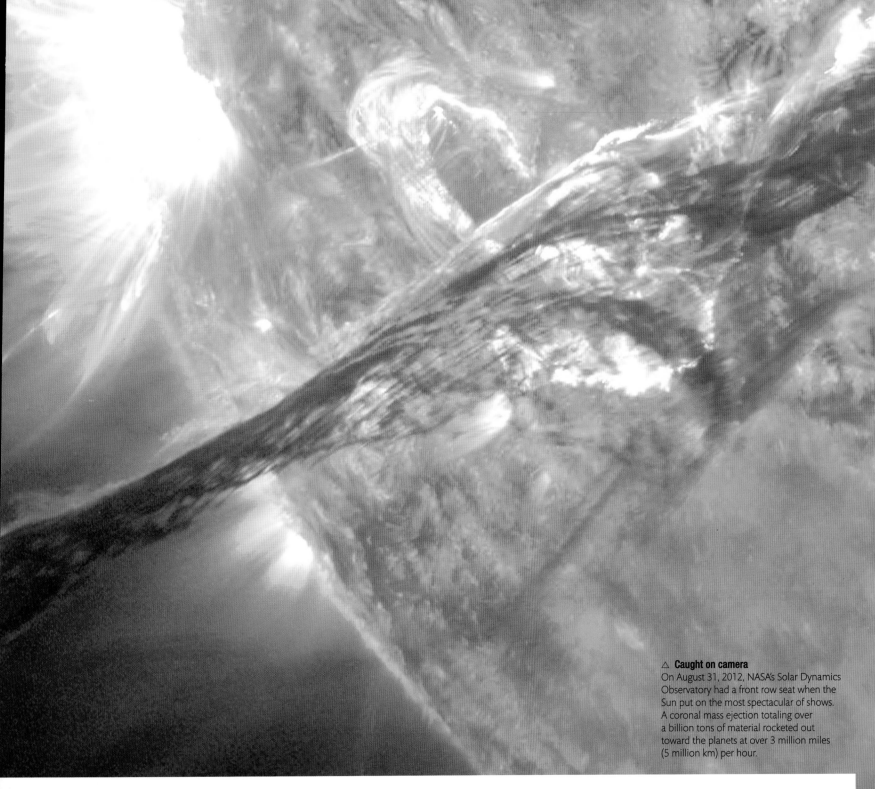

△ **Caught on camera**
On August 31, 2012, NASA's Solar Dynamics Observatory had a front row seat when the Sun put on the most spectacular of shows. A coronal mass ejection totaling over a billion tons of material rocketed out toward the planets at over 3 million miles (5 million km) per hour.

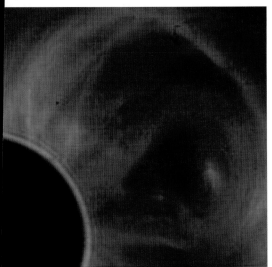

◁ **Coronal mass ejection**
The most sizable and impressive explosive events anywhere in the solar system occur when the Sun throws out a mighty eruption of plasma known as a coronal mass ejection (CME). As the name suggests, the plasma is spat out from the Sun's atmosphere (corona). The sheer violence of the explosion can accelerate solar particles toward the speed of light. When CME material reaches the Earth, it may trigger a geomagnetic storm. In the ultraviolet photograph on the left, a CME is seen swelling out from the Sun's corona like a giant bubble.

△ **Northern lights**
A geomagnetic storm caused by a CME can overwhelm the Earth's magnetic field, channeling energy poleward and producing spectacular aurorae like the one above, photographed over Thingvellir National Park in Iceland. The shimmering curtains of light are the result of oxygen atoms glowing from energy injected into the atmosphere. Normally seen in polar latitudes, aurorae can extend all the way to the tropics after a major CME.

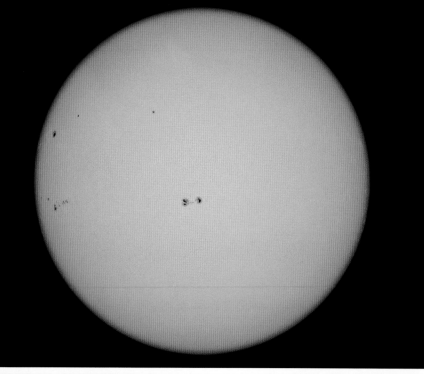

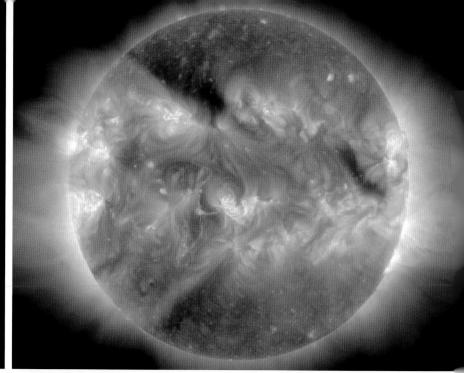

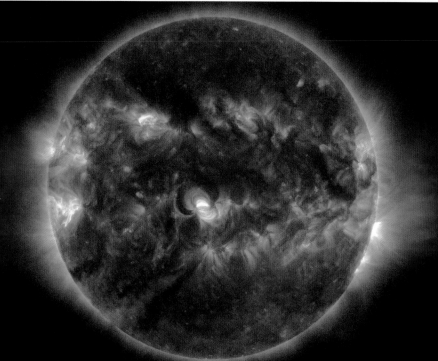

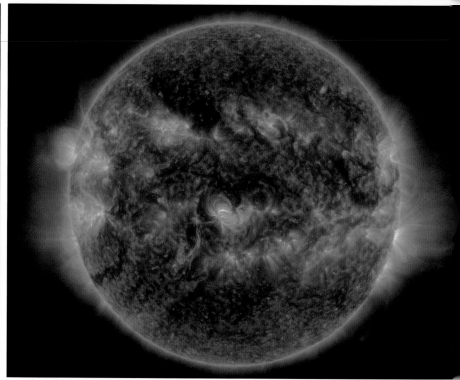

SUN RAYS

As well as producing light in the visible part of the spectrum, the Sun emits wavelengths our eyes cannot see, from radio waves and infrared to ultraviolet radiation. By capturing these rays, solar observatories can image parts of the Sun that are normally invisible. NASA's space-based Solar Dynamics Observatory (SDO) produces new images of the Sun every second; those shown here were all taken in a single hour in April 2014. The first one shows what the human eye would see if a direct glance were possible—the Sun's brilliant photosphere is reduced to a smooth yellow disk, with dark sunspots where magnetic disturbances have cooled the surface. For most of the images that follow, SDO used filters to select various wavelengths of ultraviolet light, revealing solar flares high in the Sun's outer atmosphere above sunspot regions. The final two photographs are composites that combine several wavelengths.

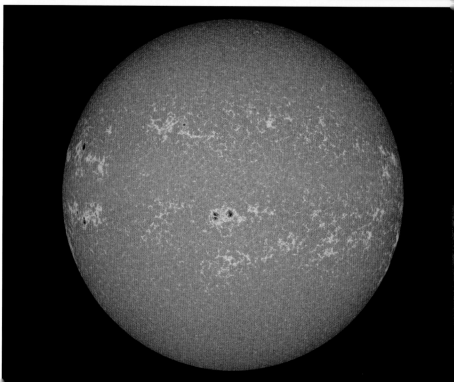

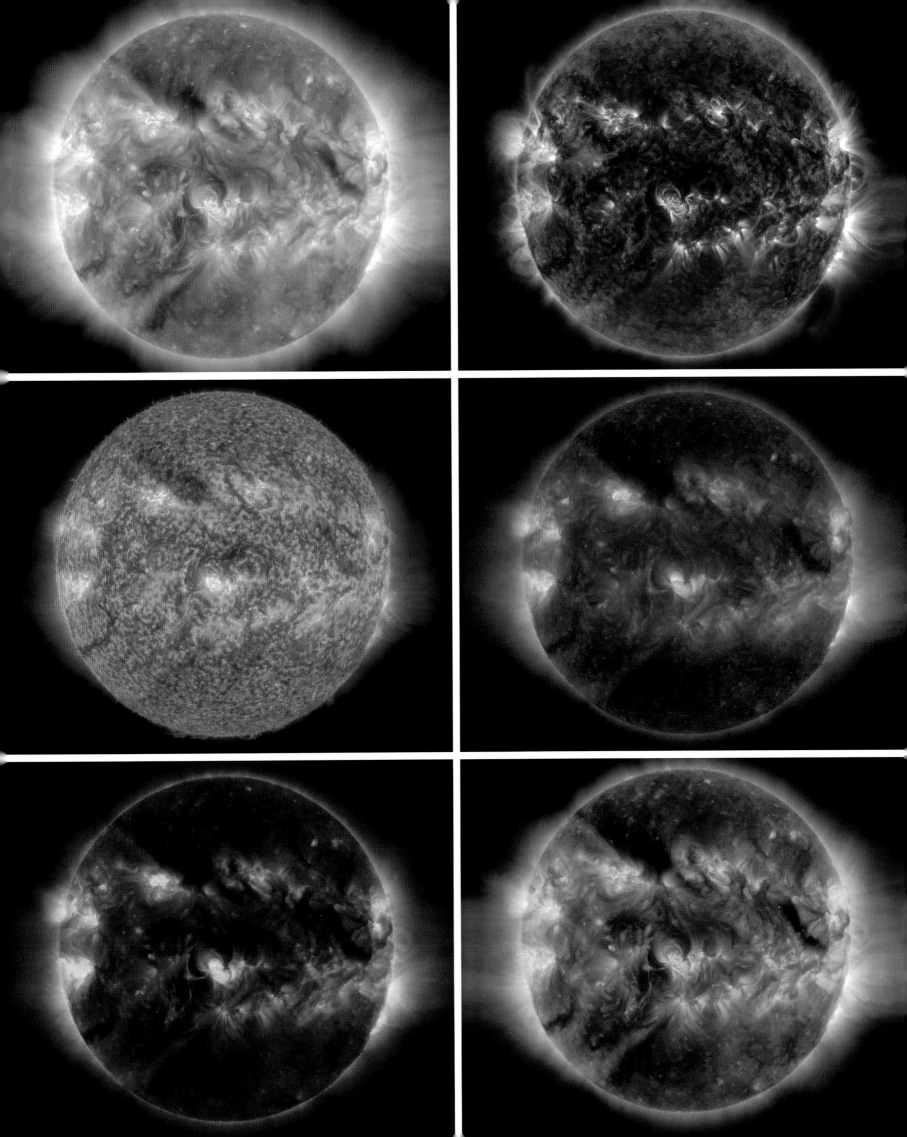

THE **SOLAR** CYCLE

THE SUN IS A CHANGEABLE STAR, SOMETIMES CALM AND PEACEFUL, SOMETIMES ERUPTING WITH GREAT VIOLENCE. THESE CHANGES FOLLOW A CLEAR PATTERN, WITH A CYCLICAL RISE AND FALL OF SOLAR ACTIVITY EVERY 11 YEARS OR SO.

For the last four centuries, scientists have kept records of the Sun's activity. During the early 19th century, German apothecary-turned-astronomer Samuel Heinrich Schwabe spent 17 years trying to spot a planet that he believed existed closer to the Sun than Mercury. He failed to see the silhouette of a new planet against the Sun, but he did keep accurate records of sunspots. Looking back over his observations, he noticed that the number of sunspots varied in a regular way, and the idea of the solar cycle was born. Today's orbiting and ground-based solar telescopes constantly scrutinize the Sun, revealing further details of this recurring pattern.

Sunspots

Once thought to be storms in the atmosphere of the Sun, we now know that sunspots are merely cooler regions of the solar surface. Typically lasting a few weeks, they are caused by intense, local magnetic activity and often appear in pairs. Records of sunspot observations date from the early 17th century, though sunspots were probably seen earlier. Scientists can trace sunspot activity further back by studying tree rings: carbon-14 levels in tree rings are lower during times of sunspot abundance, and greater when there are few sunspots.

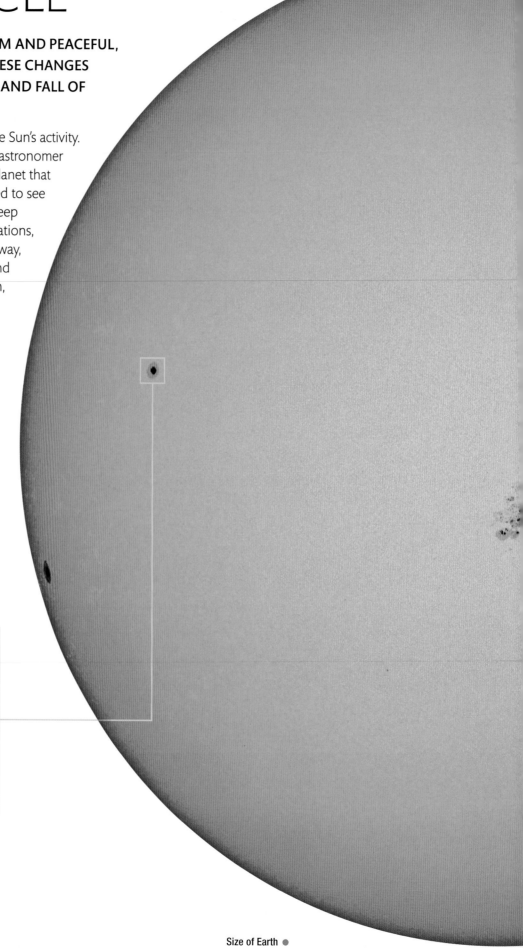

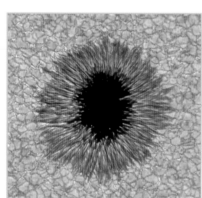

▷ **Sunspot structure**
A sunspot is usually split into two parts: an inner umbra and an outer penumbra. The dark umbra is the cooler part, with temperatures of around 4,500°F (2,500°C). By contrast, the penumbra can reach 6,300°F (3,500°C) and often exhibits streaky filaments called fibrils. The sunspots in a pair tend to have opposite magnetic polarity, akin to the poles of a magnet.

The **Great Sunspot** of 1947 was easily visible to the naked eye at sunset.

Size of Earth ●

Sunspots are frequently seen in pairs and sometimes form larger clusters.

Solar cycle

The 11-year solar cycle progresses from solar minimum (fewest sunspots) to solar maximum (most sunspots) and back again. It is linked to changes in the Sun's magnetic field, which becomes twisted during the cycle, before breaking down and renewing itself; every 22 years, the Sun's magnetic poles reverse. Solar maximum is associated not only with greater sunspot activity but also with solar flares, coronal mass ejections, and brighter aurorae on Earth.

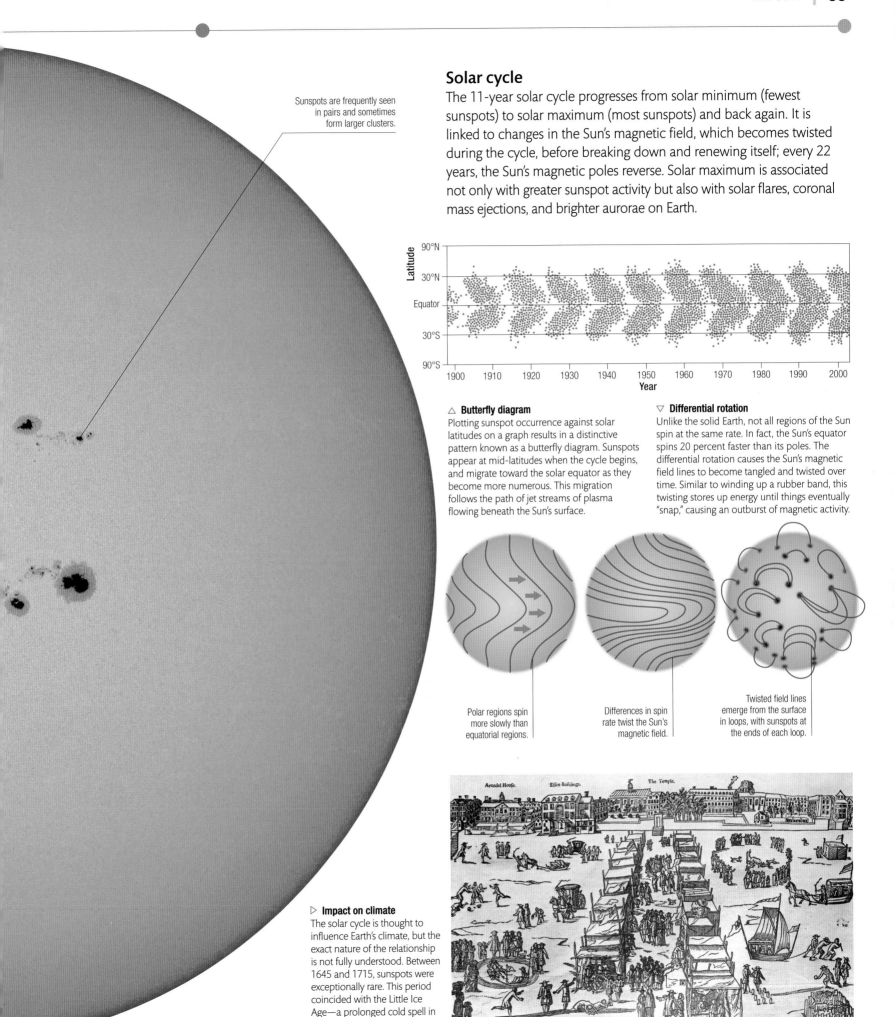

△ **Butterfly diagram**
Plotting sunspot occurrence against solar latitudes on a graph results in a distinctive pattern known as a butterfly diagram. Sunspots appear at mid-latitudes when the cycle begins, and migrate toward the solar equator as they become more numerous. This migration follows the path of jet streams of plasma flowing beneath the Sun's surface.

▽ **Differential rotation**
Unlike the solid Earth, not all regions of the Sun spin at the same rate. In fact, the Sun's equator spins 20 percent faster than its poles. The differential rotation causes the Sun's magnetic field lines to become tangled and twisted over time. Similar to winding up a rubber band, this twisting stores up energy until things eventually "snap," causing an outburst of magnetic activity.

Polar regions spin more slowly than equatorial regions.

Differences in spin rate twist the Sun's magnetic field.

Twisted field lines emerge from the surface in loops, with sunspots at the ends of each loop.

▷ **Impact on climate**
The solar cycle is thought to influence Earth's climate, but the exact nature of the relationship is not fully understood. Between 1645 and 1715, sunspots were exceptionally rare. This period coincided with the Little Ice Age—a prolonged cold spell in Europe during which normally ice-free rivers froze over.

Frost fair on the Thames River during the Little Ice Age

SOLAR ECLIPSES

WHEN SOMETHING AS CONSTANT AND UNERRING AS THE SUN'S LIGHT IS SUDDENLY INTERRUPTED DURING THE DAY, WE CANNOT FAIL TO NOTICE. FOR A FEW MINUTES, IT SEEMS AS IF THE WORLD STANDS STILL.

History books are littered with tales of the Sun disappearing; today we call these events solar eclipses. Every so often, during its steady crawl around Earth, the Moon occupies the exact same part of daytime sky as the Sun. Since the Moon is closer, its presence obscures our view of the Sun, causing an eclipse.

Total solar eclipses

During a total solar eclipse, the Sun is completely hidden by the Moon's disk for a few minutes. A total solar eclipse is perhaps nature's ultimate spectacle: the sky darkens, the temperature drops, and birds stop singing.

If the Moon orbited exactly on the line between the Sun and Earth, we would get an eclipse every month. However, because the Moon's orbit is tilted by five degrees, eclipses happen only every 18 months or so. Each is visible from only a small part of Earth's surface, where the Moon's shadow falls.

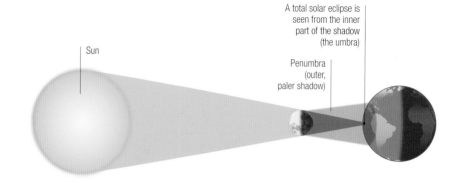

Sun

A total solar eclipse is seen from the inner part of the shadow (the umbra)

Penumbra (outer, paler shadow)

Totality

△ **How total eclipses work**
Despite being 400 times smaller than the Sun, the Moon is able to block our view of the Sun because it is 400 times closer. Where the darker part of the Moon's shadow—the umbra—falls on Earth, a total eclipse is seen; from the penumbra, a partial eclipse is visible. The umbra's path across Earth is typically 10,000 miles (16,000 km) long but only 100 miles (160 km) wide.

▽ **Totality**
Eclipse watchers view the Sun and Moon from Ellis Beach in Australia in November 2012. Totality—the stage during which the Sun is completely hidden—is a fleeting event, lasting a maximum of 7.5 minutes. During the eclipse of 2012 it lasted only two minutes.

△ Diamond ring
The Moon's surface is not perfectly smooth. Mountains and valleys allow sunlight to break through, creating an effect known as Baily's beads. A solitary bead appears as a spectacular "diamond ring," marking the beginning or end of totality.

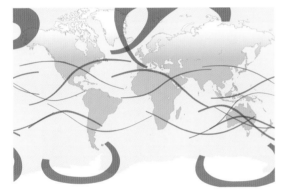

Map of total solar eclipses 2014–2040

△ Solar corona
The Sun's vast and tenuous outer atmosphere—the corona—is normally outshone by its brilliant photosphere, rendering the corona all but invisible. But when the Moon covers the Sun's face, the corona is spectacularly revealed. To study the corona through solar telescopes, astronomers use a coronagraph—an opaque disk that obscures the Sun.

Annular solar eclipses

Sometimes the Moon fails to cover the entire solar disk, allowing us to see the edge of the Sun as a ring around the Moon's silhouette. This is called an annular solar eclipse, from the Latin *annulus*, meaning "little ring." A hybrid solar eclipse—a very unusual event—appears as total from some locations on Earth and as annular from others.

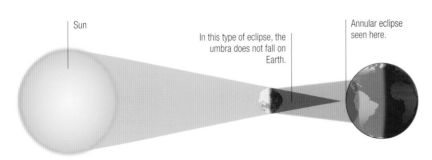

Sun

In this type of eclipse, the umbra does not fall on Earth.

Annular eclipse seen here.

△ How annular eclipses work
The Moon's orbit is elliptical rather than circular, so its distance from Earth varies. If a solar eclipse occurs when the Moon is at its farthest from Earth, it is too small in the sky to block out the Sun and causes an annular eclipse.

▷ Ring of fire
An annular eclipse is visible where the antumbra—the extension of the umbra—passes over Earth's surface. During the eclipse, the Moon leaves a spectacular "ring of fire" around it. The ring is so bright that the corona cannot be seen.

STORY OF **THE SUN**

OVER THE CENTURIES, THE SUN'S PLACE IN OUR CULTURE HAS CHANGED DRAMATICALLY—SCIENCE AND EXPERIMENTATION HAVE OVERSEEN ITS TRANSITION FROM ALL-POWERFUL GOD TO HOT, GAS-FILLED STAR.

The Sun's movements have been tracked for thousands of years, and were used by many ancient civilizations as the basis for their calendars. However, the same people still believed the Sun circled Earth; it was not until 1543 that Copernicus suggested the Sun was at the center of the solar system. Later, Newton's theory of gravity allowed the Sun's enormous mass to be calculated, and Einstein's work in the early 20th century explained how the Sun can shine for billions of years without running out of fuel. Modern spacecraft allow us to study the Sun in intimate detail and predict the storms that rage on its surface.

Stonehenge

Sun god worship in ancient Egypt

3000–2000 BCE

Astronomical calendar
Stonehenge monument is built in southwest England. Although its function remains unclear, the alignment of its stones with the sunrise and sunset in midsummer and midwinter suggests it was used as an astronomical calendar.

1350 BCE

Sun god Apollo
The Sun is worshipped as a god by the ancient Egyptians, Greeks, and later Romans. The Romans celebrate the death and rebirth of the Sun god Apollo in midwinter. When Rome converts to Christianity, this festival becomes known as Christmas.

Coronal mass ejection

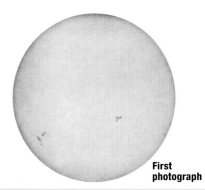

First photograph

J. Norman Lockyer

1868

Discovery of helium
English astronomer J. Norman Lockyer discovers an unknown element in the spectrum of the Sun. He names it helium after Helios, the Greek Sun god. The element is not discovered on Earth until 1895. We now know the Sun is 25 percent helium.

1859

Solar storm recorded
English astronomer Richard Carrington observes the first solar flare. It is followed by the biggest Earth-bound coronal mass ejection ever recorded. The solar storm hits Earth within days, causing aurorae as far south as Hawaii and the Caribbean.

1845

First photograph of the Sun
The new technology of photography allows the first image of the Sun to be taken by French astronomers Louis Fizeau and Lion Foucault. The pair use the daguerreotype technique to capture the image, which includes clearly visible sunspots.

Butterfly diagram

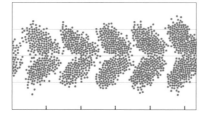

Einstein and Eddington

Nuclear fusion

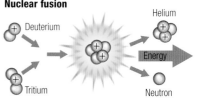

1904

Sunspots plotted
English astronomer Edward Maunder plots sunspot locations during the solar cycle, creating his famous "butterfly diagram." It shows that sunspots increase in number and move toward the solar equator as the solar cycle approaches its peak.

1919

Theory of relativity
British physicist Arthur Eddington photographs a solar eclipse from Principe in west Africa. His shots capture the positions of stars near the Sun and confirm Albert Einstein's general theory of relativity by showing that the Sun bends light.

1920

Nuclear fusion in the Sun's core
In his presidential address to the British Association for the Advancement of Science, Arthur Eddington correctly proposes that the Sun's energy is created by nuclear processes at its core. He goes on to publish a detailed account of his ideas in 1926.

Solar eclipse

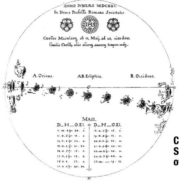

Copernicus's drawing of the solar system

364 BCE

Earliest record of sunspots

Chinese astronomer Shi Shen makes the earliest record of sunspot observation. He believes the phenomenon is due to a form of eclipse. Today we know that sunspots are cooler regions of the Sun's photosphere.

968

Sun's corona

Byzantine historian Leo Diaconus gives the first reliable description of the Sun's corona, as seen from Constantinople (now Istanbul) during a solar eclipse. He describes a "dim and feeble glow like a narrow band shining in a circle around the edge of the disk."

1543

Center of the solar system

Copernicus's *On the Revolutions of the Heavenly Spheres* is printed in Nuremburg in modern-day Germany. Previously, Ptolemy's view that Earth was at the center of the solar system prevailed. Copernicus's work places the Sun at the heart of the solar system.

Absorption lines

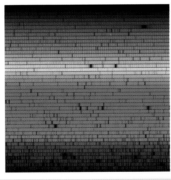

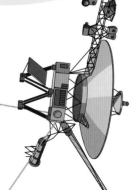

Christoph Scheiner's drawing of sunspots

1843

Sunspot cycle

German astronomer Heinrich Schwabe publishes his work on sunspots after studying them for 17 years in an attempt to find a hypothetical planet, Vulcan. He notes that sunspot numbers rise and fall over a decade or so. We now know this as the solar cycle.

1802

Discovery of absorption lines

English chemist William Wollaston discovers absorption lines in the spectrum of light from the Sun. These are later found to be caused by chemical elements in the Sun and are used to determine its composition.

1609

First telescope view of sunspots

The invention of the telescope leads to the first clear observations of sunspots by Italian scientist Galileo, German physicist Christoph Scheiner, and other astronomers. Galileo's observations of Jupiter and Venus support Copernicus's ideas about the solar system.

Comet Hale–Bopp

SDO image of Sun

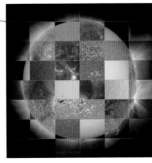

Voyager 1

1951

Discovery of the solar wind

German astronomer Ludwig F. Biermann discovers the solar wind by observing comets. He notices the tail of a comet always points away from the Sun no matter which way it is traveling, and concludes that something must be blowing it in that direction.

1995

SOHO mission

NASA and ESA's Solar and Heliospheric Observatory (SOHO) launches. It provides spectacular images and unprecedented scientific analysis of the Sun. By 2012, it will discover over 2,000 sun-grazing comets.

2010

Solar Dynamics Observatory

NASA's Solar Dynamics Observatory (SDO) launches, using high-definition technology to observe the Sun. Taking multiple wavelength images every ten seconds, it sends back data equivalent to half a million music tracks every day.

2012

Voyager 1 leaves heliosphere

The Voyager 1 spacecraft becomes the first human-made object to leave the heliosphere, the vast region of space around the Sun in which the solar wind flows.

EARTH ORBIT

LAGRANGIAN POINT ORBIT

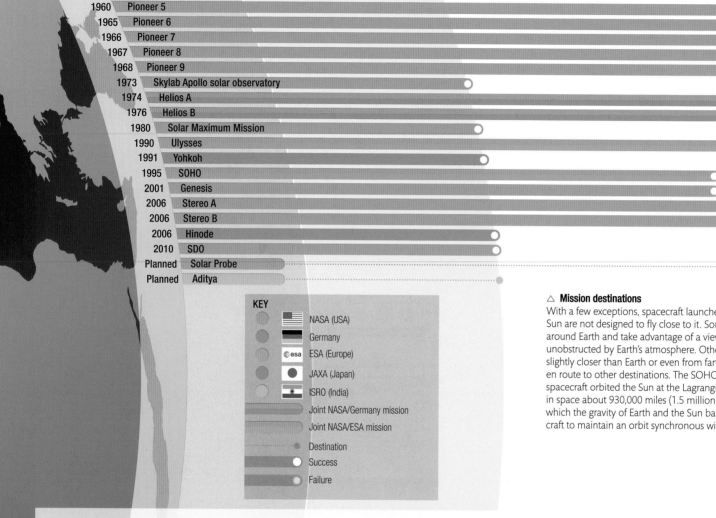

Year	Mission
1960	Pioneer 5
1965	Pioneer 6
1966	Pioneer 7
1967	Pioneer 8
1968	Pioneer 9
1973	Skylab Apollo solar observatory
1974	Helios A
1976	Helios B
1980	Solar Maximum Mission
1990	Ulysses
1991	Yohkoh
1995	SOHO
2001	Genesis
2006	Stereo A
2006	Stereo B
2006	Hinode
2010	SDO
Planned	Solar Probe
Planned	Aditya

KEY

- 🇺🇸 NASA (USA)
- 🇩🇪 Germany
- esa ESA (Europe)
- JAXA (Japan)
- ISRO (India)
- Joint NASA/Germany mission
- Joint NASA/ESA mission
- Destination
- ◯ Success
- ◯ Failure

△ **Mission destinations**

With a few exceptions, spacecraft launched to observe the Sun are not designed to fly close to it. Some stay in orbit around Earth and take advantage of a view of the Sun unobstructed by Earth's atmosphere. Others orbit the Sun slightly closer than Earth or even from farther away, sometimes en route to other destinations. The SOHO and Genesis spacecraft orbited the Sun at the Lagrangian point—a point in space about 930,000 miles (1.5 million km) from Earth at which the gravity of Earth and the Sun balance, allowing the craft to maintain an orbit synchronous with Earth's.

▷ **Pioneer 5**

This early mission lacked a camera and so could not return images. It was, however, the first true interplanetary spacecraft. Launched on a path that took it between Earth and Venus, Pioneer 5 confirmed the existence of an interplanetary magnetic field for the first time and studied how this field is affected by solar flares.

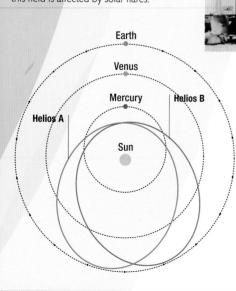

◁ **Helios A and B**

The two Helios spacecraft studied the solar wind and magnetism. They hold the records for making the closest approach to the Sun (slightly nearer than Mercury) and being the fastest human-made objects in history: they reached a top speed of 44 miles (70 km) per second. No longer functional but still in orbit, they follow elliptical paths, swooping close to the Sun at top speed and then flying back out toward Earth's orbit.

▷ **Ulysses**

Designed to observe the Sun at high latitudes, Ulysses flew to Jupiter and used the planet's gravity to fling it into an orbit that would take it over the poles of the Sun. On its travels, it discovered that 30 times more dust enters our solar system than had previously been thought. Contact with Ulysses was terminated in 2009.

▽ **SOHO**

Launched in 1995, the Solar and Heliospheric Observatory (SOHO) was the first of the modern generation of solar observatories. Still at work today, SOHO has returned many spectacular images of the Sun's violent weather, the chromosphere, and the corona, all of which it monitors from a solar orbit. While studying the Sun, SOHO has discovered 2,000 Sun-grazing comets.

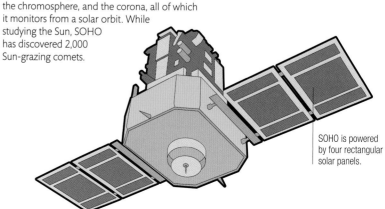

SOHO is powered by four rectangular solar panels.

ORBITER

MISSIONS TO THE **SUN**

TO LEARN MORE ABOUT THE SUN, WE NEED TO OBSERVE IT FROM ABOVE EARTH'S ATMOSPHERE. OVER THE YEARS, A NUMBER OF COUNTRIES HAVE LAUNCHED MISSIONS TO PEER AT OUR NEAREST STAR FROM SPACE.

A suite of spacecraft have revolutionized our understanding of all things solar, from the Sun's dynamic magnetic field to the way its solar wind interacts with the planets. Today, the Sun is under constant surveillance from space, 24 hours a day. Such close scrutiny is important—the more we understand about the Sun, the better we are able to predict when a potentially dangerous solar storm will be unleashed in our direction.

▽ **Genesis**
This was a mission to capture material from the solar wind. It did so in 2005, making it the first NASA mission to return a space sample to Earth since Apollo 17 astronauts returned with lunar rocks in 1972. The mission wasn't without incident, however; Genesis crash-landed, but the sample was successfully salvaged.

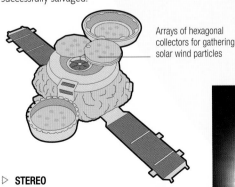

Arrays of hexagonal collectors for gathering solar wind particles

Optical telescope

UV imaging spectrometer

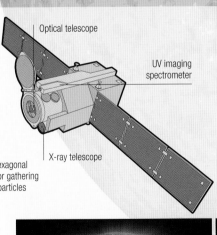

X-ray telescope

◁ **Hinode**
Armed with instruments that can see visible light, ultraviolet, and X-rays, Hinode monitors the Sun's magnetic activity, providing valuable insights into sunspots and solar flares. It is also investigating how magnetic energy moves from the photosphere to the corona.

▷ **SDO**
The Solar Dynamics Observatory (SDO) beams back high-definition images of the Sun every ten seconds. SDO was designed to study space weather. This SDO image shows loops of hot gas following the Sun's twisted magnetic field.

▷ **STEREO**
The Solar Terrestrial Relations Observatory (STEREO) consists of two identical craft orbiting the Sun in tandem. They take stereo (3-D) images of the Sun and study events such as coronal mass ejections. In 2007, STEREO photographed the Moon passing in front of the Sun (right)— a lunar transit not visible from Earth.

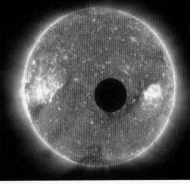

ROCKY **WORLDS**

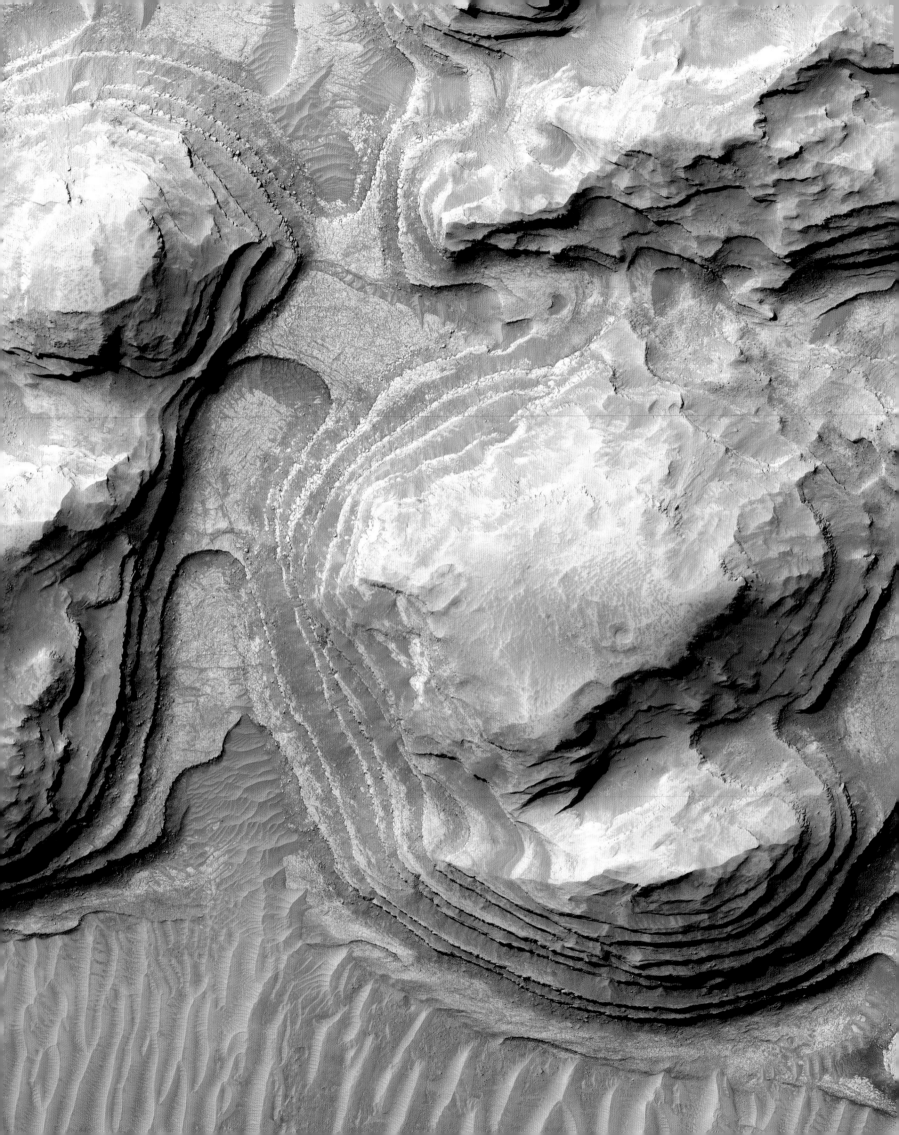

The four planets closest to the Sun—Mercury, Venus, Earth, and Mars—are a diverse group. Our own planet is the biggest of the foursome, with Venus a near-twin in size. Formed in a cloud of solar dust and gas, battered by collisions, and reduced to rocky balls by heat and gravity, they all began life in the same way. But over time, they became very different. Mercury, the smallest of the

NEIGHBORING **WORLDS**

inner planets, is closest to the Sun and has little atmosphere to protect it from our star's searing heat. The craters on Mercury's hot, dark surface— scars of long bombardment by cosmic material— resemble the craters on the Moon. While all the inner planets are thought to have an iron core, Mercury's is unusually large, perhaps because the young planet was stripped of its outer layers in a catastrophic collision. Venus, though beautiful in the twilight sky, is obscured by choking clouds of sulfuric acid and may be actively volcanic. The hottest planet in the solar system, it is the victim of a runaway greenhouse effect. Mars is the coldest rocky planet. Once, it may have been warm, with rivers flowing on its surface, but now the planet is an arid wasteland, its remaining water locked in frost and ice. Earth is a world between extremes. The right distance from the Sun for water to exist in liquid form on the surface, our planet has vast oceans, an oxygen-rich atmosphere, and a huge diversity of life forms.

◁ **Traces of the past**
More has been discovered about Mars than any other of our close planetary neighbors. NASA's Mars Reconnaissance Orbiter opened a window into the Red Planet's past when it captured this image. The layers of rock within a crater in the Arabia Terra region, caused by fluctuations in the amount of sediment deposited, reveal that the Martian climate has changed repeatedly over millions of years.

MERCURY

THE NEAREST PLANET TO THE SUN, MERCURY CAN BE SEEN CLEARLY FROM EARTH ONLY FOR SHORT PERIODS EACH YEAR. IT IS USUALLY VISIBLE IN SPRING AND FALL AS A BRIGHT GLIMMER JUST ABOVE THE HORIZON AT DUSK AND DAWN.

Mercury is a tiny, dense, deeply cratered world, so close to the Sun that it is continually scorched and blasted by solar emissions. Temperatures during its long daytime period reach a roasting 800°F (430°C)—hot enough to melt lead. Yet because there is only a thin atmosphere, heat escapes quickly enough for nighttime temperatures to drop to −290°F (−180°C). No other planet in the solar system experiences such extremes.

Mercury spins on its axis very slowly, with one rotation taking almost 59 Earth days. Yet it is also the fastest-orbiting of all the planets, completing its circuit of the Sun in just 88 days. By the time the sunny side begins to turn away, the whole planet has been swept around to face the Sun from the opposite side. So once the Sun comes up, it takes a long time to set again—there are 176 days from one sunrise to the next, during which time the planet orbits the Sun more than twice. Despite the long Mercurial days, Mercury's sky always looks black due to the incredibly thin atmosphere, which is not thick enough to reflect light.

Mercury has a **maximum orbital speed** of 30 miles (50 km) per second.

MERCURY DATA

Average diameter	3,032 miles (4,879 km)
Mass (Earth = 1)	0.055
Gravity at equator (Earth = 1)	0.38
Mean distance from Sun (Earth = 1)	0.38
Axial tilt	0.01°
Rotation period (day)	58.6 Earth days
Orbital period (year)	87.97 Earth days
Minimum temperature	−290°F (−180°C)
Maximum temperature	800°F (430°C)
Moons	0

▷ **Northern hemisphere**
At the north pole are huge expanses of smooth plains some 1.5 million square miles (4 million km²) in area—half the size of the United States. A feature known as Goethe Basin contains ghost craters that have been flooded and buried by lava flows.

▷ **Western hemisphere**
Until NASA's MESSENGER made its first flyby in 2008, this half of Mercury was unknown. The flyby revealed that 40 percent of the planet's surface is covered by smooth, volcanic plains. Its crust is more like that of Mars than the Moon's, despite the similar appearance.

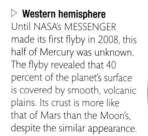

▷ **Southern hemisphere**
Around Mercury's poles, in shadowy places that are permanently shielded from the Sun's heat, such as in Chao Meng-Fu crater, NASA's MESSENGER spacecraft found radar-bright patches that could be a mix of frozen water and organic materials.

Brahms Crater is a large, old, complex crater with a prominent central peak over 1.9 miles (3 km) high.

Tyagaraja Crater

Named after the Hungarian composer, Bartok Crater is about 45 miles (73 km) across with a central peak.

Michelangelo Crater

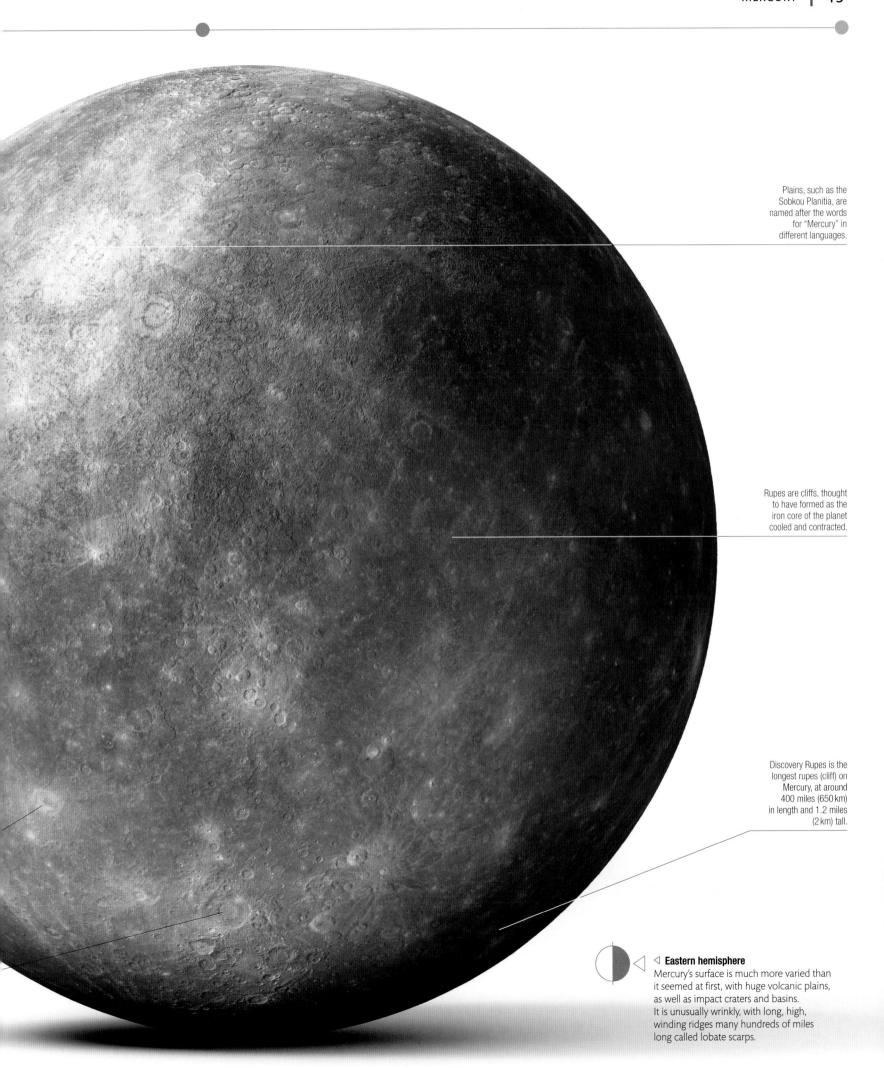

Plains, such as the Sobkou Planitia, are named after the words for "Mercury" in different languages.

Rupes are cliffs, thought to have formed as the iron core of the planet cooled and contracted.

Discovery Rupes is the longest rupes (cliff) on Mercury, at around 400 miles (650 km) in length and 1.2 miles (2 km) tall.

◁ **Eastern hemisphere**
Mercury's surface is much more varied than it seemed at first, with huge volcanic plains, as well as impact craters and basins. It is unusually wrinkly, with long, high, winding ridges many hundreds of miles long called lobate scarps.

MERCURY STRUCTURE

MERCURY IS ONE OF THE FOUR SOLID, TERRESTRIAL PLANETS MADE OF ROCK AND METAL. IT IS SMALLER THAN SOME MOONS, YET MORE DENSE THAN ANY OTHER PLANET APART FROM EARTH.

For such a small planet to be so dense, Mercury must have a very large iron core. This suggests that it has lost rock from its outer layers. If so, one explanation might be that early in its history Mercury was struck by a planetesimal, one of the many protoplanets that whirled through the solar system as it formed. The devastating impact caused by this planetesimal—which was probably about one-sixth the size of the planet itself—blasted away much of Mercury's rocky exterior.

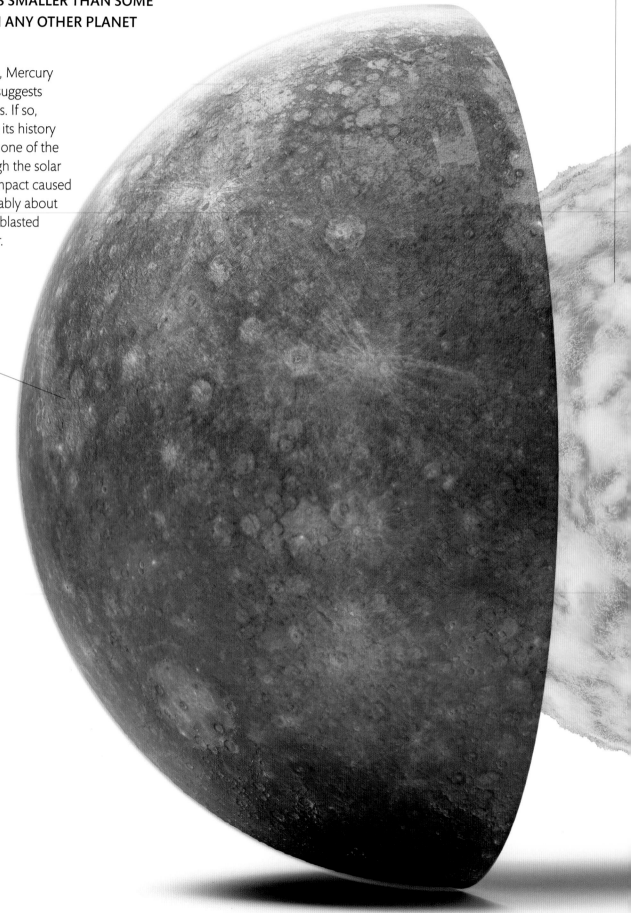

Liquid core
The iron core is 2,240 miles (3,600 km) wide. Researchers discovered its liquid state by bouncing radio waves off the planet to measure how much it wobbles as it rotates. A solid core would have rigid rotation, but Mercury's wobbly spin indicates liquid sloshing around inside.

In places, Mercury's surface is tinged yellow by sulfur. There is more of this element in Mercury than any other planet.

Mercury's **iron core** takes up 61% of its volume— Earth's takes up only 17%.

▷ **Layer by layer**
Data from MESSENGER and the forthcoming BepiColombo missions may support the theory that Mercury's unusual structure—with a vast core and narrow outer layers—is due to the loss of rock when it was struck by a planetesimal. Another explanation is that rock vaporized in the hot conditions of the early solar system, before the Sun stabilized. A third idea is that rock was stripped away by the drag of the solar nebula early on.

Mantle
The semi-molten rock of the mantle layer is around 400 miles (600 km) thick. Like Earth's mantle, Mercury's consists of silicate rocks and is far less dense than the planet's core. It is a relatively thin layer, accounting for approximately 20 percent of the planet's radius.

Crust
Mercury's crust is likely to be made of magnesium-rich basalt and other silicate rocks. It is 70–190 miles (100–300 km) thick. The surface is stable, with no moving plates, meaning that features such as impact craters remain undisturbed on the planet for billions of years.

▽ **Atmosphere**
Mercury has an atmosphere, but it is very thin, which is why it is called an exosphere. Some astronomers believe Mercury once had a thick atmosphere like Earth, but because the planet is small, its gravity could not stop the atmosphere from being blown away by the solar wind. The gases that remain include hydrogen, oxygen, helium, water vapor, sodium, and potassium.

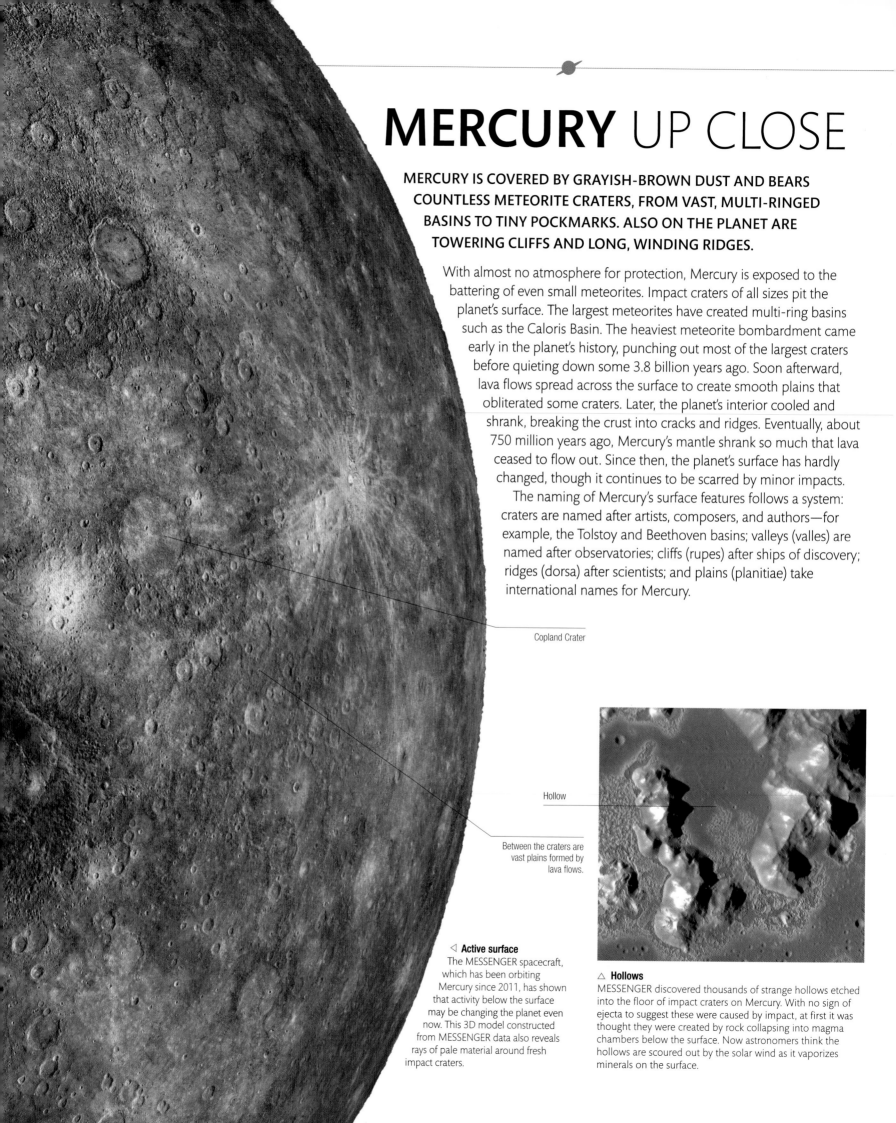

MERCURY UP CLOSE

MERCURY IS COVERED BY GRAYISH-BROWN DUST AND BEARS COUNTLESS METEORITE CRATERS, FROM VAST, MULTI-RINGED BASINS TO TINY POCKMARKS. ALSO ON THE PLANET ARE TOWERING CLIFFS AND LONG, WINDING RIDGES.

With almost no atmosphere for protection, Mercury is exposed to the battering of even small meteorites. Impact craters of all sizes pit the planet's surface. The largest meteorites have created multi-ring basins such as the Caloris Basin. The heaviest meteorite bombardment came early in the planet's history, punching out most of the largest craters before quieting down some 3.8 billion years ago. Soon afterward, lava flows spread across the surface to create smooth plains that obliterated some craters. Later, the planet's interior cooled and shrank, breaking the crust into cracks and ridges. Eventually, about 750 million years ago, Mercury's mantle shrank so much that lava ceased to flow out. Since then, the planet's surface has hardly changed, though it continues to be scarred by minor impacts.

The naming of Mercury's surface features follows a system: craters are named after artists, composers, and authors—for example, the Tolstoy and Beethoven basins; valleys (valles) are named after observatories; cliffs (rupes) after ships of discovery; ridges (dorsa) after scientists; and plains (planitiae) take international names for Mercury.

Copland Crater

Hollow

Between the craters are vast plains formed by lava flows.

◁ **Active surface**
The MESSENGER spacecraft, which has been orbiting Mercury since 2011, has shown that activity below the surface may be changing the planet even now. This 3D model constructed from MESSENGER data also reveals rays of pale material around fresh impact craters.

△ **Hollows**
MESSENGER discovered thousands of strange hollows etched into the floor of impact craters on Mercury. With no sign of ejecta to suggest these were caused by impact, at first it was thought they were created by rock collapsing into magma chambers below the surface. Now astronomers think the hollows are scoured out by the solar wind as it vaporizes minerals on the surface.

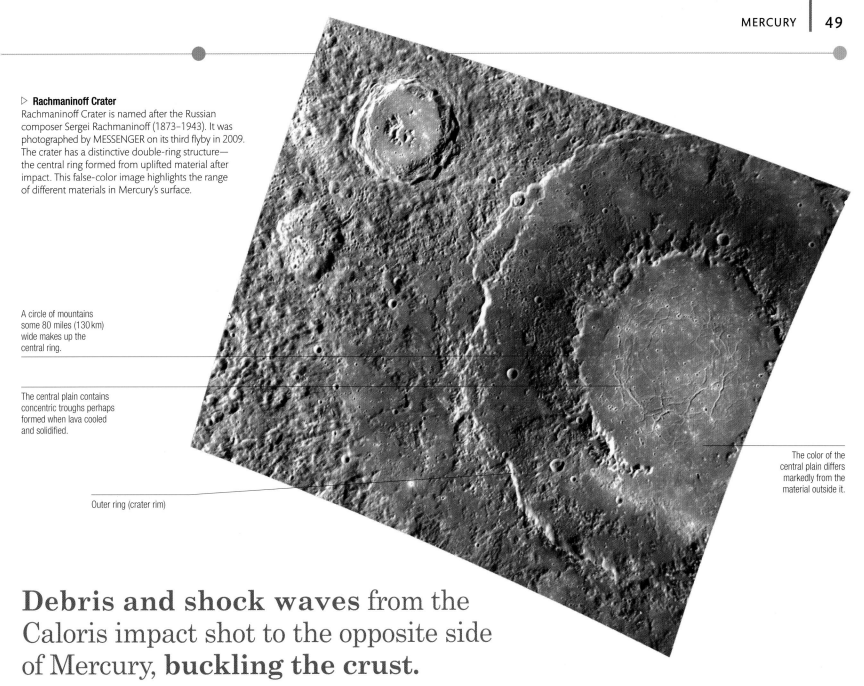

▷ **Rachmaninoff Crater**
Rachmaninoff Crater is named after the Russian composer Sergei Rachmaninoff (1873–1943). It was photographed by MESSENGER on its third flyby in 2009. The crater has a distinctive double-ring structure—the central ring formed from uplifted material after impact. This false-color image highlights the range of different materials in Mercury's surface.

A circle of mountains some 80 miles (130 km) wide makes up the central ring.

The central plain contains concentric troughs perhaps formed when lava cooled and solidified.

Outer ring (crater rim)

The color of the central plain differs markedly from the material outside it.

Debris and shock waves from the Caloris impact shot to the opposite side of Mercury, **buckling the crust.**

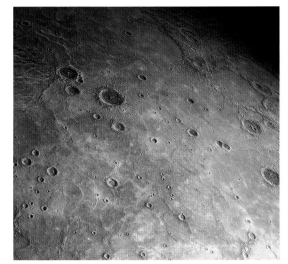

△ **Plains**
Much of Mercury's surface comprises vast, empty plains, or planitiae. Most are ancient and heavily scarred with craters. Other, gently rolling plains called intercrater plains are pitted with only small craters; these plains probably formed when lava buried older terrain. There are even younger smooth lava plains, like those around Caloris Basin, laid down too recently to show many craters.

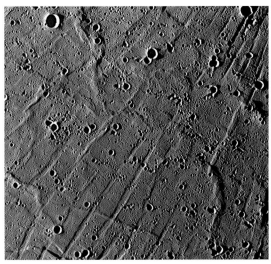

△ **Spider troughs**
One of the most striking discoveries of MESSENGER's first Mercury flyby in 2008 was this extraordinary series of troughs, or fossae. They radiate around the center of Caloris Basin like the threads of a spiderweb, which is why the feature was originally called the Spider. Now it is officially named Pantheon Fossae, after its resemblance to the sunken panels that radiate around the dome of the Pantheon in Rome.

△ **Basins**
The large yellow patch in this false-color image is Caloris Basin, one of the largest impact craters in the solar system, some 800 miles (1,300 km) across and now partially filled by a lava flow. It probably formed when a huge asteroid crashed into Mercury early in its history, creating a vast basin, sending ejecta 600 miles (1,000 km) from the crater's rim, and fracturing rocks around it into troughs.

MERCURY MAPPED

NASA's MESSENGER spacecraft, which has been orbiting Mercury since 2011, has built up a global map of the planet. The probe's imaging systems continue to seek details of regions in permanent shadow.

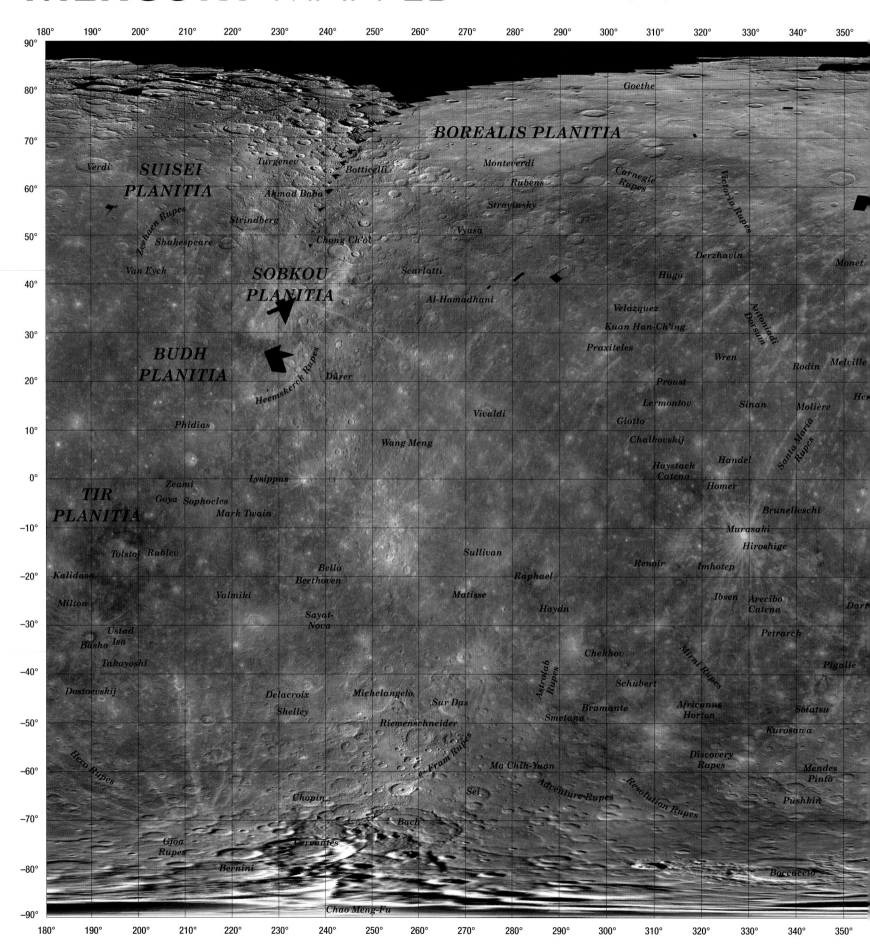

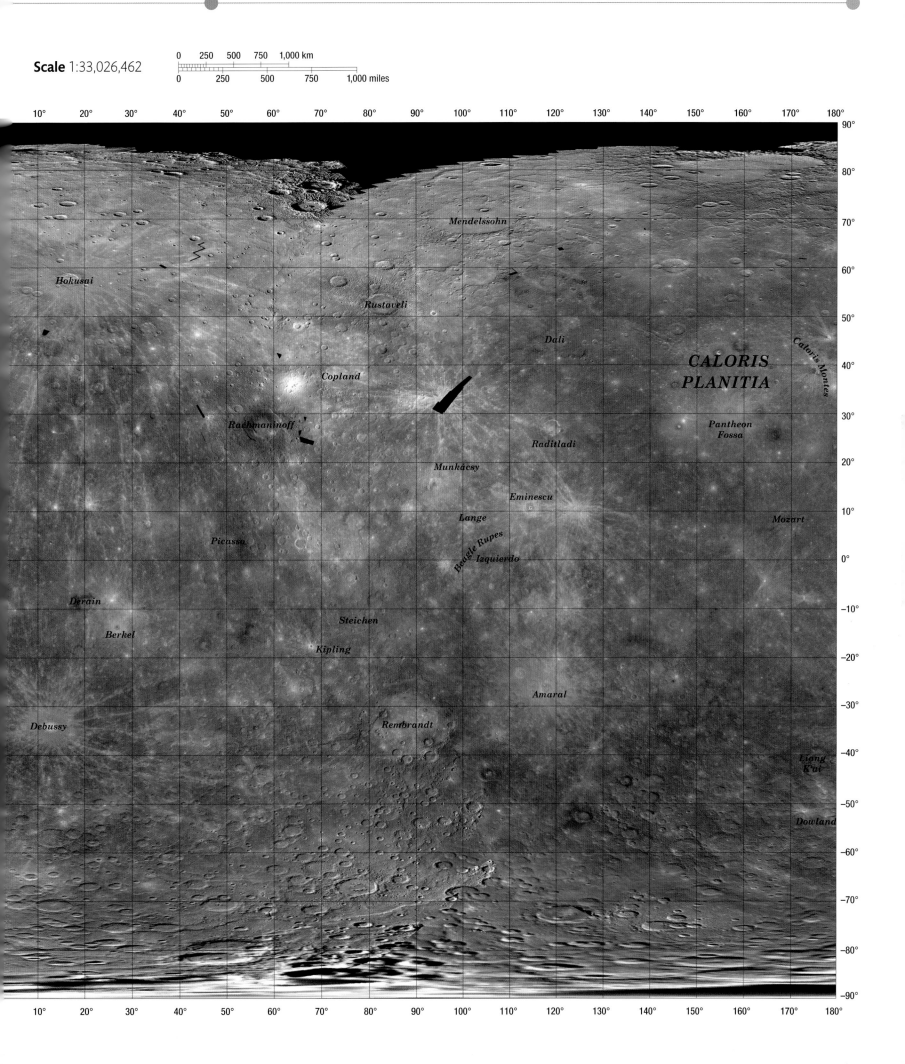

Scale 1:33,026,462

| 0 | 250 | 500 | 750 | 1,000 km |

| 0 | 250 | 500 | 750 | 1,000 miles |

Hokusai

Mendelssohn

Rustaveli

Dali

CALORIS PLANITIA

Caloris Montes

Copland

Rachmaninoff

Pantheon Fossa

Raditladi

Munkácsy

Eminescu

Lange

Mozart

Picasso

Beagle Rupes

Izquierdo

Derain

Steichen

Berkel

Kipling

Amaral

Debussy

Rembrandt

Liang K'ai

Dowland

DESTINATION CARNEGIE RUPES

AMONG MERCURY'S MOST DISTINCTIVE FEATURES ARE HUNDREDS OF LONG CLIFFS, KNOWN AS RUPES, THAT WIND FOR MILES ACROSS THE LANDSCAPE, CUTTING THROUGH ANCIENT CRATERS. CARNEGIE RUPES IS 166 MILES (267 KM) LONG AND REACHES 6,600 FT (2,000 M) TALL IN PLACES.

Carnegie Rupes is located in Mercury's northern hemisphere. Like all the planet's rupes, it is the steep face of a long ridge that slopes gently away on the opposite side. Geologists describe such cliffs as lobate scarps, because of their curved shapes. Rupes are thought to have formed at least 3 billion years ago as the planet contracted, cracking the surface. The shrinkage was slight, but enough to thrust up blocks of crust along the cracks, or faults. Some scientists believe the planet's contraction was the result of cooling. Others think the drag of the Sun's gravity slowed Mercury's rotation, a phenomenon known as tidal despinning, reducing the equatorial bulge. Following an established theme, rupes are named after ships of discovery—in the case of Carnegie, a research vessel that mapped the Earth's magnetic field in the early 20th century.

Artist's impression based on images from MESSENGER spacecraft

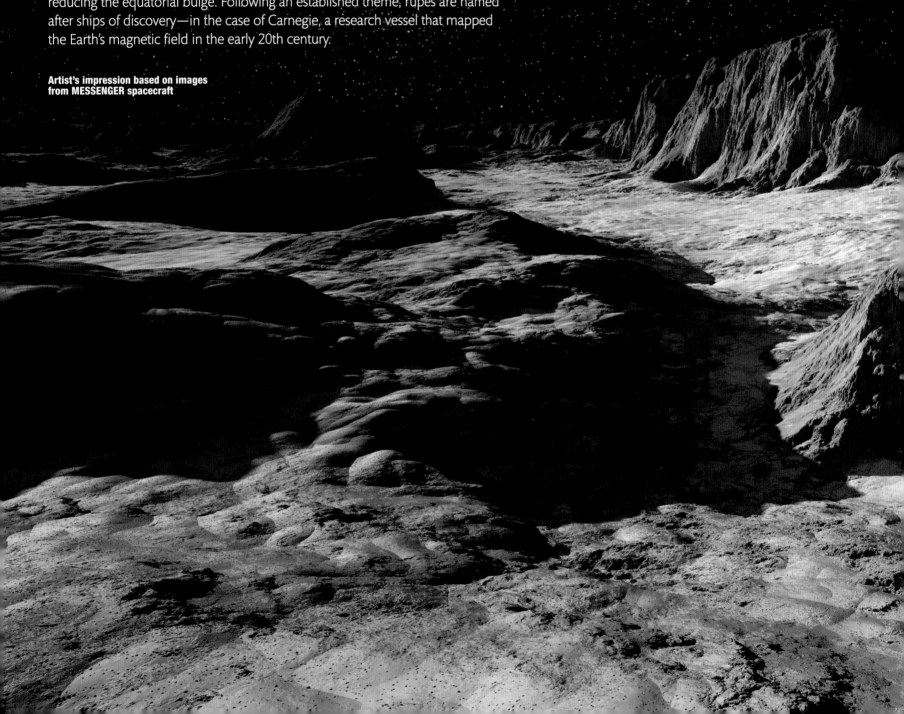

LOCATION

Latitude 59°N; longitude 53°E

TOPOGRAPHY

This image taken by NASA's MESSENGER spacecraft looks northwest across Carnegie Rupes, as it slices across an unnamed crater about 60 miles (100 km) in diameter. The colors indicate elevation, with low areas in blue and high areas in red. The line of the cliff shows a sharp change in elevation—over 1.4 miles (2 km) in places.

Unnamed crater

Impact crater

Carnegie Rupes

THE WINGED MESSENGER

VISIBLE TO THE NAKED EYE AROUND SUNRISE AND SUNSET, MERCURY WAS WELL KNOWN IN ANCIENT TIMES. IT IS THE FASTEST-MOVING OF ALL THE PLANETS AND WAS NAMED AFTER THE WINGED MESSENGER OF ROMAN MYTHOLOGY.

Mercury is so small and distant—and circles so close to the Sun—that it is difficult to see from Earth. Consequently, very little was known about the planet until comparatively recently. Although Mercury was first observed through a telescope by Italian scientist Galileo Galilei in the early 17th century, it was not until the late 20th century that telescopes with sufficient power to resolve surface details were developed.

The big breakthrough in understanding the Sun's closest neighbor came when spacecraft began to beam back close-ups from the planet. The first was Mariner 10, which made flybys in 1974 and 1975. This was followed, after a gap of more than 30 years, by the MESSENGER spacecraft, which remains in orbit today.

Hermes, Greek messenger god

1000 BCE

Babylon tablets
The earliest known record of the observation of Mercury is on the Mul.Apin tablets—catalogs of celestial bodies from ancient Babylon. The Babylonians call the planet Nabu, after their messenger god.

c. 350 BCE

Apollo and Hermes
At first, the ancient Greeks believe that Mercury is two planets: they call it Apollo when it appears in the morning sky, and Hermes when they see it after sunset. In the 4th century BCE, they realize it is actually a single planet and name it Hermes.

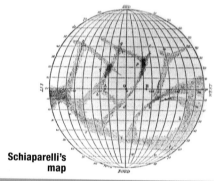

Schiaparelli's map

1962

Mercury by radar
Soviet scientists led by Vladimir Kotelnikov at Moscow's Institute of Radio-engineering and Electronics become the first to bounce a radar signal off Mercury and receive its echo, enabling them to make the first radar observations of the planet.

1880s

Schiaparelli's map
Italian astronomer Giovanni Schiaparelli observes Mercury and creates the most accurate map yet. He believes, wrongly, that Mercury is locked in its orbit—the same side always faces the Sun, and the planet takes 88 days to orbit the Sun and make one rotation.

1800–1808

Clouds on Mercury
German astronomer Johann Schröter claims, wrongly, to have seen features such as clouds and mountains on Mercury. Using Schröter's drawings, the astronomer Friedrich Bessel estimates (wrongly) that Mercury spins at the same speed as Earth and tilts strongly.

Arecibo radio telescope

Mariner 10 mosaic of Mercury

1965

Rotation speed
American astronomers Gordon Pettengill and Rolf Dyce use the radio telescope dish at Arecibo, Puerto Rico, to measure Mercury's spin rate. From radar pulses reflected by the planet's surface, they calculate that Mercury's rotation is not tidally locked, as Schiaparelli had thought, but takes just 59 days—about two-thirds of its orbital period of 88 days. Most of Mercury has now been mapped by the Arecibo dish.

1975

Mariner 10
NASA's Mariner 10 is the first spacecraft to visit Mercury and photograph it up close. In three separate flybys, starting March 29, 1975, Mariner 10 images almost half of the planet's surface, revealing a landscape similar to that of the Moon.

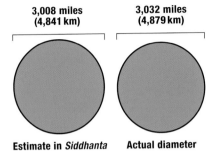

**3,008 miles
(4,841 km)**

**3,032 miles
(4,879 km)**

Estimate in *Siddhanta*

Actual diameter

5th century CE

Mercury's diameter
By unknown means and without a
telescope, an Indian astronomer
estimates Mercury's diameter with
99 percent accuracy—an astonishing
achievement or a lucky guess. The result
is recorded in the book *Surya Siddhanta*.

1611

Galileo's observations
Galileo makes the first observations of Mercury
through a telescope. He guesses it is a planet,
but his telescope is not powerful enough to
reveal that Mercury has phases, just like Venus
and the Moon, and that these phases depend
on how much we see of Mercury's sunlit half.

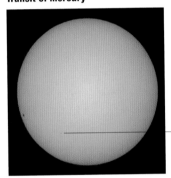

**Galileo
Galilei**

Transit of Mercury

Phases of Mercury

Mercury passes in
front of the Sun

1737

Occultation by Venus
Occultations—when one planet passes in
front of another, as seen from Earth—are
rare events. English astronomer John Bevis
sees the occultation of Mercury by Venus
on May 28—the only time in history that
this has been witnessed.

1639

Phases
Italian astronomer Giovanni Zupi observes
through a powerful telescope that Mercury
has phases similar to those of Earth's Moon.
This proves that Mercury orbits the Sun,
revealing varying amounts of its surface
as it catches the Sun at different angles.

1631

Gassendi observes transit
French astronomer Pierre Gassendi sees
Mercury pass in front of the Sun. This is the
first time the transit of a planet has been
observed through a telescope. It enables
Gassendi to make the first reliable
measurement of a planet's diameter.

**MESSENGER image
of Mercury**

**MESSENGER
launch**

2002

Skinakas Basin
Astronomers at the Skinakas Astrophysical
Observatory in Crete believe that they have
found a giant crater missed by Mariner 10.
Subsequently, the MESSENGER spacecraft
shows that the crater, dubbed the Skinakas
Basin, is in fact an illusion.

2008

MESSENGER flyby
Launched on August 3, 2004, NASA's
MESSENGER makes the first of its three
flybys of Mercury in January 2008. During
the flybys, MESSENGER maps most of the
planet's surface in color and studies the
atmosphere and magnetosphere.

2011

MESSENGER in orbit
On March 18, MESSENGER goes
into long-term orbit around Mercury.
The craft completes its mapping of
Mercury, discovers water at the planet's
north pole, and continues to send valuable
data about Mercury back to Earth.

LAUNCH   EARTH ORBIT   JOURNEY TO MERCURY

1973   Mariner 10
2004   MESSENGER
Planned   BepiColombo

MISSIONS TO **MERCURY**

MERCURY IS THE LEAST EXPLORED OF THE ROCKY PLANETS, VISITED BY JUST TWO MISSIONS TO DATE: MARINER 10 IN THE MID-1970S AND THE MORE RECENT MESSENGER SPACECRAFT, WHICH STUDIED MERCURY FROM ORBIT.

One reason for the lack of missions to Mercury is the sheer technical difficulty. Spacecraft have to travel extremely fast to get to Mercury, and when they reach the planet, they must suddenly slow down enough to get into orbit just as the Sun's gravity is trying to accelerate them even more. In addition, the Sun's pull is so strong near Mercury that orbits around the planet are unstable, and proximity to the Sun makes it hard for spacecraft to maintain a stable temperature. Nonetheless, Mariner and MESSENGER have reached the planet successfully and studied its features and properties. A third major mission, the joint European–Japanese BepiColombo, may reveal more about this intriguing planet.

KEY

NASA (USA)

JAXA (Japan)

ESA (Europe)

Joint ESA/JAXA mission

Destination

▽ Mariner 10

Mariner 10's first Mercury flyby took place on March 29, 1974. Because getting a craft into orbit around the planet was so difficult, Mariner 10 was designed to orbit the Sun instead, enabling it to fly past Mercury three times. These flybys revealed a highly cratered surface and, to the great surprise of astronomers, a magnetic field around the planet.

Mariner 10 image of Mercury's cratered surface

▷ MESSENGER

MESSENGER (Mercury surface, space environment, geochemistry, and ranging) left Earth in 2004 but took over six years to achieve orbit around Mercury—the first craft ever to do so. On March 29, 2011, it sent the first photo from Mercury orbit. Since then, MESSENGER's cameras and other instruments have returned a flood of data about the planet. Its investigations have discovered water ice and organic compounds in shadowed craters near Mercury's north pole.

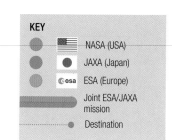

Magnetometer for studying Mercury's magnetic field

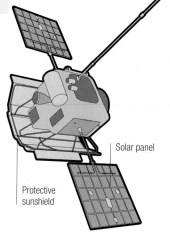

Solar panel

Protective sunshield

▽ MESSENGER's journey

MESSENGER had to circle the Sun seven times to get into its orbit around Mercury. It passed Earth a year after launch and then Venus twice, using both planets' gravity to slingshot itself onward. It then made three flybys of Mercury to slow down before entering orbit. Its orbit is very eccentric: its lowest point is just 124 miles (200 km) above the surface, while the highest is at an altitude of over 9,300 miles (15,000 km).

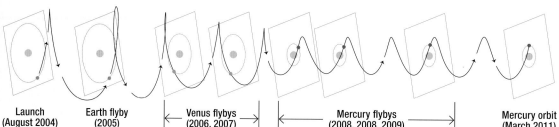

Launch (August 2004)   Earth flyby (2005)   Venus flybys (2006, 2007)   Mercury flybys (2008, 2008, 2009)   Mercury orbit (March 2011)

FLYBY   ORBITER

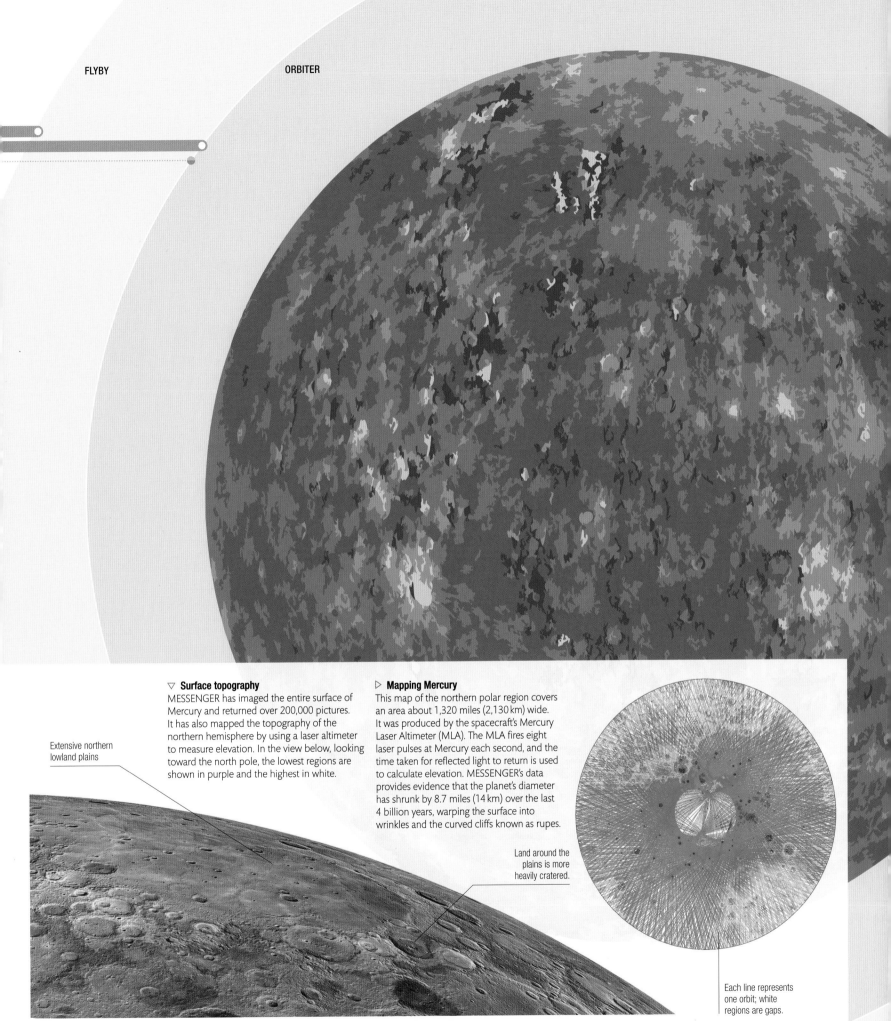

Extensive northern
lowland plains

▽ **Surface topography**

MESSENGER has imaged the entire surface of
Mercury and returned over 200,000 pictures.
It has also mapped the topography of the
northern hemisphere by using a laser altimeter
to measure elevation. In the view below, looking
toward the north pole, the lowest regions are
shown in purple and the highest in white.

▷ **Mapping Mercury**

This map of the northern polar region covers
an area about 1,320 miles (2,130 km) wide.
It was produced by the spacecraft's Mercury
Laser Altimeter (MLA). The MLA fires eight
laser pulses at Mercury each second, and the
time taken for reflected light to return is used
to calculate elevation. MESSENGER's data
provides evidence that the planet's diameter
has shrunk by 8.7 miles (14 km) over the last
4 billion years, warping the surface into
wrinkles and the curved cliffs known as rupes.

Land around the
plains is more
heavily cratered.

Each line represents
one orbit; white
regions are gaps.

VENUS

VENUS IS THE SECOND PLANET FROM THE SUN AND OUR NEAREST NEIGHBOR. IT IS A ROCKY WORLD, SIMILAR IN SIZE TO EARTH, BUT THE TWO PLANETS COULD HARDLY BE MORE DIFFERENT IN CHARACTER.

From Earth, Venus is visible at dusk or dawn, just above the horizon, shining brighter than anything in the sky but the Sun and Moon. Through a telescope, the planet can be seen passing through phases, like our Moon's, from crescent to nearly full as it orbits the Sun, revealing different amounts of its sunlit side. Seen from space, Venus is swathed in pale yellow clouds that hide its surface, but radars and probes carried by spacecraft have looked beneath to discover a hellish world.

Venusian clouds are flooded with droplets of sulfuric acid, and the planet's thick atmosphere weighs down so heavily that the pressure on the surface is 90 times that on Earth. The surface comprises either flat, barren rock, or volcanoes, both dormant and possibly active. Beneath a deep orange sky, a runaway greenhouse effect traps the Sun's heat, sending temperatures soaring to a blistering 880°F (470°C) and making Venus the hottest planet in the solar system.

> Venus rotates in the **opposite direction** to most of the planets. It turns so **slowly** on its axis that its **day lasts longer** than its year.

VENUS DATA

Average diameter	7,520 miles (12,104 km)
Mass (Earth = 1)	0.82
Gravity at equator (Earth = 1)	0.9
Mean distance from Sun (Earth = 1)	0.72
Axial tilt	2.6°
Rotation period (day)	243 Earth days
Orbital period (year)	224.7 Earth days
Average surface temperature	880°F (470°C)
Moons	0

▷ **Northern hemisphere**
The north pole is a scorched landscape of bare rock and rubble. Nearby are the rugged ridges of Atalanta Planitia, and also Ishtar Terra, Venus's highest mountain range.

▷ **Uplands**
Besides the three main highland regions, or terrae, Venus has 20 or so smaller upland areas called regio. These include Alpha Regio, the bright patch seen bottom center, which is highly deformed and probably of ancient origin.

▷ **Southern hemisphere**
Venus's southern regions are as hot and bare as the north. Lada Terra, the second biggest of Venus's three raised land masses, or terrae, is near the south pole. It has more volcanic upwellings, or coronae, than the other terrae.

The 57-mile- (92-km-) wide double-ring crater Greenaway has a rough, radar-bright base, suggesting volcanic activity after the impact that formed it.

The Diana Chasma gorge, four times as long as the Grand Canyon, contains perhaps the deepest point on the planet, where temperatures soar to 930°F (500°C).

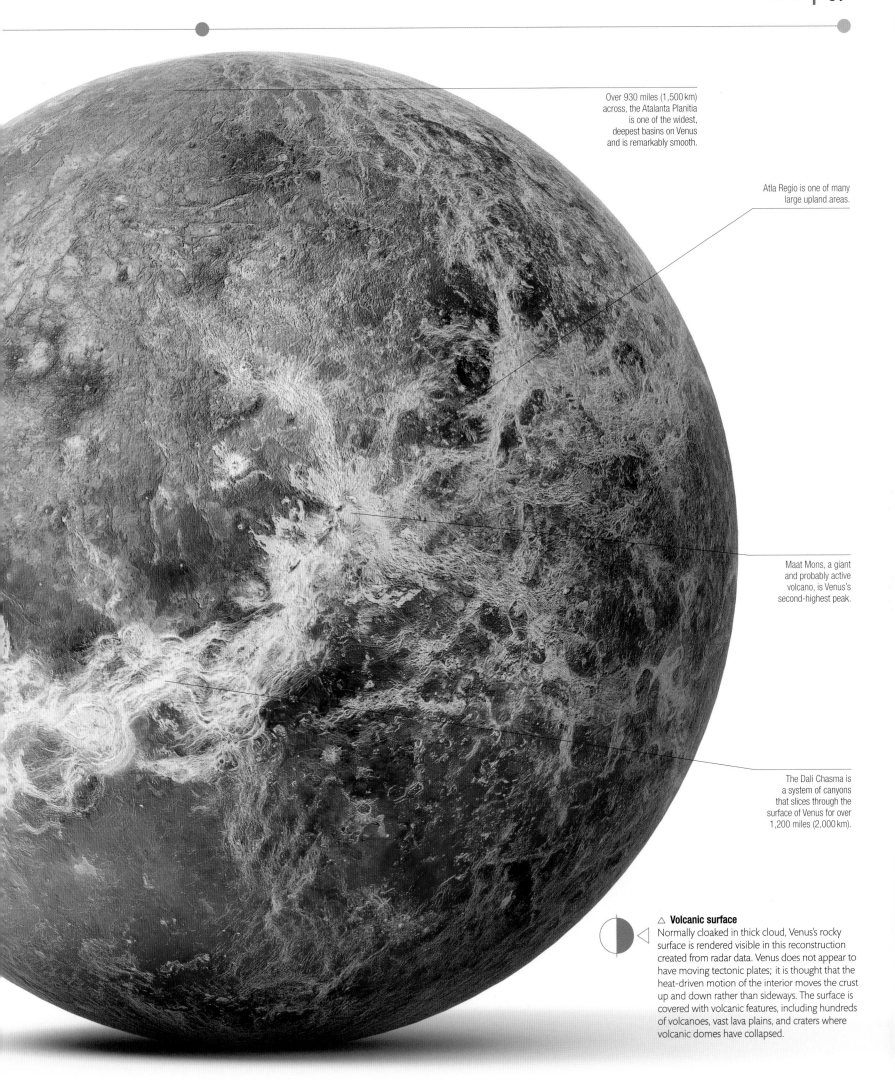

Over 930 miles (1,500 km) across, the Atalanta Planitia is one of the widest, deepest basins on Venus and is remarkably smooth.

Atla Regio is one of many large upland areas.

Maat Mons, a giant and probably active volcano, is Venus's second-highest peak.

The Dali Chasma is a system of canyons that slices through the surface of Venus for over 1,200 miles (2,000 km).

△ **Volcanic surface**
Normally cloaked in thick cloud, Venus's rocky surface is rendered visible in this reconstruction created from radar data. Venus does not appear to have moving tectonic plates; it is thought that the heat-driven motion of the interior moves the crust up and down rather than sideways. The surface is covered with volcanic features, including hundreds of volcanoes, vast lava plains, and craters where volcanic domes have collapsed.

VENUS STRUCTURE

ALTHOUGH FORMED FROM THE SAME SOLAR DEBRIS AS EARTH, VENUS IS UNDENIABLY DIFFERENT FROM OUR PLANET ON THE OUTSIDE. HOWEVER, SCIENTISTS BELIEVE THAT THE INTERIORS OF THE TWO WORLDS MAY BE SIMILAR.

Almost equal in size and density to Earth, Venus probably has much the same internal structure and chemistry. At the heart of the planet there is thought to be a metal core with a solid center and a molten outer layer. Surrounding this is a deep mantle of hot rock and a thin, brittle crust that shows abundant evidence of volcanic activity.

Although Venus has a metal core like Earth's, it has no detectable magnetic field. This may be because it rotates too slowly—taking eight months to turn once—to produce the circulations within the outer core that would generate a dynamo effect.

Venus has the thickest, most dense atmosphere of all the rocky planets. Its air is 96.5 percent carbon dioxide and contains small amounts of other chemicals, including sulfuric acid; a thick blanket of sulfuric acid clouds covers the entire planet.

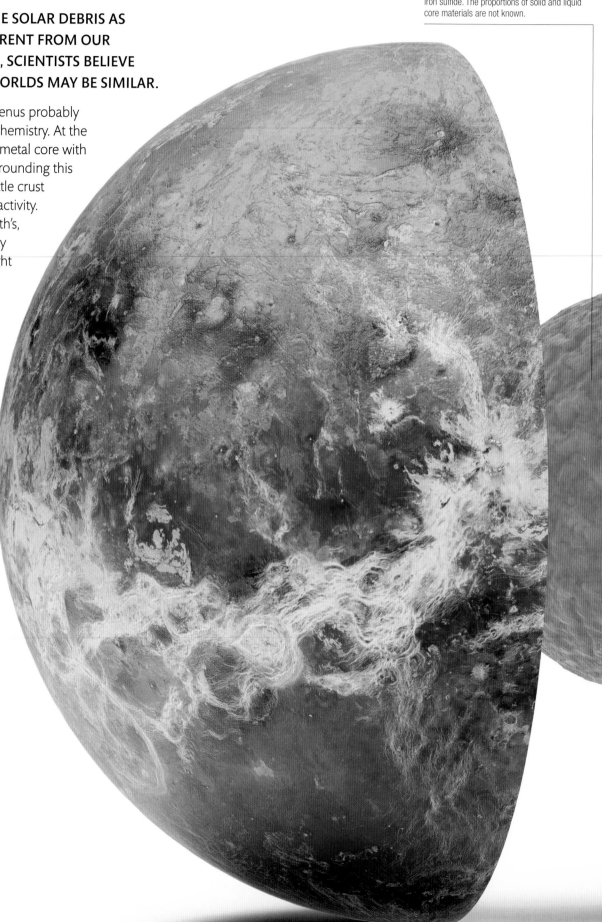

Core
The center of Venus is a core of mostly solid iron with perhaps a trace of sulfur. There is probably also a semi-liquid outer core of molten iron sulfide. The proportions of solid and liquid core materials are not known.

Venus on the inside
While Venus's core is likely to be mostly iron and nickel—the same as Earth's—its slightly lower density suggests there may be a lighter element, such as sulfur, in there too. Again, like Earth, Venus has a mantle of rock that is made fluid enough by interior heat to creep slowly up and down in convection currents. These currents push molten rock through the crust to create volcanoes on the surface. Layers in this 3D model are not shown to scale: the crust, surface relief, and atmosphere are exaggerated for clarity.

Mantle
The mantle is hot, plastic rock, churned by convection currents that move slowly over thousands of years. Similar in composition to Earth's, Venus's mantle may contain rocks rich in iron and magnesium.

Crust
The thin outer layer above the mantle is made of basalt and other silicate rocks. In places the surface of the crust bulges outward, lifted by tremendous volcanic forces in the upper part of the mantle.

Atmosphere
Venus's unbroken cloud deck extends from 20 to 55 miles (32 to 90km) above the surface. At surface level, the carbon dioxide "air" is clear and slow moving, but it is so dense that it acts like a liquid, forming a kind of sea and dragging dust and stones across the ground as it flows.

The clouds are made from droplets, and perhaps solid crystals, of sulfuric acid.

The lower layer of the atmosphere is clear, dense, and extremely hot.

A thin hazy layer lies between the bottom of the clouds and the lower atmosphere.

The atmosphere above the cloud deck thins out into space.

VENUS UP CLOSE

THE SURFACE OF THIS INHOSPITABLE WORLD IS ALMOST ENTIRELY VOLCANIC. MORE THAN 1,600 VOLCANOES HAVE BEEN IDENTIFIED ON VENUS, A GREATER NUMBER THAN ON ANY OTHER PLANET IN THE SOLAR SYSTEM.

The first detailed maps of Venus were made in the early 1990s, when the Magellan spacecraft used radar to penetrate the thick cloud hiding the planet. What Magellan's images revealed was a world covered in volcanoes.

The Venusian landscape is characterized by vast plains covered by lava flows, and mountain or highland regions deformed by geological activity. No ongoing eruptions have yet been confirmed, but there are many signs of recent volcanic activity, including ash flows, impact craters partially covered by lava flows, and fluctuating levels of sulfur dioxide in the atmosphere that could be caused by eruptions.

The surface of the planet is young. It is thought that Venus was entirely resurfaced by a cataclysmic volcanic event that created a single gigantic tectonic plate that now wraps around the entire planet—very different from Earth's surface, which is broken into nearly 50 plates. From the estimated rate of asteroid strikes on the planet and slow weathering of the craters, it is thought that this happened some 300–500 million years ago.

Weather on Venus

Venus is cloaked in clouds of sulfuric acid that block out 80 percent of all sunlight. The atmosphere glides rapidly around the planet on winds of up to 220 mph (360 km/h)—cloud systems can sail completely around the planet in under four days. Venus's clouds rain sulfuric acid, but the lower atmosphere is so hot that the raindrops evaporate before reaching the ground. The heavy cloud layer in Venus's atmosphere appears to shield it from most meteorite bombardments.

▽ The greenhouse effect

Most sunlight is reflected back into space from the tops of Venus's thick clouds. However, some penetrates the clouds to reach the surface, and is then reemitted as heat (infrared radiation). This heat cannot escape back into space, and is trapped by carbon dioxide in the atmosphere. Carbon dioxide and other gases give Earth a similar "greenhouse effect," but on Venus the huge quantities of carbon dioxide in the air mean that this effect is extreme—so much heat is trapped that the planet's surface is hot enough to melt lead.

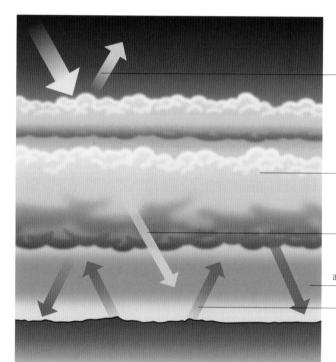

Roughly 80 percent of sunlight is reflected off the cloud deck.

Thick layers of gas and cloud prevent heat from escaping.

About 20 percent of sunlight reaches Venus's surface.

Carbon dioxide in the atmosphere holds in heat.

Infrared radiation from the Sun-warmed ground is absorbed by carbon dioxide and cannot escape into space.

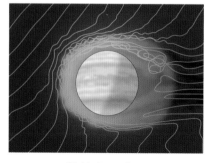

Stable ionosphere

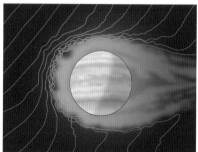

Drifting ionosphere

◁ The ionosphere

Like Earth, Venus is enveloped in a cloud of charged particles (ions) called the ionosphere. While Earth's ionosphere is shaped and stabilized by its magnetic field, Venus has almost no magnetic field and its ionosphere is shaped instead by the solar wind—a stream of charged particles flowing from the Sun. When a lull in the solar wind occurs, Venus's ionosphere balloons outward on the downwind side, forming a teardrop shape like the tail of a comet.

△ **Coronae**
Venus's surface features enormous, crown-shaped depressions called coronae. They are thought to be the result of hot magma in the mantle moving upward. This pushes the surface up, only for it to then collapse when the magma cools.

Artemis Corona is Venus's largest corona, measuring 1,600 miles (2,600 km) in diameter.

Dali Chasma is a system of deep troughs.

Venusian volcanoes

Venus does not have steep-sided, explosive volcanoes like typical volcanoes on Earth. Instead, most are shield volcanoes (shallow, gently sloping structures made from multiple layers of lava flows). On the lowland plains are types of volcanoes called pancake domes, formed by very thick lava, and tick volcanoes, with a central body and radiating leglike valleys. Other volcanic features include circular depressions called coronae and spiderlike arachnoids.

▽ **Volcanic hot spots**
Unlike Earth, Venus's surface is not broken into tectonic plates that create volcanoes as they move. Instead, Venusian volcanoes form above hot spots where plumes of hot magma well up from the interior. The result is runny lava that forms volcanoes of various sizes and shapes.

Pancake dome—formed when thick lava erupts very slowly.

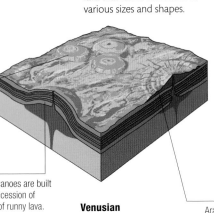

Venusian volcanoes

Shield volcanoes are built from a succession of eruptions of runny lava.

Arachnoid volcanoes look like a series of oval shapes, surrounded by a complex network of fractures.

▽ **Maat Mons**
The second-highest mountain and the highest volcano on Venus, Maat Mons—named after the Egyptian goddess of truth and justice—rises nearly 3 miles (5 km) above the surrounding plains. It is a huge shield volcano with a caldera (crater) about 20 miles (30 km) across at the summit, and may be active.

△ **Mead Crater**
Most features on Venus are named after historical or mythological women. Mead Crater, for example, is named after cultural anthropologist Margaret Mead (1901–78) and is the largest impact crater on Venus, over 174 miles (280 km) across. It has two distinct, concentric rings. The bright inner ring is a cliff formed by the initial impact. The darker outer ring is crossed by streaks made by ejecta and probably formed when the whole structure later collapsed.

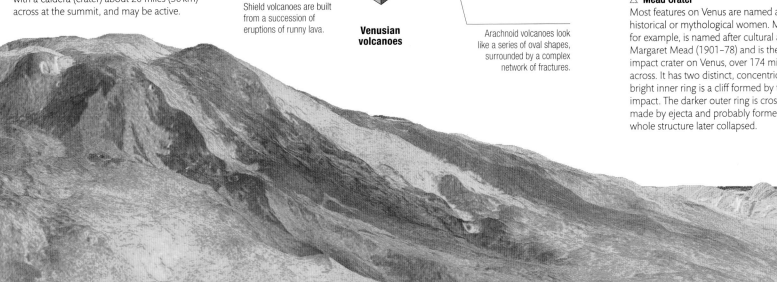

VENUS MAPPED

While most of Venus's surface comprises undulating plains, there are two significant highland areas: Ishtar Terra, where the planet's highest mountains are found; and Aphrodite Terra, near the equator.

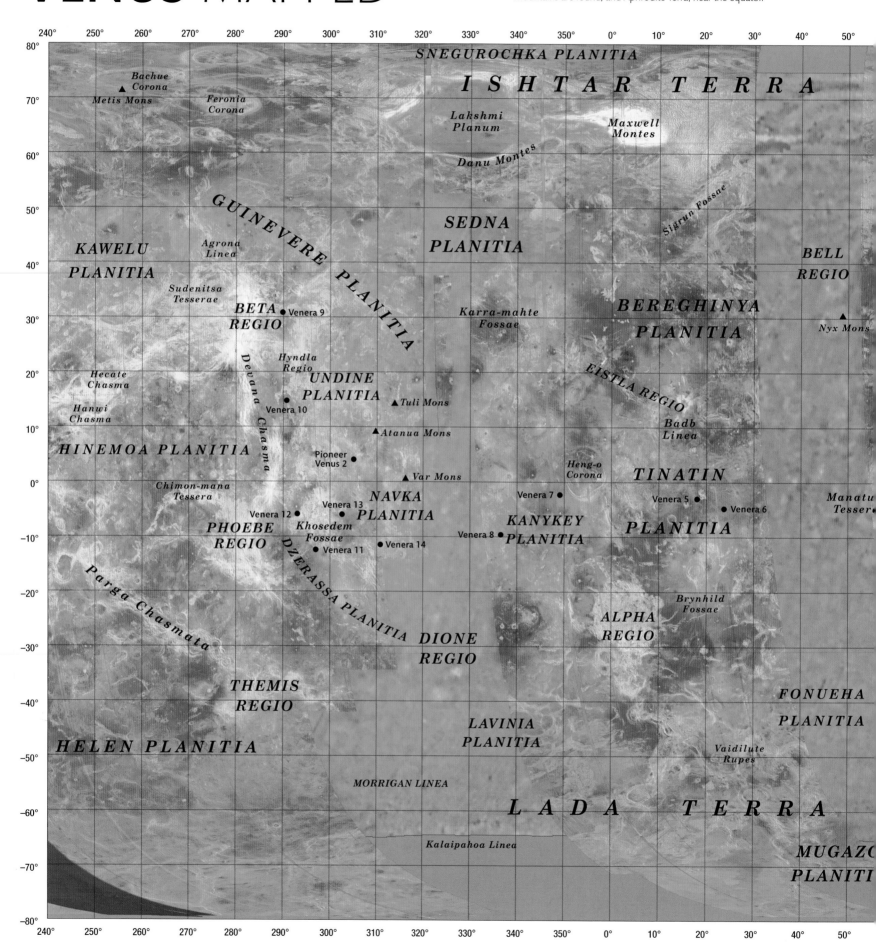

SNEGUROCHKA PLANITIA

ISHTAR TERRA

Bachue Corona ▲

Metis Mons

Feronia Corona

Lakshmi Planum

Maxwell Montes

Danu Montes

Sigrun Fossae

GUINEVERE PLANITIA

KAWELU PLANITIA

Agrona Linea

SEDNA PLANITIA

BELL REGIO

Sudenitsa Tesserae

BETA REGIO ● *Venera 9*

Karra-mahte Fossae

BEREGHINYA PLANITIA

▲ *Nyx Mons*

Hecate Chasma

Hyndla Regio

UNDINE PLANITIA

EISTLA REGIO

Devana Chasma

● *Venera 10*

▲ *Tuli Mons*

Hanwi Chasma

Badb Linea

▲ *Atanua Mons*

HINEMOA PLANITIA

Heng-o Corona

TINATIN

Pioneer Venus 2 ●

▲ *Var Mons*

Chimon-mana Tessera

NAVKA PLANITIA

Venera 7 ●

Venera 5 ●

● *Venera 6*

Manatu Tesser

Venera 13 ●

Venera 12 ●

KANYKEY PLANITIA

PLANITIA

PHOEBE REGIO

Khosedem Fossae

Venera 8 ●

Venera 11 ●

● *Venera 14*

Parga Chasmata

DZERASSA PLANITIA

DIONE REGIO

ALPHA REGIO

Brynhild Fossae

THEMIS REGIO

FONUEHA PLANITIA

HELEN PLANITIA

LAVINIA PLANITIA

Vaidilute Rupes

MORRIGAN LINEA

LADA TERRA

Kalaipahoa Linea

MUGAZO PLANITI

240° 250° 260° 270° 280° 290° 300° 310° 320° 330° 340° 350° 0° 10° 20° 30° 40° 50°

80° 70° 60° 50° 40° 30° 20° 10° 0° -10° -20° -30° -40° -50° -60° -70° -80°

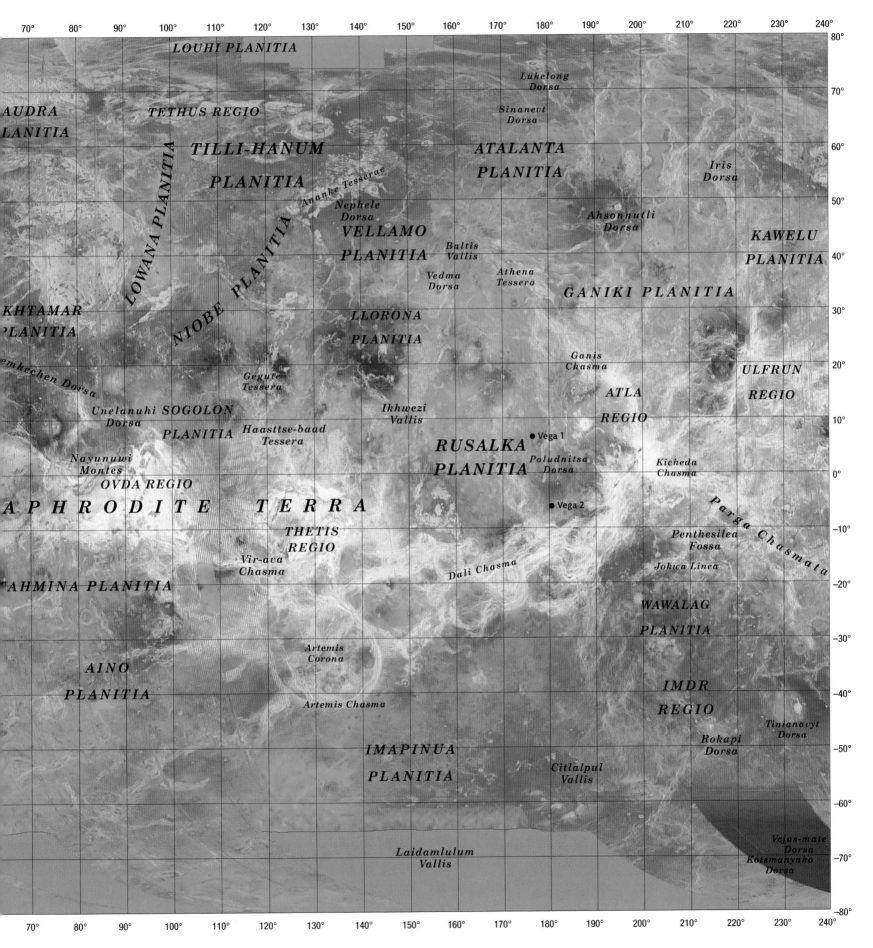

Scale 1:81,956,988

0 500 1000 1500 2000 km

0 500 1000 1500 2000 miles

70° 80° 90° 100° 110° 120° 130° 140° 150° 160° 170° 180° 190° 200° 210° 220° 230° 240°

80° 70° 60° 50° 40° 30° 20° 10° 0° -10° -20° -30° -40° -50° -60° -70° -80°

LOUHI PLANITIA

Lukelong Dorsa

Sinanevt Dorsa

AUDRA LANITIA

TETHUS REGIO

TILLI-HANUM PLANITIA

ATALANTA PLANITIA

Iris Dorsa

LOWANA PLANITIA

Ananke Tesserae

Nephele Dorsa

VELLAMO PLANITIA

Baltis Vallis

Ahsonnutli Dorsa

KAWELU PLANITIA

KHTAMAR PLANITIA

NIOBE PLANITIA

Vedma Dorsa

Athena Tessera

GANIKI PLANITIA

mkechen Dorsa

Gegute Tessera

LLORONA PLANITIA

Ganis Chasma

ULFRUN REGIO

Unelanuhi Dorsa SOGOLON

Ikhwezi Vallis

ATLA REGIO

PLANITIA

Haasttse-baad Tessera

RUSALKA PLANITIA

● Vega 1

Nayunuwi Montes

Poludnitsa Dorsa

Kicheda Chasma

OVDA REGIO

● Vega 2

A P H R O D I T E   T E R R A

Parga Chasmata

THETIS REGIO

Penthesilea Fossa

Vir-ava Chasma

Dali Chasma

Jokwa Linea

AHMINA PLANITIA

WAWALAG PLANITIA

Artemis Corona

IMDR REGIO

AINO PLANITIA

Artemis Chasma

Tinianavyt Dorsa

Rokapi Dorsa

IMAPINUA PLANITIA

Citlalpul Vallis

Vejas-mate Dorsa

Laidamlulum Vallis

Kotsmanyako Dorsa

DESTINATION **MAXWELL MONTES**

MAXWELL MONTES IS THE HIGHEST MOUNTAIN RANGE ON VENUS, TOWERING NEARLY 7 MILES (11 KM) INTO THE PLANET'S SCORCHING CLOUDS. ALTHOUGH THE TEMPERATURE HERE IS SLIGHTLY COOLER THAN IN VENUS'S LOWLANDS, AND REFLECTIVE MINERALS EVEN GIVE THE ILLUSION OF SNOW-CAPPED PEAKS, THE GROUND IS HOT ENOUGH TO MELT LEAD.

The sharp ridges of Maxwell Montes rise above the vast Lakshmi Planum, the volcanic plain that forms the western edge of Ishtar Terra, a continent-sized plateau near Venus's north pole. Just how the mountains formed is uncertain. Compression may have produced folding and faulting in the same process that created big mountains on Earth. Another theory suggests that volcanic action above a hot spot of molten magma in the planet's interior uplifted the range. At certain elevations, radar instruments have detected bright surfaces on Maxwell, shining with not snow but frosted metal. In Venus's fierce heat, minerals vaporize, forming a mist that condenses and freezes, and may even fall as metal snowflakes.

Artist's impression based on Magellan radar data with topography exaggerated.

LOCATION

Latitude 65°N; longitude 3°E

LAND PROFILE

Maxwell Montes is the highest volcanic range on Venus. It is slightly taller than Mauna Kea in Hawaii, the tallest volcano on Earth, but is dwarfed by Mars's Olympus Mons.

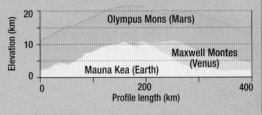

495 MILES (797 KM)—THE LENGTH OF THE MAXWELL MONTES RANGE.

MINERAL SNOW

The shining metal snow on Maxwell's peaks comprises tiny crystals of minerals, including lead sulfide (galena) and bismuth sulfide (bismuthinite). The rock samples shown below are from Earth.

**LEAD SULFIDE
(GALENA)**

**BISMUTH SULFIDE
(BISMUTHINITE)**

THE **PLANET** OF LOVE

NAMED BY THE ROMANS AFTER THEIR GODDESS OF LOVE, VENUS HAS A BRIGHT, JEWEL-LIKE APPEARANCE THAT HAS MADE IT AN OBJECT OF CURIOSITY FOR ASTRONOMERS SINCE ANCIENT TIMES.

Venus was the first planet scrutinized through a telescope, when Italian scientist Galileo Galilei observed in 1610 that it had phases like the Moon. However, its thick cloud cover meant that nothing was known of the surface until recently. Venus's proximity to Earth and similar size led to speculation that its clouds hid dense jungles and even civilizations. In the 1970s, the clouds were finally penetrated: Earth-based radar and a succession of spacecraft revealed a surface that is entirely barren and ferociously hot. Since then, the stark Venusian landscape has been mapped in detail.

Venus at twilight

Venus Tablet

c. 10,000 BCE

Venus in the night sky
Venus has been familiar since prehistoric times. Its proximity to the Sun and the high reflectivity of its thick cloud cover make it the brightest object in the night sky after the Moon—so bright it can even cast shadows on Earth when the Moon is hidden.

c. 1600 BCE

Venus Tablet of King Ammisaduqa
The clay Venus Tablet of Babylonian King Ammisaduqa is one of the most ancient of all astronomical records. It dates from about 1600 BCE and records in cuneiform writing the time that Venus appears on the horizon in the evening and morning over 21 years.

Emperor Napoleon I of France

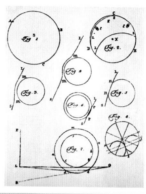

Lomonosov's drawings of atmospheric refraction

1812

Napoleon and Venus
When advancing with his armies on Moscow, the French Emperor Napoleon sees Venus in the daytime sky—said to be a lucky sign. He views it as an omen of victory. But what follows is his worst military setback as his armies retreat from Russia in disarray.

1761

Venus's atmosphere
Russian astronomer Mikhail Lomonosov observes a transit of Venus and notices that the Sun's light creates a bulge around Venus. He believes that this bulging is evidence that Venus has an atmosphere, and that the atmosphere is refracting light from the Sun.

1667

Cassini's spot
Italian-French astronomer Giovanni Cassini tracks the movement of a spot on the face of Venus, leading him to estimate, wrongly, that the planet rotates every 24 hours. In 1877, Italian astronomer Giovanni Schiaparelli correctly calculates the rotation period as 225 days.

Richard Proctor

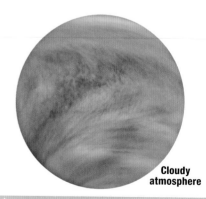

Cloudy atmosphere

1813

Polar spots
German physician and astronomer Franz von Gruithuisen—a keen observer of Venus—observes bright spots at Venus's poles. He believes these spots might be polar ice caps, but in fact they turn out to be shifting vortexes of bright cloud in Venus's atmosphere.

1875

Life on Venus
English astronomer Richard Proctor believes it is highly likely that life exists elsewhere in the universe. He suggests that Venus, so similar to Earth in size, could be inhabited, and that its thick clouds may hide an advanced Venusian civilization.

1920s

Carbon dioxide identified
Spectroscopy—analysis of the spectrum of light emitted by objects—allows astronomers to identify the chemical elements in celestial bodies. In the 1920s they discover that Venus's cloudy atmosphere consists of unbreathable carbon dioxide.

El Caracol, Chichen Itza

Venus attacks an ocelot warrior in the Dresden Codex

c. 6th century BCE

Phosphorus and Hesperus
Originally the ancient Greeks believe the morning and evening star are two planets: Phosphorus and Hesperus. They later agree with the Babylonians that it is actually a single planet, which the Babylonians called Ishtar, after their goddess of love.

906 CE

Mayan observatory
The remarkable structure of El Caracol in the ancient Mayan city of Chichen Itza, Mexico, is an astronomical observatory for Mayan priests. It is designed, in particular, for the observation of Venus. To the Mayans, Venus is Kulkulkán—Earth's twin and a god of war.

12th century

The Dresden Codex
The Codex, possibly found by Spanish conquistador Hernán Cortés in 1519, is the oldest book written in the Americas. It is believed to be a copy of an 8th century Mayan text, and contains accurate tables charting the arrival of Venus in the sky.

Venusian phases, sketched by Galileo

Horrocks's diagram of the 1639 transit of Venus

1643

Ashen light
The mysterious glow on Venus's night side—the ashen light—is first observed by Italian astronomer-priest Giovanni Battista Riccioli. In 1812, German astronomer Franz von Gruithuisen asserts that ashen light is smoke from the fires of the Venusian emperor.

1639

Transit of Venus
English astronomers Jeremiah Horrocks and William Crabtree are the first to observe a transit of Venus (when the planet passes between Earth and the Sun). It enables astronomers to make the first accurate calculations of the Earth–Sun distance.

1610

Galileo and Venus's phases
While studying Venus through a telescope, Galileo discovers that the planet has phases, as our changing point of view reveals varying amounts of its sunlit face. This supports the idea of Polish astronomer Copernicus that Venus travels around the Sun, not Earth.

Magellan radar map of Venus

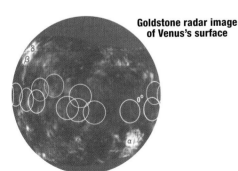

Goldstone radar image of Venus's surface

Surface of Venus from Venera 3

1961

Radar exploration
Venus's thick clouds prevent conventional telescopes from observing its surface. But from 1961, radar images—supplied first by radio telescopes at Goldstone, California, then by the Arecibo dish in Puerto Rico—reveal the Venusian surface for the first time.

1962

First visit: Mariner 2
NASA's Mariner 2 becomes the first spacecraft to fly past another planet when it swings within 22,000 miles (35,000 km) of Venus on December 14. Mariner 2's investigations confirm that Venus has cool clouds and a scorching surface.

1966

First landing: Venera 3
The Soviet Venera 3 is the first spacecraft to reach another planet, crashing onto Venus on March 1. The first successful landings are made by the Venera 7 and 8 probes of 1970 and 1972, which reveal extreme surface temperatures of 851–887°F (455–475°C).

1990

Magellan mission
NASA's Magellan goes into orbit around Venus, aerobraking to reduce speed. Using radar, it maps 98 percent of the surface. After completing its mission in 1994, it plunges into Venus's atmosphere.

LAUNCH   **EARTH ORBIT**   **JOURNEY TO VENUS**

1961	Sputnik 7
1961	Venera 1
1962	Mariner 1
1962	Sputnik 19
1962	Mariner 2
1962	Sputnik 20
1962	Sputnik 21
1963	Kosmos 21
1964	Venera 1964A
1964	Venera 1964B
1964	Kosmos 27
1964	Zond 1
1965	Venera 2
1965	Venera 3
1965	Kosmos 96
1965	Venera 1965A
1967	Venera 4
1967	Mariner 5
1967	Kosmos 167
1969	Venera 5
1969	Venera 6
1970	Venera 7
1970	Kosmos 359
1972	Venera 8
1972	Kosmos 482
1973	Mariner 10
1975	Venera 9
1975	Venera 10
1978	Pioneer Venus 1
1978	Pioneer Venus 2
1978	Venera 11
1978	Venera 12
1981	Venera 13
1981	Venera 14
1983	Venera 15
1983	Venera 16
1984	Vega 1
1984	Vega 2
1989	Magellan
1989	Galileo
1997	Cassini
2004	MESSENGER
2005	Venus Express
2010	Akatsuki
Planned	Venus Orbiter
Planned	BepiColombo
Planned	Solar Probe+
Planned	Venera-D

KEY

- RFSA (USSR/Russia)
- NASA (USA)
- ESA (Europe)
- JAXA (Japan)
- ISRO (India)
- Joint ESA/JAXA mission
- Destination
- Success
- Failure

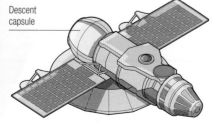

Descent capsule

Venera 7, the first craft to survive a landing on Venus

First surface image, from Venera 9

◁ ▽ **Venera**

In 1966, the Soviet probe Venera 3 crashed into Venus's surface and became the first spacecraft to reach another planet. Over the next 17 years, the Soviet Union sent another 13 craft to Venus, revealing a huge amount about the planet. On October 22, 1975, Venera 9 landed and beamed back the first pictures from the surface, showing a landscape littered with broken rock. Visibility was surprisingly good considering the thickness of Venus's atmosphere; a Soviet scientist described it as like "a cloudy day in Moscow."

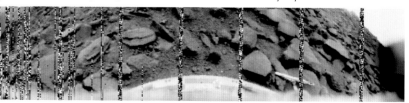

FLYBY   ORBITER   PROBE   LANDER

MISSIONS TO **VENUS**

VENUS WAS THE FIRST PLANET TO BE VISITED BY SPACECRAFT, IN 1962. SINCE THEN, THERE HAVE BEEN NEARLY 40 MISSIONS, SOME OF THEM FLYBYS AND OTHERS PROBING THE ATMOSPHERE OR LANDING ON THE SURFACE.

After a string of failures, the first spacecraft to reach Venus was NASA's Mariner 2, which revealed the planet's scorching surface temperature during a flyby in 1962. The first soft landing was made by the Soviet craft Venera 7 in 1970, but it was able to broadcast data for only 23 minutes. Since then there have been 20 Venus landings, with varying degrees of success, which is not surprising considering the extreme heat and pressure on Venus. Proposals for future missions include a robust rover vehicle that can explore the planet's surface in the same way that rovers investigate Mars.

Air pressure on Venus is about **90 times** greater than on Earth.

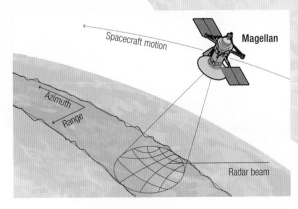

▷ **Mapping the surface**
Much of our knowledge of Venus comes from the NASA spacecraft Magellan (named after Portuguese explorer Ferdinand Magellan), which reached the planet on August 10, 1990. It spent four years in orbit and mapped 98 percent of the surface, peering through the thick clouds by firing a radar beam at the ground and capturing the echo. It imaged craters, hills, ridges, and a wide range of volcanic formations in incredible detail. Its mission complete, Magellan was sent into Venus's atmosphere, where it was vaporized, though some wreckage may have reached the surface.

Spacecraft motion · Magellan · Azimuth · Range · Radar beam

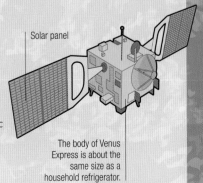

▷ **Venus Express**
Launched in 2005, Europe's Venus Express spacecraft was sent to the planet to study its atmosphere and climate in detail. It arrived in April 2006 and has since beamed back a vast amount of data. It has found evidence of oceans in Venus's past, captured flashes of lightning, and revealed a huge double atmospheric vortex at the south pole.

Solar panel

The body of Venus Express is about the same size as a household refrigerator.

EARTH

SITUATED AROUND 93 MILLION MILES (150 MILLION KM) FROM THE SUN, EARTH IS ALONE AMONG THE PLANETS IN HAVING VAST OCEANS OF LIQUID WATER ON ITS SURFACE AND THE UNDISPUTED PRESENCE OF LIFE.

When the solar system formed, Earth was the largest predominantly solid object to take shape and acquired the most internal heat energy of the rocky planets. As a result, Earth was the most susceptible to the development of internal heat flows and to the breaking up of its surface into large slabs, or plates, which slowly grind past each other.

Through a combination of plate movements known as plate tectonics, volcanic activity, and comet impacts, large amounts of water accumulated on Earth's surface. The planet's distance from the Sun, its gravity, and an insulating atmosphere combined to create conditions for this water to exist in each of its three physical states, including liquid water, which was essential to the development of life. As a result, Earth today appears unique, with its swirling water clouds, vast oceans, and continents colored green in parts by the presence of plants.

Earth is the only planet known to have large amounts of **water** in all three states: **solid, liquid, and gas.**

EARTH DATA

Average diameter	7,918 miles (12,742 km)
Axial tilt	23.5°
Rotation period (day)	24 hours
Orbital period (year)	365.26 Earth days
Minimum surface temperature	−128°F (−89°C)
Maximum surface temperature	136°F (58°C)
Moons	1

▷ **Northern hemisphere**
This view of Earth is dominated by the continents of North America and Eurasia, which until about 70 million years ago were joined. They are separated by the North Atlantic Ocean and, to its north, the smaller, partly iced-over Arctic Ocean.

▷ **Eurasia and Australia**
Eurasia is Earth's biggest landmass. It sits above the planet's third-largest body of water, the Indian Ocean. Australia is the smallest of Earth's seven continents.

▷ **Southern hemisphere**
Earth's southern hemisphere is centered on a single landmass, Antarctica. The ring-shaped Southern Ocean surrounds the ice-covered continent, with Australia and parts of South America and Africa making major incursions.

The Rocky Mountains run down the west of the North American Plate for a distance of 3,000 miles (4,800 km).

Earth's equator is encircled by a persistent band of cloud, making tropical regions humid and rainy.

The Pacific Ocean has the largest surface area of any body of water on Earth at 65.4 million square miles (169.5 million square km), making up almost half the total area covered by oceans.

Clouds in the southern hemisphere spiral in a clockwise motion, while those in the northern hemisphere spiral counterclockwise.

The Atlantic Ocean is the second-largest body of water on Earth at 41.1 million square miles (106.5 million square km). It is bounded by Africa and Europe to the east and the Americas to the west.

The west coast of Africa echoes the shape of the east coast of South America. The two continents were once joined but started separating around 130 million years ago.

The Amazon Basin is a densely forested region covering around 2.7 million square miles (7 million square km).

The longest mountain range on Earth at 4,300 miles (7,000 km), the Andes form the western edge of the South American Plate.

The southernmost tip of South America is known as Cape Horn. Winds below this latitude circle Earth uninterrupted by land, causing formidable waves in the Southern Ocean.

◁ **The Americas**
Looking at Earth from this angle, the amount of water that covers the planet is striking. Two vast oceans—the Pacific, covering about one-third of Earth's whole surface, and the Atlantic—are separated by the landmasses of North and South America. The continents are joined by the narrow bridge of Central America.

EARTH STRUCTURE

EARTH'S LAYERED INTERNAL STRUCTURE IS MIRRORED IN A MULTILAYERED ATMOSPHERE THAT EXTENDS FOR HUNDREDS OF MILES ABOVE THE PLANET'S SURFACE, GRADUALLY MERGING WITH SPACE.

What we know of Earth's internal structure has been learned largely through the study of earthquake waves, particularly the routes they take as they travel inside the planet. Each layer beneath the surface is progressively denser, hotter, and under increased pressure. A unique aspect of Earth is that its outer, rigid shell, the lithosphere (made up of the crust and topmost layer of the mantle), is split into chunks called tectonic plates, which move relative to each other, driven by internal heat flows. Surrounding the planet's surface, Earth's atmosphere provides important protection to the life that flourishes on the planet.

Inner core
The innermost layer of Earth consists of a solid iron–nickel alloy and has an average temperature of about 9,900°F (5,500°C). Despite the high temperature, the metals in the inner core cannot melt because of the intense pressure exerted on them.

Over a quarter of Earth's surface is covered by land. Continental crust is thicker than the oceanic crust that occurs under Earth's oceans.

▷ **Earth layer by layer**
Earth has three primary layers—core, mantle, and crust—each with a unique chemical composition. The core has two distinct parts, inner and outer. There are also two types of crust—the thinner oceanic and the thicker continental crust. The layers of the mantle increase in density with depth, and the topmost layer is fused to the crust, forming the lithosphere. Layers in this 3D model are not shown to scale: the crust, surface relief, and atmosphere are exaggerated for clarity.

In Earth's early history, the **planet was hot** and **liquid**—heavy **iron sank,** forming the **core.**

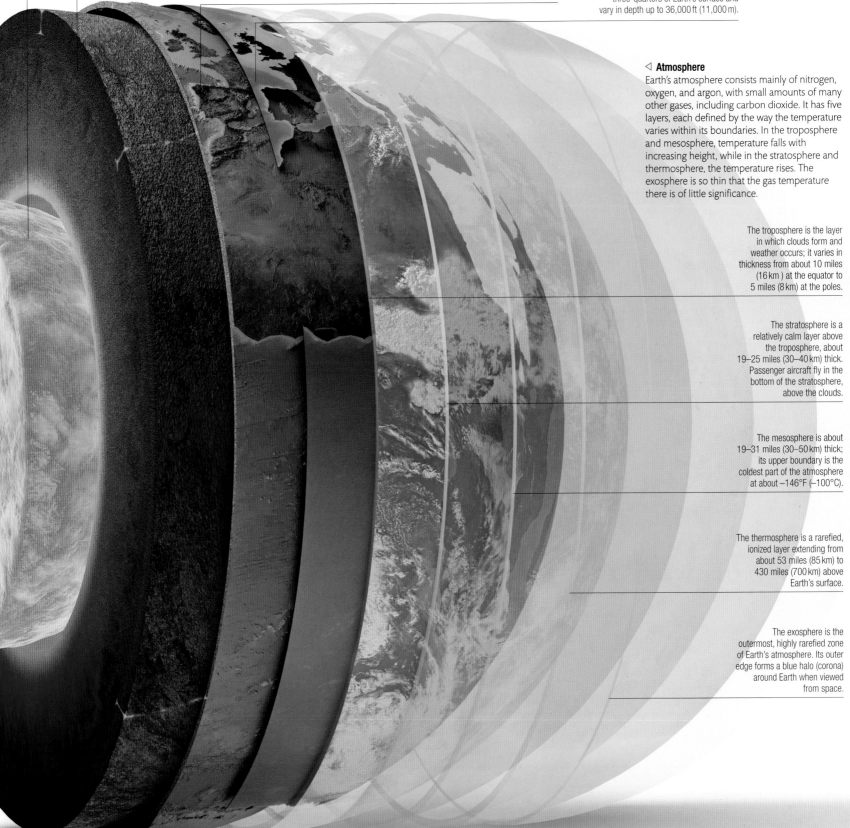

Outer core
The outer core is liquid iron with some nickel and has an average temperature of about 9,000°F (5,000°C). Currents in the outer core are thought to generate Earth's magnetic field and cause the magnetic poles to wander.

Mantle
The largest of Earth's internal layers is basically solid, consisting of rocks such as peridotite. However, it can slowly deform, allowing heat to enter from the core and cause convection currents over geological time scales. These currents drive crustal movements.

Crust
Oceanic crust consists of dark volcanic rocks such as basalt and is 4–5 miles (7–8 km) thick. Continental crust consists of many types of relatively light rock and is 16–45 miles (25–70 km) thick.

Ocean
Saltwater oceans cover almost three-quarters of Earth's surface and vary in depth up to 36,000 ft (11,000 m).

◁ **Atmosphere**
Earth's atmosphere consists mainly of nitrogen, oxygen, and argon, with small amounts of many other gases, including carbon dioxide. It has five layers, each defined by the way the temperature varies within its boundaries. In the troposphere and mesosphere, temperature falls with increasing height, while in the stratosphere and thermosphere, the temperature rises. The exosphere is so thin that the gas temperature there is of little significance.

The troposphere is the layer in which clouds form and weather occurs; it varies in thickness from about 10 miles (16 km) at the equator to 5 miles (8 km) at the poles.

The stratosphere is a relatively calm layer above the troposphere, about 19–25 miles (30–40 km) thick. Passenger aircraft fly in the bottom of the stratosphere, above the clouds.

The mesosphere is about 19–31 miles (30–50 km) thick; its upper boundary is the coldest part of the atmosphere at about −146°F (−100°C).

The thermosphere is a rarefied, ionized layer extending from about 53 miles (85 km) to 430 miles (700 km) above Earth's surface.

The exosphere is the outermost, highly rarefied zone of Earth's atmosphere. Its outer edge forms a blue halo (corona) around Earth when viewed from space.

TECTONIC **EARTH**

EARTH'S OUTER ROCKY SHELL IS SPLIT INTO MANY HUGE FRAGMENTS CALLED TECTONIC PLATES. THESE SLOWLY MOVING PLATES INTERACT, CAUSING VARIOUS GEOLOGICAL EVENTS AND CREATING CHANGES ON THE PLANET'S SURFACE.

Earth's tectonic plates are irregularly shaped and fit together like a jigsaw puzzle. Their movements relative to each other, caused by convective heat flows deep within the planet, occur at a rate of just a few inches each year, but over millions of years, plate movement has shifted continents. A variety of features have formed at or near plate boundaries. These include mountain ranges, deep-sea trenches, volcanoes where two plates move toward each other, and mid-ocean ridges where they move apart. Earthquakes are more common at plate boundaries.

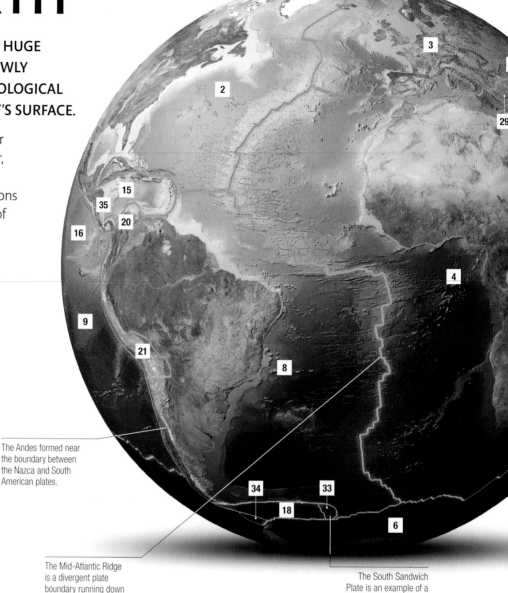

The Andes formed near the boundary between the Nazca and South American plates.

The Mid-Atlantic Ridge is a divergent plate boundary running down the Atlantic.

The South Sandwich Plate is an example of a microplate.

KEY

1	Pacific
2	North American
3	Eurasian
4	African (Nubian)
5	African (Somalian)
6	Antarctic
7	Australian
8	South American
9	Nazca
10	Indian
11	Sunda
12	Philippine Sea
13	Arabian
14	Okhotsk
15	Caribbean
16	Cocos
17	Yangtze
18	Scotia
19	Caroline
20	North Andes

21	Altiplano
22	Anatolian
23	Banda Sea
24	Burma
25	Okinawa
26	Woodlark
27	Mariana
28	New Hebrides
29	Aegean Sea

30	Timor
31	Bird's Head
32	North Bismarck
33	South Sandwich
34	South Shetland
35	Panama
36	South Bismarck
37	Maoke
38	Solomon Sea

▷ **Earth's plates**
There are seven major plates—for example, the Pacific and Eurasian plates—as well as a dozen or so medium-sized plates, such as the Arabian Plate, and numerous much smaller microplates. Listed here are most of the recognized plates, in approximate order of decreasing size. The plates are also numbered on the globes shown on the right. A few of the microplates are sometimes considered just parts of larger plates.

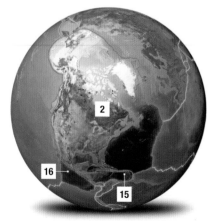

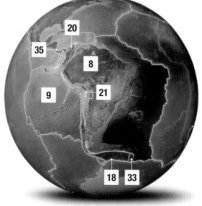

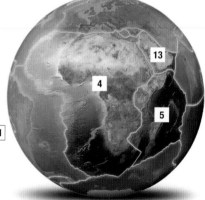

△ **North American**
The North American Plate (2) makes up just under one-sixth of Earth's surface. It contains parts of the Arctic and Atlantic oceans and a section of Siberia. A notable volcanic hot spot has existed for millions of years under this plate and is currently the cause of vigorous geyser activity in Yellowstone National Park, Wyoming.

△ **South American**
With the neighboring Nazca (9), Scotia (18), and other smaller plates, the South American Plate (8) accounts for about one-eighth of Earth's surface. The Andes mountain range in South America rises where the eastward-moving Nazca Plate is pushed under the edge of the South American Plate.

△ **Eurasian**
This plate (3) includes Europe and most of the landmass of Asia. A number of medium-sized plates to the east and southeast, such as the Sunda Plate (11), were formerly considered part of the Eurasian Plate. Millions of years ago, the Indian Plate (10) crashed into the Eurasian Plate, creating the Himalayas.

△ **African**
The two African plates (4 and 5) include the African continent and large parts of the Atlantic and Indian oceans. Africa is believed to be in the process of splitting into two parts along the East African Rift—a gigantic split in Earth's crust that runs for about 2,500 miles (4,000 km) through East Africa.

Siberia is an example of an ancient, tectonically stable chunk of continental crust.

A transform boundary in northern Turkey is a source of frequent earthquakes.

Convergent boundaries, associated with deep trenches, exist all around the Pacific.

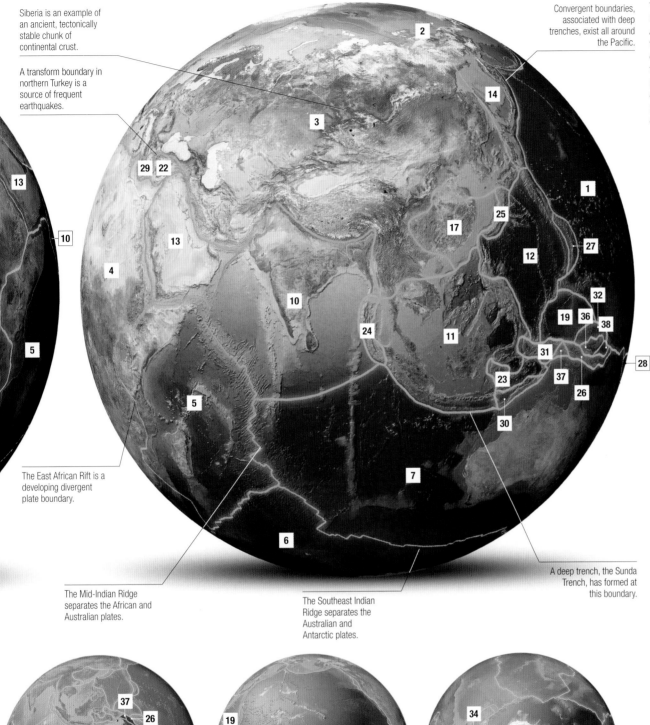

The East African Rift is a developing divergent plate boundary.

The Mid-Indian Ridge separates the African and Australian plates.

The Southeast Indian Ridge separates the Australian and Antarctic plates.

A deep trench, the Sunda Trench, has formed at this boundary.

▽ Plate boundaries

Boundaries between plates are of three types. At convergent boundaries, two plates move toward one another; one may dip beneath the other, often causing volcanoes or mountains to form. At transform boundaries, plates grind past each other. At divergent boundaries, which are either mid-ocean ridges or continental rifts, plates move apart and new plate is created along the boundary.

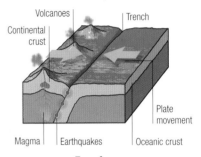

Convergent

Volcanoes | Trench
Continental crust
Plate movement
Magma | Earthquakes | Oceanic crust

Transform

Earthquakes | Plate movement

Divergent

Plate movement | New plate created at boundary
Magma

▽ Swimming between plates

Divers and snorkelers visiting Thingvallavatn Lake in southwestern Iceland can swim in a gap between the North American and Eurasian plates. At the lake floor, the plate boundary is visible in the clear water at a deep rift, known as Silfra. In one section, the fissure is 200 ft (63 m) deep at the bottom, too narrow and steep-sided for most divers to risk exploring.

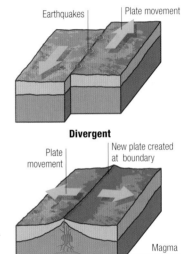

△ Australian

This plate (7) comprises Australia, parts of New Zealand and New Guinea, and parts of the Indian and Southern oceans. Major features include Australia's deserts, the Great Dividing Range, and the Great Barrier Reef. The whole plate is moving in a northeastern direction at a rate of about 2.5 in (6.5 cm) per year.

△ Pacific

The largest tectonic plate, the Pacific Plate (1) covers about one-fifth of Earth. It contains no large landmasses, but many volcanic islands and subsea volcanoes occur where plumes of magma burst through the surface. The Pacific Plate is moving northwest at a rate of about 4 in (10 cm) per year.

△ Antarctic

Making up about one-eighth of Earth's surface, the Antarctic Plate (6) includes the Antarctic continent at its center, together with most of the encircling Southern Ocean. Over millions of years, this plate has become larger, as all around its edges new plate is continually created at divergent plate boundaries.

EARTH'S CHANGING SURFACE

UNLIKE THE FACE OF OUR MOON, WHICH HAS CHANGED LITTLE IN BILLIONS OF YEARS, EARTH'S SURFACE IS DYNAMIC. OUR PLANET IS SHAPED BY MANY PROCESSES, FROM RESTLESS TECTONIC PLATES TO CORROSIVE WATER IN THE ATMOSPHERE.

Some of the main drivers of change are internal, such as convection within Earth's mantle, which drives the movement of tectonic plates and builds new landscapes. On the surface, rock is subjected to continual weathering and erosion, processes driven ultimately by the Sun's energy. Over millions of years, these processes wear down entire mountain ranges, reducing rock to rubble, sand, and silt. Some of these processes are unique to Earth, helping to explain why the planet's surface changes so rapidly compared to other rocky worlds.

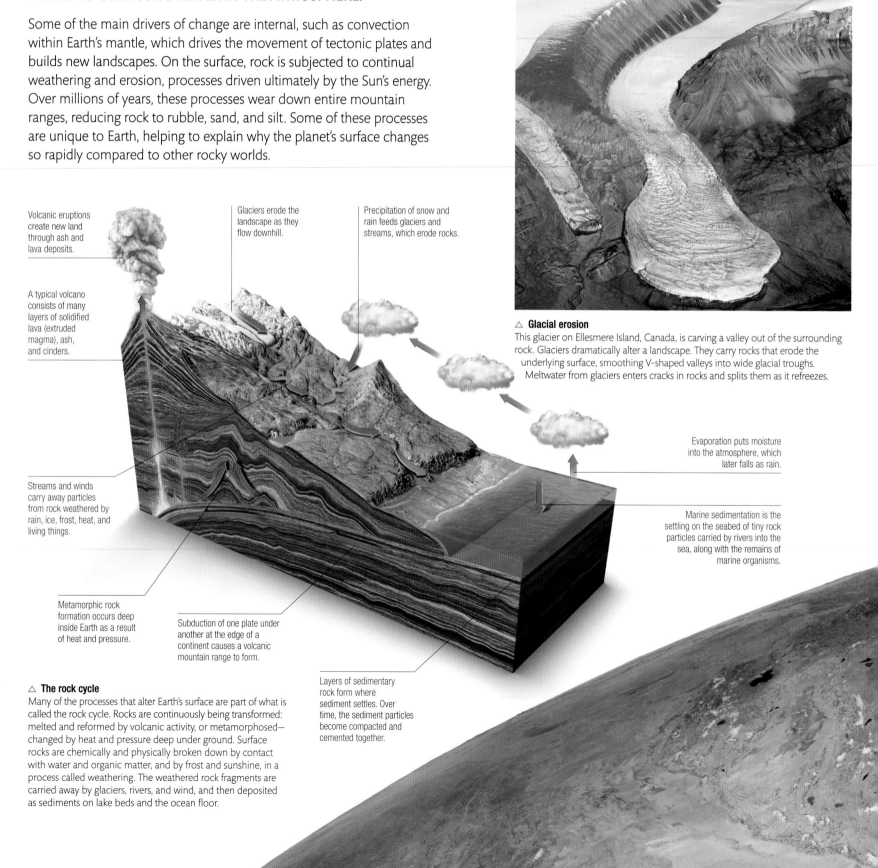

Volcanic eruptions create new land through ash and lava deposits.

Glaciers erode the landscape as they flow downhill.

Precipitation of snow and rain feeds glaciers and streams, which erode rocks.

A typical volcano consists of many layers of solidified lava (extruded magma), ash, and cinders.

Streams and winds carry away particles from rock weathered by rain, ice, frost, heat, and living things.

Metamorphic rock formation occurs deep inside Earth as a result of heat and pressure.

Subduction of one plate under another at the edge of a continent causes a volcanic mountain range to form.

△ **Glacial erosion**
This glacier on Ellesmere Island, Canada, is carving a valley out of the surrounding rock. Glaciers dramatically alter a landscape. They carry rocks that erode the underlying surface, smoothing V-shaped valleys into wide glacial troughs. Meltwater from glaciers enters cracks in rocks and splits them as it refreezes.

Evaporation puts moisture into the atmosphere, which later falls as rain.

Marine sedimentation is the settling on the seabed of tiny rock particles carried by rivers into the sea, along with the remains of marine organisms.

△ **The rock cycle**
Many of the processes that alter Earth's surface are part of what is called the rock cycle. Rocks are continuously being transformed: melted and reformed by volcanic activity, or metamorphosed—changed by heat and pressure deep under ground. Surface rocks are chemically and physically broken down by contact with water and organic matter, and by frost and sunshine, in a process called weathering. The weathered rock fragments are carried away by glaciers, rivers, and wind, and then deposited as sediments on lake beds and the ocean floor.

Layers of sedimentary rock form where sediment settles. Over time, the sediment particles become compacted and cemented together.

A mountain range as high as the **Himalayas** could be worn flat in less than **20 million years.**

△ **Eroded landscape**
These stunning rock formations, called hoodoos, in Bryce Canyon, USA, were formed primarily by a process called frost wedging. The canyon area experiences numerous freeze–thaw cycles each year. In the winter, water produced by melting snow seeps into cracks in the layers of limestone and other sedimentary rocks and then freezes at night. Water expands as it freezes, prying open the cracks and causing the rocks to fracture.

▷ **Mountain building**
A major surface-changing process affecting Earth's land areas is mountain-building. This typically occurs where tectonic plates are pushed together. A lateral squeezing of multilayered sedimentary rock causes thrust faulting, in which the rock layers break along gently inclined planes called faults. Mountains gradually rise as the rock layers stack on top of each other. This is how most of Earth's major mountain ranges have been built at different times in the past.

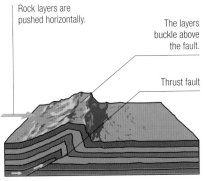

Rock layers are pushed horizontally.

The layers buckle above the fault.

Thrust fault

Initial break along thrust fault

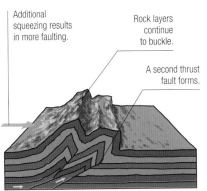

Additional squeezing results in more faulting.

Rock layers continue to buckle.

A second thrust fault forms.

Further faulting and buckling

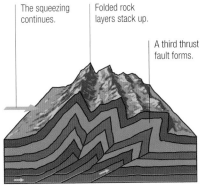

The squeezing continues.

Folded rock layers stack up.

A third thrust fault forms.

Complex of fractured and buckled rock layers

▽ **Himalayas**
Earth's highest mountain range, the Himalayas, began forming about 50–70 million years ago, when the Indian Plate crashed into the Eurasian Plate. If the neighboring Karakoram Range is included, this mountain belt includes Earth's 14 highest peaks, each over 5 miles (8 km) high.

1

WATER AND ICE

1 | Coral atoll
Sculpted by water, many of Earth's features are unique in the solar system. The Great Blue Hole off the coast of Belize is a submerged sinkhole surrounded by a coral atoll. Stretching 1,000 ft (300 m) across, the sinkhole sits in the Mesoamerican reef system that runs 600 miles (1,000 km) along the coast of Central America. Coral reefs form when coral larvae attach to underwater rocks along the edge of a landmass.

2 | Delta
A tangling network of channels, the great Ganges River creates an abstract jigsaw puzzle of islands as it meets the Bay of Bengal. Mature rivers deposit sediment as they slow on their course to the sea. As the sediment builds up, it forms low-lying land—a delta. The soil in a delta is often very fertile because it is rich in organic matter and minerals carried from upstream.

3 | Waves
Oceans cover more than two-thirds of Earth's surface, giving our planet its unique blue-jewel appearance from space. Seawater is constantly moving, and the pattern of currents is governed by the continents, sunlight, Earth's rotation, and the pull of the Moon. Whipped into waves by the wind, the sea in turn shapes the land. Tides, waves, and currents erode, deposit, and transport material, carving out coastlines.

4 | Meltwater
Around 80 percent of Greenland is covered by an ice sheet; it holds about 10 percent of the world's ice. When temperatures rise, it begins to melt, and meltwater carves deep channels in the remaining ice. This huge ice canyon is 150 ft (45 m) deep. Ice caps, glaciers, and permanent snow hold almost 70 percent of Earth's fresh water. Satellite imaging and data are used to track the rate at which the ice is melting.

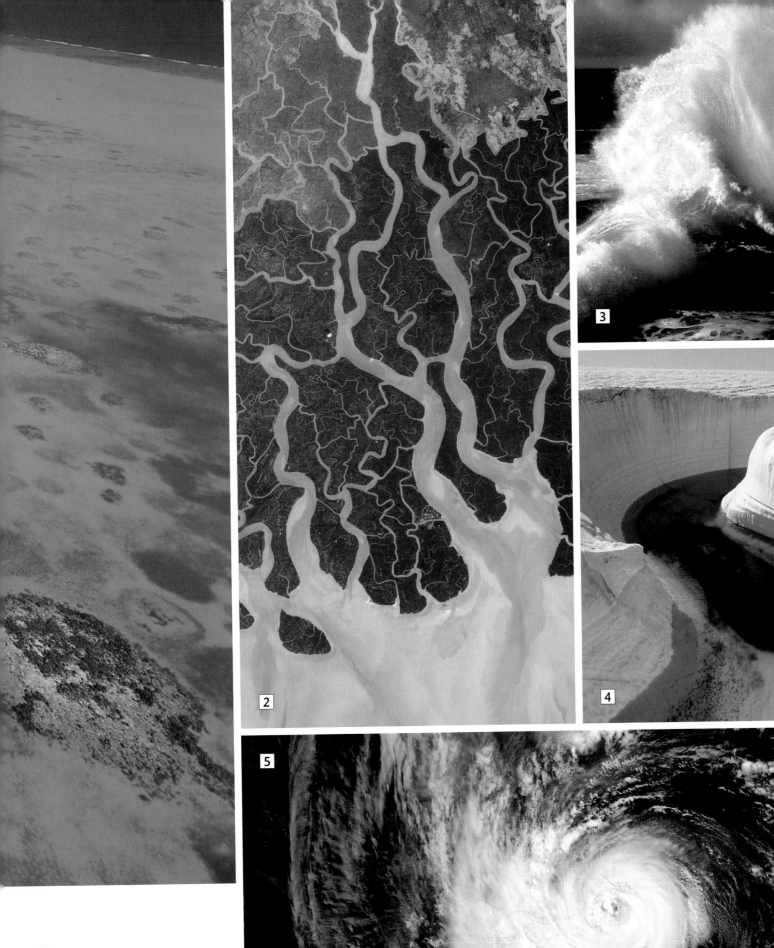

2

3

4

5

5 | **Weather**

Satellites can track storms from beginning to end. Hurricane Isabel, pictured here over North Carolina, formed over East Africa, and satellites recorded its growth into a tropical cyclone with a wind speed of 166 mph (267 km/h). Cyclones are areas of low atmospheric pressure. Winds rush in to fill a gap and spiral up, pulling energy and moisture from warm seas as they travel across the ocean.

LIFE ON **EARTH**

EARTH IS THE ONLY PLACE IN THE UNIVERSE KNOWN TO HARBOR LIFE. OTHER SOLAR SYSTEM LOCATIONS, SUCH AS THE SUBSURFACE OCEAN OF JUPITER'S MOON EUROPA, MIGHT THEORETICALLY SUPPORT LIFE, BUT FOR NOW OUR HOME PLANET APPEARS TO BE UNIQUE.

Life has existed on Earth for at least 3.7 billion years. We cannot be sure that it originated on our planet—it might have developed elsewhere and spread to Earth in objects such as comets. The presence on Earth of extremophiles—organisms that can exist in extremely challenging conditions—supports the idea that life may be able to survive in seemingly hostile environments elsewhere in the universe. However, at present, the consensus is that the life we see on Earth originated here from non-living matter.

Why Earth?

Life on Earth began almost as soon as the young planet ceased to be bombarded by asteroids, allowing its surface to cool. Since then, our planet has continually provided conditions conducive for living organisms to thrive and evolve. Earth's distance from the Sun puts it in the solar system's habitable Goldilocks zone. Here, surface temperature and atmospheric pressure conspire to allow water to exist as a liquid on the surface—a prerequisite for life as we know it. Earth also benefits from a rich supply of energy (solar radiation and heat generated from Earth's interior), a protective electromagnetic field (generated by currents in the liquid-iron core), and a large moon, which maintains climatic stability by minimizing swings in the planet's tilt.

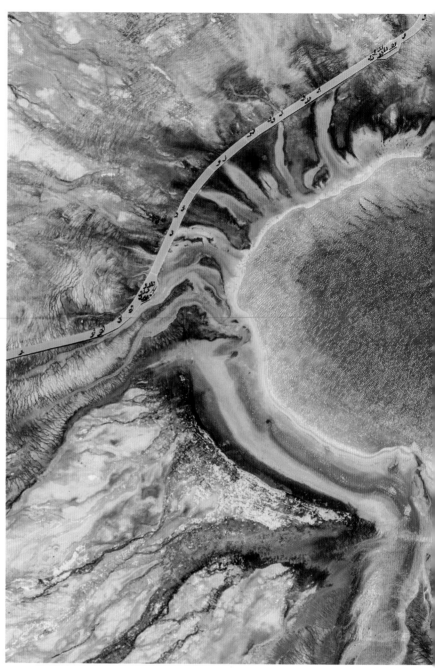

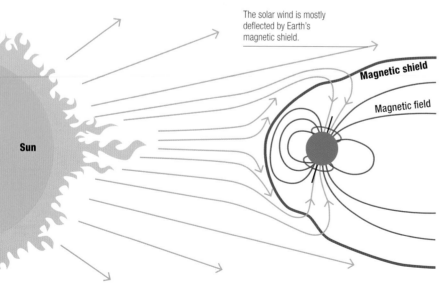

The solar wind is mostly deflected by Earth's magnetic shield.

Magnetic shield

Magnetic field

Sun

△ **Magnetic field**
Earth's strong magnetic field protects life on our planet by creating a shield that prevents most of the potentially harmful particles in the solar wind from reaching the planet's surface. The solar wind consists of a stream of high-energy charged particles, mainly electrons and protons, released from the Sun's upper atmosphere.

How life developed

The precursors of the first life forms were probably complex organic (carbon-containing) molecules that arose by chance in Earth's surface water. At some point, an organic molecule with a unique property appeared: it had the ability to catalyze production of copies of itself. This self-replicating organic molecule was the earliest ancestor of DNA. Through the process of evolution by natural selection, its descendants became more sophisticated, acquiring the ability to manufacture protective structures and substances that helped them survive and multiply—thus becoming primitive cells. Later, the single cells formed intimate, cooperative colonies, giving rise to multicellular organisms. Around a billion years ago, the advent of sexual reproduction set off a diversification of these organisms into more complex forms, such as plants and animals, that has continued to the present day.

△ **Extremophiles**
Extremophiles are organisms that thrive in conditions normally inhospitable to life, such as scalding or acidic water, or inside rocks. In Grand Prismatic Hot Spring, the green, yellow, and orange areas are mats of pigmented extremophile bacteria. Different-colored species favor different temperatures, which reach up to 188°F (87°C) at the center of the hot, mineral-rich pool.

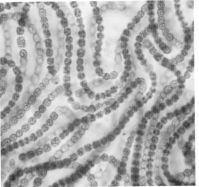

◁ **Cyanobacteria**
Thought to have existed for 3.5 billion years, cyanobacteria are among Earth's oldest life forms. These microbes obtain energy by photosynthesis. By releasing oxygen as a waste product of photosynthesis, ancient cyanobacteria altered Earth's atmosphere, creating a protective blanket of ozone that made the surface habitable and triggering the evolution of oxygen-breathing life forms.

▷ **Hydrothermal vent**
Among the sites where life possibly originated are hydrothermal vents—cracks in the seabed from which gush plumes of hot water containing minerals and some simple dissolved gases, such as ammonia and carbon dioxide. The minerals might have catalyzed chemical reactions between the gas molecules, thereby forming the basic building blocks of life.

◁ **Changing life**
Through evolution by natural selection, life on Earth developed from simple early forms, limited in number, to the enormous diversity seen today. Most species that have existed are now extinct, but traces remain preserved in Earth's rocks as fossils. The fossil record reveals that on several occasions in Earth's past, catastrophic events caused mass extinctions that wiped out hundreds of species at once.

Ammonite fossil

▽ **Biodiversity**
The diversity of life within a region is known as its biodiversity. Africa's grasslands are well known for their great diversity of animal species. The Serengeti plains of East Africa, for example, support around 45 mammal species and 500 bird species, to name just a small fraction of the total number present.

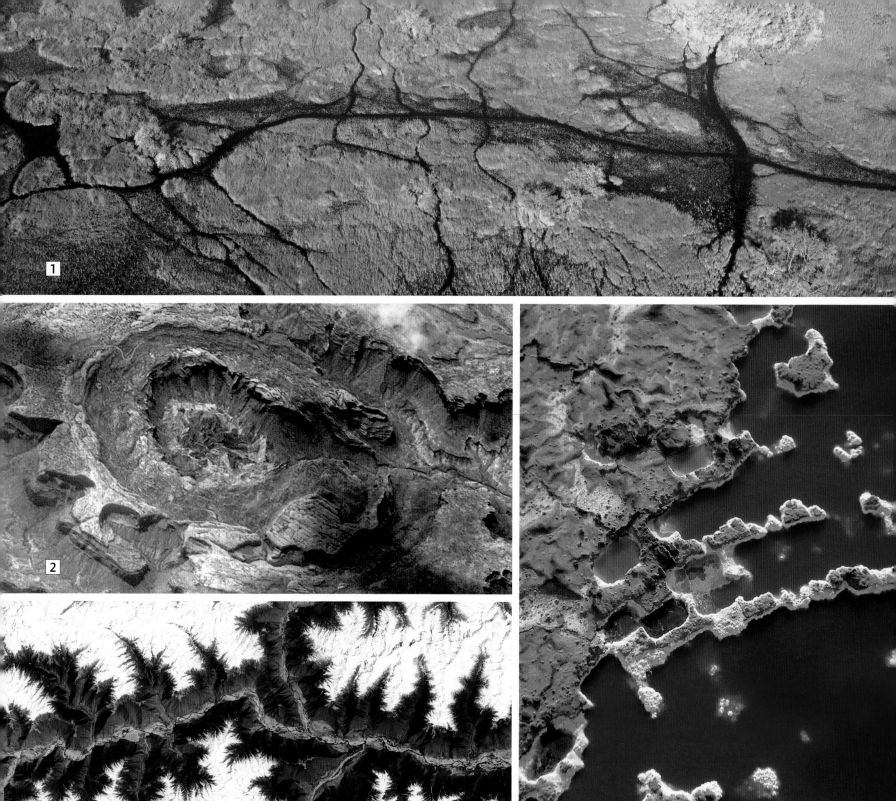

EARTH FROM ABOVE

1 Wetlands
Earth's abundance of surface water creates unique habitats such as wetlands. Formed wherever water is slowed in its progress, such as where a mature river meets the sea, wetlands provide a rich source of nutrients for diverse wildlife. This aerial view reveals the flat plains of the Okavango Delta in Botswana as a canvas of lakes, islands, and channels cutting through lush green vegetation.

2 Impact crater
This photograph of Upheaval Dome in Canyonlands National Park in Utah was taken from the International Space Station. The origin of the circular structure is uncertain, but the discovery of "shocked quartz" in the rock suggests it is a deeply eroded impact crater, perhaps 60 million years old. An alternative theory is that the dome is the eroded stump of a salt deposit.

3 Mountains
Rising to a towering 16,000 ft (5,000 m), the snow-capped peaks of the eastern Himalayas in Tibet and southwest China are among the highest on Earth. In this image from the Terra satellite, vegetation on the lower slopes appears red, while rivers are blue. Thrown up by the collisions of the Eurasian and Indian plates 50–70 million years ago, the Himalayas are still rising at the rate of ¾ in (2 cm) per year.

4 Salt pans
Algae growing in salt evaporation ponds in a coastal lagoon near Alexandria, Egypt, paint the landscape vivid red in this aerial view. Seawater is trapped in ponds for salt extraction; once the water has evaporated, the salt is harvested. As the salinity of the ponds changes, different algae flourish, turning the ponds from green to orange and red.

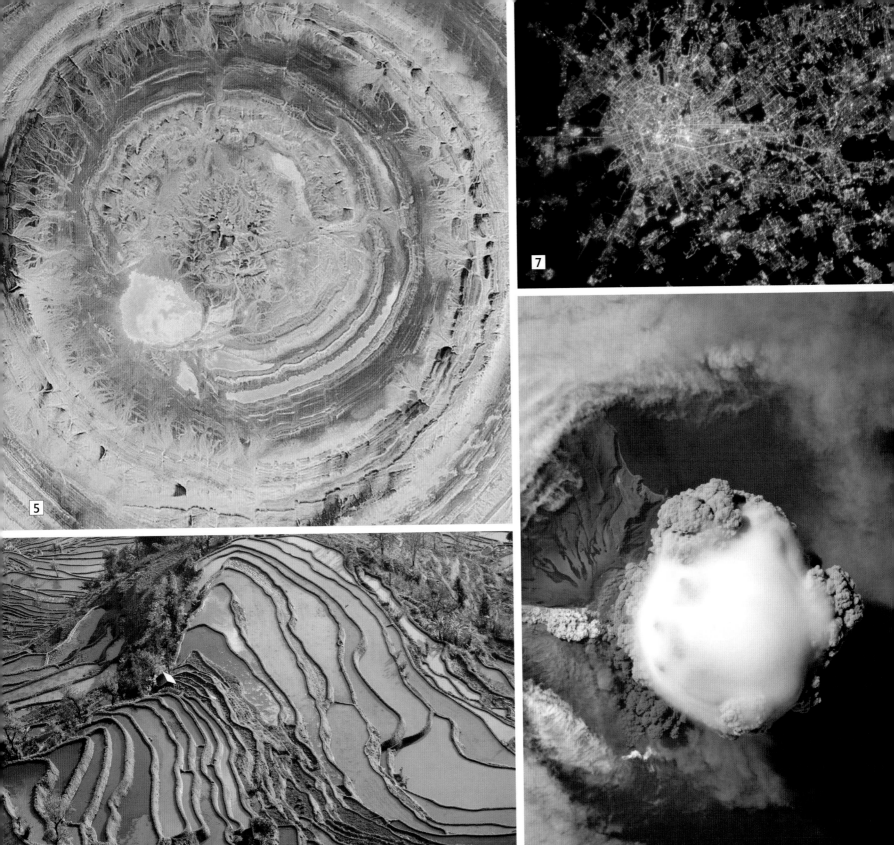

5 Desert
Known as Earth's Bull's-Eye, the Richat Structure in the Sahara Desert is a landmark for astronauts. When viewed from space, the 31-mile- (50-km-) wide rock dome stands out amid the otherwise featureless landscape. The circular structure probably formed as a result of uplift of layered sedimentary rock that became exposed to erosion.

6 Farmland
Carved into the mountainsides of Yunnan Province, China, terraced fields turn the landscape into an abstract patchwork. The fields at lower altitude are warm enough for rice to flourish, while those at higher elevations are used to grow hardier crops, such as corn. Viewed from above, the extent of human impact on the landscape becomes apparent— the mountain is a monument to agriculture.

7 City
At night, cities light up. In this photograph from the International Space Station, Milan illuminates the region of Lombardy, Italy. From space, the extent of city growth and light pollution is obvious. The brightest areas of Earth are the most urbanized. More than a century after the invention of the electric light, some regions remain unpopulated and entirely unlit— Antarctica is still in the dark.

8 Volcano
Viewed from the International Space Station, Sarychev Peak in the Kuril Islands erupts. At 4,900 ft (1,500 m), it is dwarfed by the volcano Olympus Mons on Mars, which stands nearly 74,000 ft (22,000 m) tall. With around 60 currently active volcanoes, Earth is much quieter, geologically speaking, than Jupiter's moon Io, which has over 400 and is the most volcanically active body in the solar system.

OUR **PLANET**

FOR THOUSANDS OF YEARS, PEOPLE HAVE TRIED TO UNDERSTAND EARTH'S STRUCTURE AND WORKINGS. KNOWLEDGE HAS ACCUMULATED SLOWLY, WITH MANY KEY THEORIES DEVELOPING IN THE LAST FEW DECADES.

Unlike other celestial bodies, Earth was not visible to people in its entirety until the first camera-carrying satellites were launched in the 1960s. Nevertheless, more than 2,000 years ago the proto-scientists of ancient civilizations worked out that our planet is a sphere and gained some idea of its size and the extent of its oceans. It was not until the 20th century that Earth's age and internal structure were confirmed and the tectonic plates that move continents were discovered.

The world as understood by Greek philosopher Anaximander (c. 610–546 BCE)

c. 3000–500 BCE

Flat Earth
Ancient Mediterranean societies believe that the world's landmasses sit on the flat surface of a disk, possibly surrounded by sea. The concept of a flat Earth is illustrated in some early maps.

c. 330 BCE

Aristotle claims Earth is a sphere
Greek philosopher Aristotle reasons that Earth is a sphere. In support of his theory, he points out that some stars become visible only as a person travels far to the south. If Earth were flat, he argues, the same stars would be visible everywhere.

200 million years ago

130 million years ago

70 million years ago

Today

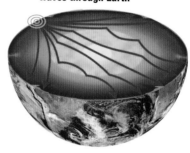

Passage of earthquake waves through Earth

1912

Wegener's continental drift
German scientist Alfred Wegener proposes that all the continents were once joined together and have since moved apart through an unknown mechanism, which he calls "continental drift." His ideas are rejected by most other scientists.

1906

Evidence for Earth's core
From the study of earthquake waves and their behavior as they pass through Earth, Irish geologist Richard Oldham concludes that Earth has a distinct core. This, he suggests, is denser than the rest of the planet and slows waves that pass through it.

1830s–40s

Ice age theory
Swiss geologist Louis Agassiz and other scientists study glacier-eroded landscapes in the Alpine regions of Europe. Agassiz is the first to propose that Earth has been through an ice age in the relatively recent past.

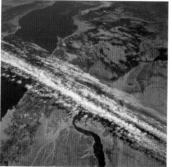

Jet stream clouds over southern Egypt

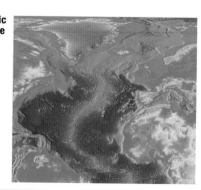

Mid-Atlantic ocean ridge

1920s–30s

Jet streams discovered
From the study of balloon and high-altitude aircraft flights, scientists in Japan, the US, and Europe understand that fast-moving, narrow air currents flow west to east in Earth's atmosphere. These currents come to be known as jet streams.

1955

Earth's age established
US geochemist Clair Patterson establishes that Earth is 4.55 billion years old, by measuring the ratios of lead isotopes in meteorites that formed in the early solar system. This is known as radiometric dating.

1960

Seafloor spreading
US geologist Harry Hess proposes that new seafloor is constantly created at mid-ocean ridges, from which it then slowly spreads away. This concept, which is quickly accepted, is key to the later development of the theory of plate tectonics.

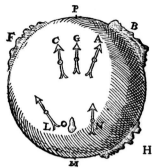

Gilbert's model of magnetic Earth

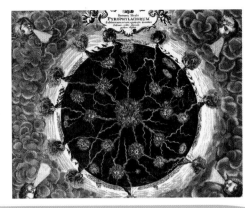

Earth's interior as imagined by Kircher

c. 240 BCE

Circumference calculated
Greek scholar Eratosthenes makes the first accurate calculation of Earth's circumference. To do so, he compares the Sun's elevation above the horizon at two different places, one far south of the other, at the same date and time.

1600 CE

Magnetic Earth
After studying the behavior of compass needles, English scientist William Gilbert, in his book *De Magnete*, suggests that Earth is a giant, spherical magnet. He correctly proposes that its center is made mostly of iron.

1600s

Looking inward
Many ideas are advanced about Earth's internal structure. In England, Edmond Halley claims our planet contains concentric, gas-filled spheres, while German scholar Athanasius Kircher thinks it might hold enormous, interlinked, fiery chambers.

Georges Cuvier (1769–1832)

James Hutton (1726–97)

1810s–20s

Cuvier's catastrophism
French naturalist Georges Cuvier champions the theory of catastrophism—the idea that many catastrophes have occurred over history, altering our planet suddenly rather than gradually, and killing off large groups of animal species.

1798

Cavendish weighs Earth
English scientist Henry Cavendish calculates Earth's average density by means of gravity-measuring experiments. Using his findings, Earth's mass can also be calculated, so Cavendish is said to have "weighed the world."

1785

Hutton's geological theory
In his book, *Theory of the Earth*, Scottish scientist James Hutton—considered the father of modern geology—proposes that Earth was shaped entirely by slow-moving forces still in operation today, acting over long periods of time.

Earth's tectonic plates

Impact causes extinction event

Late 1960s

Theory of plate tectonics
Building on the ideas of seafloor spreading and continental drift, researchers explore the possibility that Earth's outer shell is split into a dozen or so moving plates. This theory of plate tectonics revolutionizes Earth science.

1980

Explaining the extinction of dinosaurs
US physicist Luis Alvarez and others propose that a large asteroid or comet impacting Earth 65.5 million years ago (at the end of the Cretaceous period) caused extinction of the dinosaurs and many other animal and plant groups.

Late 20th century

Anthropocene
A new name, Anthropocene, is proposed for the modern era, during which human activity has begun to have a profound effect on the planet, its climate, and its natural ecosystems.

THE **MOON**

THE MOON, EARTH'S COMPANION IN SPACE, IS OUR PLANET'S LONE SATELLITE. IT IS THE LARGEST AND BRIGHTEST OBJECT IN THE NIGHT SKY AND THE ONLY ONE WHOSE SURFACE FEATURES CAN BE EASILY SEEN WITH THE NAKED EYE.

With a diameter one-quarter of Earth's, the Moon is the largest satellite in the solar system compared to its parent planet. Earth and the Moon exert a powerful influence on each other through their gravity. Tidal forces have slowed the Moon's rotation so that it spins once on its axis in the same time it takes to orbit Earth (27.32 days), and consequently keeps one face permanently toward our planet.

The Moon is a barren ball of rock that lacks sufficient gravity to hold on to a substantial atmosphere. Exposed alternately to the heat of the Sun and the emptiness of space, the lunar surface experiences wild temperature swings, from 248°F (120°C) at local noon to –274°F (–170°C) in the middle of the long lunar night. The floors of permanently shadowed craters get even colder.

With no weather or tectonic activity to erase craters, much of the Moon's battered landscape preserves a barely altered record of conditions in our part of the solar system over the last 4 billion years.

Hermite Crater, near the lunar north pole, is one of the coldest places in the solar system.

MOON DATA

Average diameter	2,159 miles (3,474 km)
Mass (Earth = 1)	0.012
Gravity at equator (Earth = 1)	0.167
Mean distance from Earth	239,000 miles (385,000 km)
Axial tilt	1.5°
Rotation period (day)	27.32 Earth days
Orbital period (year)	27.32 Earth days
Minimum temperature	–413°F (–247°C)
Maximum temperature	248°F (120°C)

▷ **Northern hemisphere**
The Moon orbits bolt upright in relation to the Sun, so its polar regions receive horizontal sunlight. As a result, crater floors near the poles can be permanently shadowed and may contain water ice.

▷ **Far side**
The hemisphere that faces away from Earth is more densely cratered than the near side. It has fewer of the dark lava plains, or maria, that dominate the near side, and those it does have are relatively small.

▷ **Southern hemisphere**
The south pole is located at the edge of a vast impact crater called the South Pole–Aitken Basin. Smaller craters within it contain areas of permanent shadow and ice from collisions with comets.

At 698 miles (1,123 km) across, the Mare Imbrium (Sea of Rains) is one of the largest lunar maria. It is ringed by mountains thrown up by the meteor strike that formed the Imbrium impact basin.

Oceanus Procellarum (Ocean of Storms)

The Aristarchus Crater is a relatively young impact crater (450 million years old) and one of the Moon's brightest features.

Grimaldi Crater

Copernicus Crater has high central peaks and terraced walls.

Mare Humorum (Sea of Moisture)

Mare Nubium (Sea of Clouds)

The Clavius Crater is a huge ancient crater in the southern highlands. It is 140 miles (225 km) in diameter.

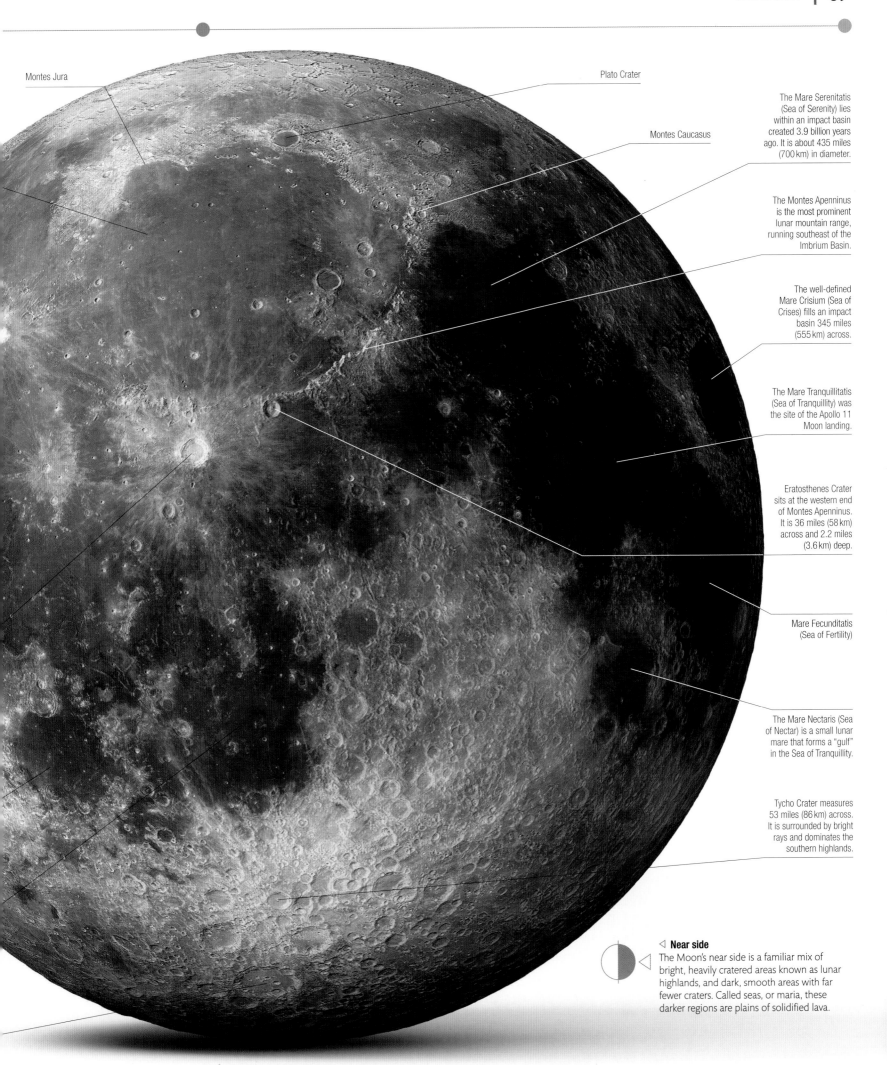

Montes Jura

Plato Crater

Montes Caucasus

The Mare Serenitatis (Sea of Serenity) lies within an impact basin created 3.9 billion years ago. It is about 435 miles (700 km) in diameter.

The Montes Apenninus is the most prominent lunar mountain range, running southeast of the Imbrium Basin.

The well-defined Mare Crisium (Sea of Crises) fills an impact basin 345 miles (555 km) across.

The Mare Tranquillitatis (Sea of Tranquillity) was the site of the Apollo 11 Moon landing.

Eratosthenes Crater sits at the western end of Montes Apenninus. It is 36 miles (58 km) across and 2.2 miles (3.6 km) deep.

Mare Fecunditatis (Sea of Fertility)

The Mare Nectaris (Sea of Nectar) is a small lunar mare that forms a "gulf" in the Sea of Tranquillity.

Tycho Crater measures 53 miles (86 km) across. It is surrounded by bright rays and dominates the southern highlands.

◁ Near side

The Moon's near side is a familiar mix of bright, heavily cratered areas known as lunar highlands, and dark, smooth areas with far fewer craters. Called seas, or maria, these darker regions are plains of solidified lava.

MOON STRUCTURE

AS A RELATIVELY SMALL BODY, THE MOON HAS COOLED CONSIDERABLY IN THE 4.5 BILLION YEARS SINCE ITS FORMATION. ITS ROCKY INTERIOR HAS LARGELY SOLIDIFIED AROUND A CORE OF RED-HOT OR PARTIALLY MOLTEN IRON.

The Moon's proximity to Earth has permitted scientists to investigate its inner structure in detail. Using seismometers placed on the surface by astronauts during the Apollo Moon landings, geologists can map the lunar interior by measuring the properties of moonquakes—seismic tremors triggered when tidal forces distort the shape of the Moon, or when meteorite impacts send shock waves through its interior.

More recently, spacecraft, including NASA's twin GRAIL (Gravity Recovery and Interior Laboratory) satellites, have mapped the Moon's structure by measuring slight variations in its gravitational field.

The Moon's inner core is remarkably small, only around **150 miles (240 km)** across.

Periodic moonquakes occur 600 miles (1,000 km) or more below the surface.

◁ **Birth of the Moon**
Studies of lunar rock suggests that the Moon was formed around 4.5 billion years ago, when a Mars-sized world called Theia collided with the still-molten Earth. The impact obliterated Theia and blasted huge amounts of debris into orbit around Earth. Over time, much of this material came together to form a single large satellite—the Moon.

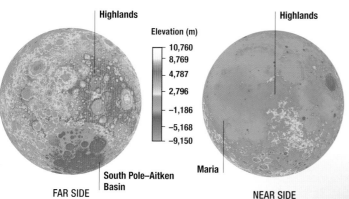

Highlands

Elevation (m)

10,760
8,769
4,787
2,796
−1,186
−5,168
−9,150

Highlands

◁ **Surface elevation**
The Moon's highest regions are on its far side, which is on average 3 miles (5 km) higher than the near side. The lowest region—the 8-mile- (13-km-) deep South Pole-Aitken Basin—is also on the far side. Low-lying lava plains called maria (seas) cover 31 percent of the Moon's near side.

South Pole–Aitken Basin

Maria

FAR SIDE

NEAR SIDE

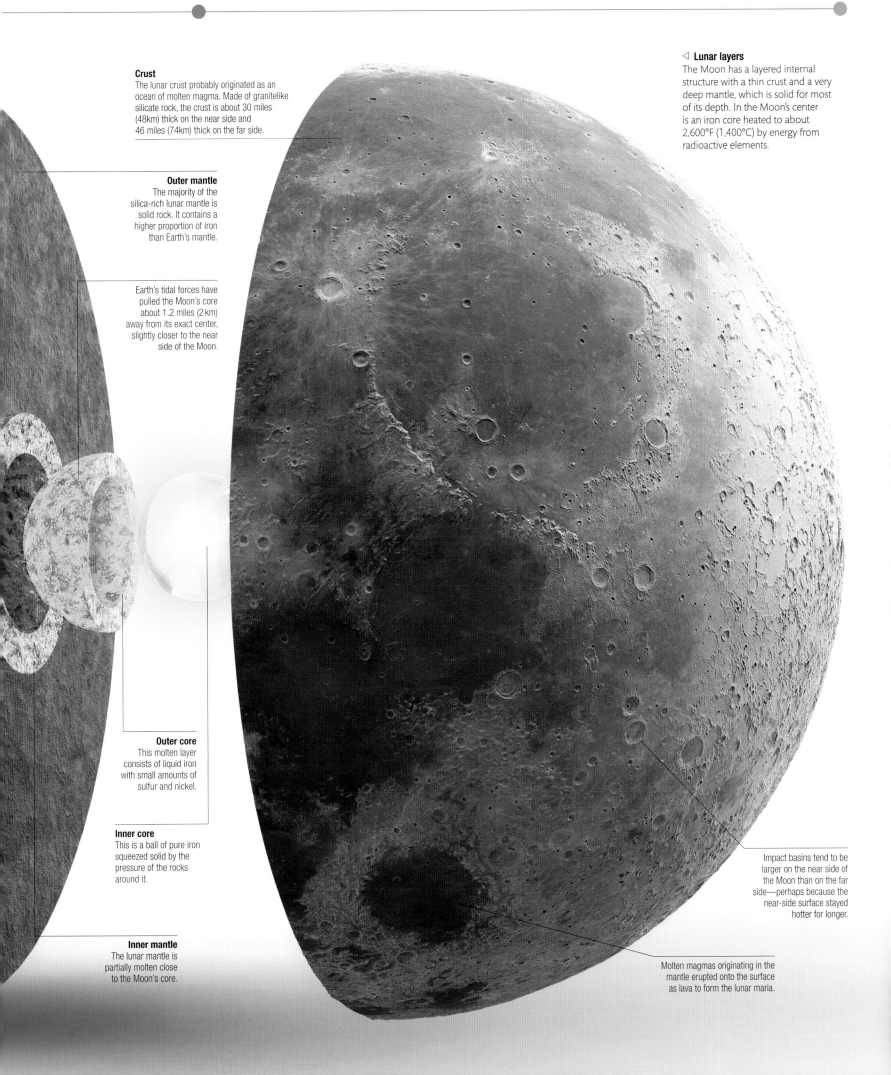

Crust
The lunar crust probably originated as an ocean of molten magma. Made of granitelike silicate rock, the crust is about 30 miles (48km) thick on the near side and 46 miles (74km) thick on the far side.

Outer mantle
The majority of the silica-rich lunar mantle is solid rock. It contains a higher proportion of iron than Earth's mantle.

Earth's tidal forces have pulled the Moon's core about 1.2 miles (2 km) away from its exact center, slightly closer to the near side of the Moon.

Outer core
This molten layer consists of liquid iron with small amounts of sulfur and nickel.

Inner core
This is a ball of pure iron squeezed solid by the pressure of the rocks around it.

Inner mantle
The lunar mantle is partially molten close to the Moon's core.

◁ **Lunar layers**
The Moon has a layered internal structure with a thin crust and a very deep mantle, which is solid for most of its depth. In the Moon's center is an iron core heated to about 2,600°F (1,400°C) by energy from radioactive elements.

Impact basins tend to be larger on the near side of the Moon than on the far side—perhaps because the near-side surface stayed hotter for longer.

Molten magmas originating in the mantle erupted onto the surface as lava to form the lunar maria.

EARTH'S **COMPANION**

THE MOON IS SO LARGE COMPARED TO EARTH, AND ORBITS SO CLOSE TO ITS PARENT PLANET, THAT THE TWO WORLDS EXERT A CONSIDERABLE INFLUENCE ON EACH OTHER THROUGH THE FORCE OF GRAVITY.

The Moon's large size relative to its parent planet is due to the unique way in which it formed. Most natural satellites in the solar system either formed from leftover debris after the new planet took shape or are small, captured objects such as asteroids. Consequently, moons are usually dwarfed by their parent planet. Earth's moon, in contrast, formed after a collision between Earth and another planet created a huge cloud of debris. Today, separated by an average distance of 238,900 miles (384,400 km), Earth and its moon exert a strong gravitational pull on one another that generates tidal forces in both worlds. These forces have slowed the Moon's period of rotation and raise substantial tides in Earth's oceans.

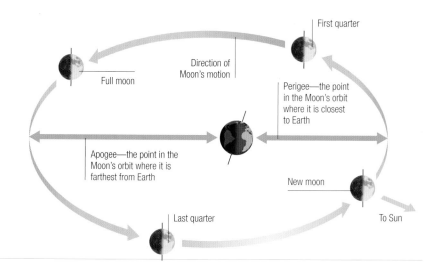

△ **Spin and orbit**
Tidal forces pulling at irregularities in the Moon's spherical shape have caused our satellite's rotation to slow and its orbital position to drift slowly outward. The Moon has developed a synchronous rotation, meaning it spins once on its axis with each orbit of Earth. As a result, one hemisphere—the near side—always faces Earth, while the far side is forever turned away.

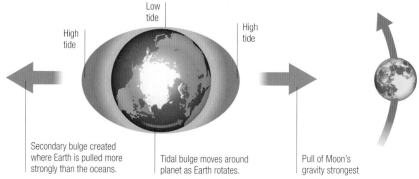

△ **Tidal forces**
Tidal forces arise because different parts of a celestial body experience different gravitational forces, depending on their distance from another object. Earth's oceans are lifted slightly on the side nearest the Moon, causing a high tide. On the opposite side of the planet, a second bulge occurs when the Moon's gravity is weakest.

△ **Lunar eclipse**
Lunar eclipses take place when the Moon passes into Earth's shadow and turns a dark reddish color as it catches sunlight scattered by Earth's atmosphere. Because Earth is much larger than the Moon and its shadow much wider, lunar eclipses are more frequent than solar eclipses.

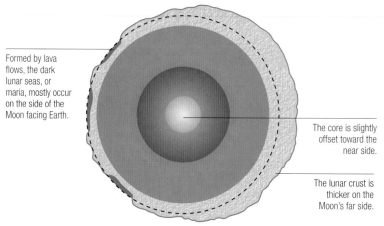

△ **Offset structure**
Early in the Moon's history, tidal forces created by Earth's gravity pulled the core about 1.2 miles (2 km) closer to the near side. The structures surrounding the core are also offset, with the mantle closer to the near side and the crust thicker on the far side. This may explain why the volcanic eruptions that formed the lunar seas were concentrated on the near side.

The lunar month

The Moon's most obvious feature from Earth is its monthly cycle of phases, waxing (growing) from a dark new moon through crescent, first quarter, and gibbous phases to a brilliant full moon drenched in sunlight, before waning (decreasing) back to another new moon. The entire cycle, known as a synodic or lunar month, takes 29.53 days and sees the Moon make a complete eastward circle around the sky relative to the Sun. The lunar month is slightly longer than the Moon's orbital period of 27.32 days because the Sun is also moving eastward through Earth's skies, and it takes a little longer for the Moon to catch up and return to the same relative position.

MOON MAPPED

The large lunar seas, or maria, in the center of the map are familiar features of the Moon's Earth-facing aspect. On the Moon's far side (right on the map), impact craters dominate the terrain.

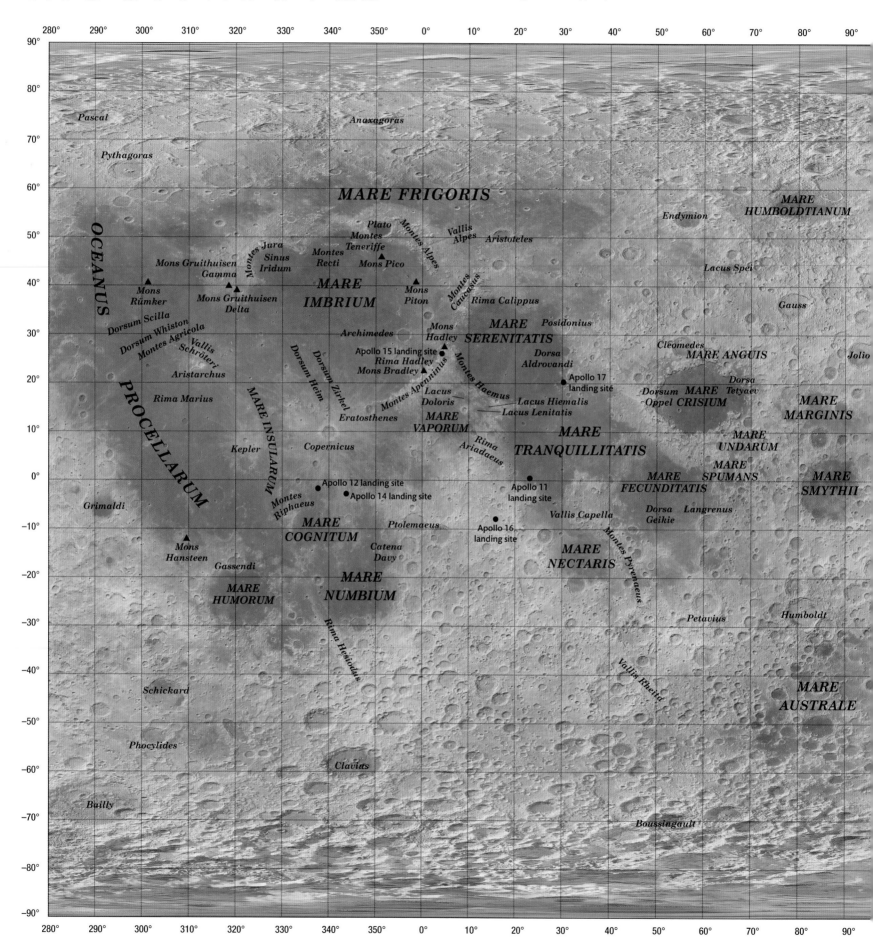

- 280° 290° 300° 310° 320° 330° 340° 350° 0° 10° 20° 30° 40° 50° 60° 70° 80° 90°

Pascal

Anaxagoras

Pythagoras

MARE FRIGORIS

Endymion

MARE HUMBOLDTIANUM

OCEANUS

Mons Gruithuisen Gamma

Montes Jura

Sinus Iridum

Plato

Montes Teneriffe

Montes Alpes

Vallis Alpes

Aristoteles

Lacus Spei

Mons Rümker

Mons Gruithuisen Delta

Montes Recti

Mons Pico

MARE IMBRIUM

Mons Piton

Montes Caucasus

Rima Calippus

MARE SERENITATIS

Posidonius

Cleomedes

MARE ANGUIS

Gauss

Jolio

Dorsum Scilla

Dorsum Whiston

Montes Agricola

Vallis Schröteri

Archimedes

Mons Hadley

Apollo 15 landing site

Rima Hadley

Mons Bradley

Dorsa Aldrovandi

Dorsum Oppel

Dorsa Tetyaev

MARE CRISIUM

MARE MARGINIS

Aristarchus

PROCELLARUM

Rima Marius

Dorsum Zirkel

Dorsum Heim

Montes Apenninus

Montes Haemus

Apollo 17 landing site

Lacus Hiemalis

Lacus Lenitatis

MARE UNDARUM

Kepler

MARE INSULARUM

Copernicus

Eratosthenes

Lacus Doloris

MARE VAPORUM

Rima Ariadaeus

MARE TRANQUILLITATIS

MARE SPUMANS

MARE SMYTHII

Grimaldi

Montes Ripbaeus

Apollo 12 landing site

Apollo 14 landing site

Ptolemaeus

Apollo 11 landing site

MARE FECUNDITATIS

Dorsa Geikie

Langrenus

Mons Hansteen

Gassendi

MARE COGNITUM

Catena Davy

Apollo 16 landing site

Vallis Capella

MARE NECTARIS

Montes Pyrenaeus

MARE HUMORUM

MARE NUMBIUM

Rima Hesiodus

Petavius

Humboldt

Schickard

Vallis Rheita

MARE AUSTRALE

Phocylides

Clavius

Bailly

Boussingault

- 280° 290° 300° 310° 320° 330° 340° 350° 0° 10° 20° 30° 40° 50° 60° 70° 80° 90°

Scale 1:23,566,109

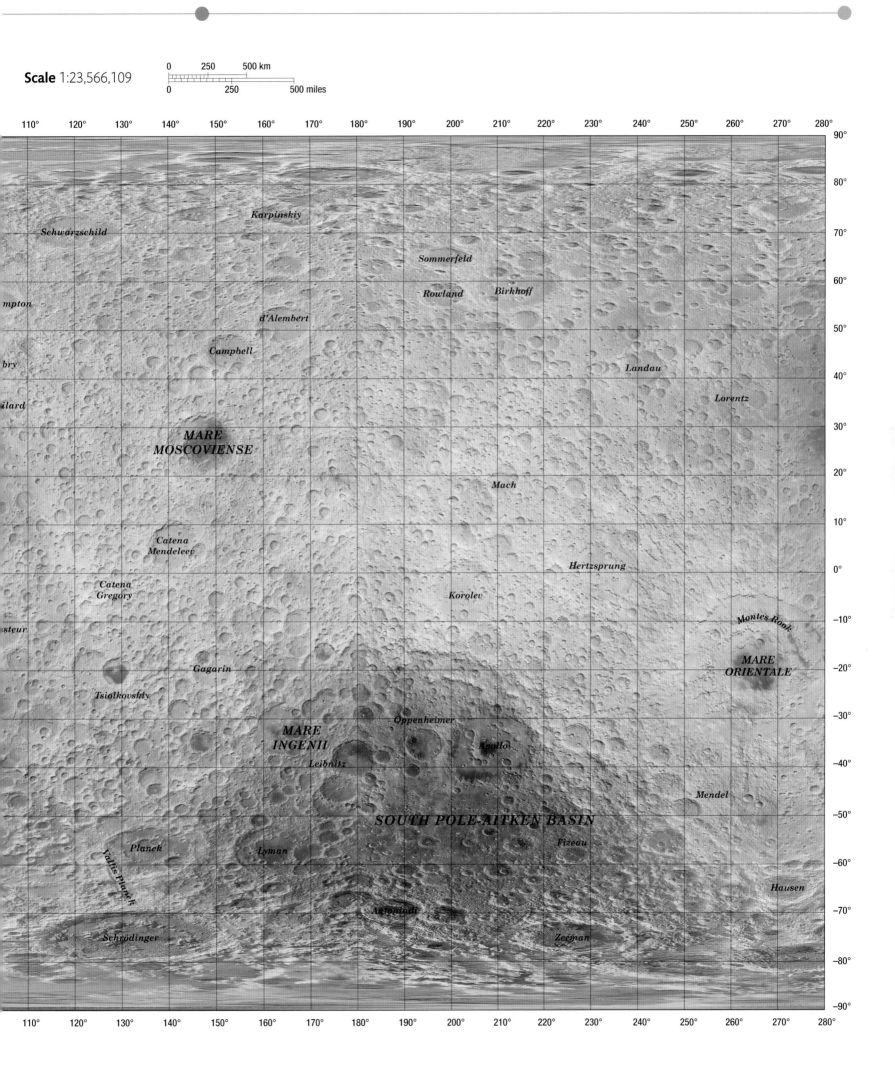

110°   120°   130°   140°   150°   160°   170°   180°   190°   200°   210°   220°   230°   240°   250°   260°   270°   280°

Schwarzschild

Karpinskiy

Sommerfeld

Rowland   *Birkhoff*

mpton

d'Alembert

bry

Campbell

Landau

ilard

Lorentz

MARE MOSCOVIENSE

Mach

Catena Mendeleev

Hertzsprung

Catena Gregory

Korolev

steur

Montes Rook

Gagarin

MARE ORIENTALE

Tsiolkovsky

Oppenheimer

MARE INGENII

Apollo

Leibnitz

Mendel

SOUTH POLE-AITKEN BASIN

Planck

Fizeau

Vallis Planck

Lyman

Hausen

Antoniadi

Schrödinger

Zeeman

110°   120°   130°   140°   150°   160°   170°   180°   190°   200°   210°   220°   230°   240°   250°   260°   270°   280°

DESTINATION **HADLEY RILLE**

A STEEP-SIDED VALLEY RUNNING FOR 60 MILES (100 KM) ACROSS THE LUNAR LANDSCAPE, HADLEY RILLE IS THE REMNANT OF AN ANCIENT LAVA STREAM THAT FLOWED ACROSS THE MOON'S SURFACE AROUND 3 BILLION YEARS AGO.

Hadley Rille lies on the edge of the Mare Imbrium impact basin at the foot of the Montes Apenninus mountain range. The valley originates at an elongated crater called Bela, from where it winds across a plain known as the Palus Putridinus. The Rille is thought to be a lava channel that formed when lava flowed across the lunar surface like a river. This image was taken in 1971, when Hadley Rille was observed during the Apollo 15 mission. Astronaut David Scott left a memorial sculpture and plaque at the site to commemorate astronauts who died in training and on missions.

David Scott with a lunar rover at the Apollo 15 landing site, photographed by James Irwin

LOCATION

Latitude 3°E; longitude 26°N

LAND PROFILE

For much of its length, Hadley Rille is about 0.9 miles (1.5 km) wide and between 600 and 900 ft (180 and 270 m) deep.

Elevation (m)

Profile width (m)

170 LB (77 KG) OF LUNAR MATERIAL WAS RETURNED TO EARTH BY THE APOLLO 15 MISSION.

APOLLO 15

On July 30, 1971, Apollo 15's Lunar Module landed near one of Hadley Rille's deepest points, where the valley plunges to 1,200 ft (370 m). Using the Lunar Roving Vehicle for the first time, the astronauts made three excursions to explore nearby craters and collect rock samples.

— LRV 1
— LRV 2
— LRV 3

Landing site

Hadley Rille

Earthlight

Dune

St. George

EARTHRISE

This awe-inspiring Earthrise—the rising of Earth over the Moon's horizon—was filmed by the Japanese spacecraft Kaguya on April 6, 2008, from an altitude of about 60 miles (100 km). For Kaguya to capture this spectacle, the orbits of the Moon, Earth, Sun, and spacecraft had to line up. Since Earth is almost stationary when viewed from the Moon, an Earthrise is most easily observed from orbiting spacecraft. To see one from the Moon's surface, an astronaut would need to be standing near one of the Moon's poles.

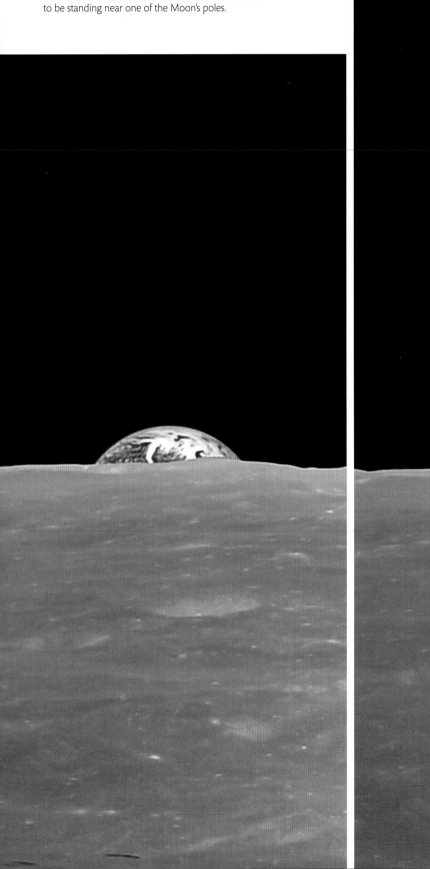
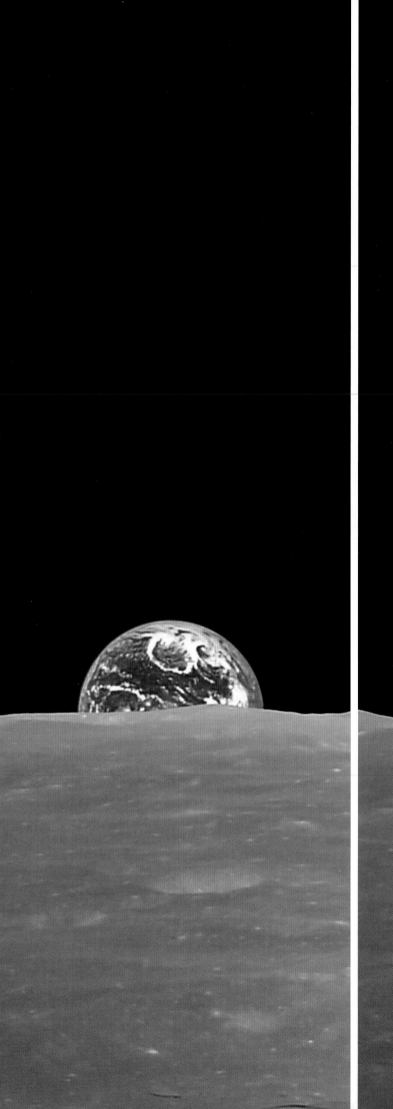

LUNAR CRATERS

THE SURFACE OF THE MOON IS COVERED IN COUNTLESS CRATERS OF ALL SIZES. THEIR FORMATION HAS BEEN THE DRIVING FORCE SHAPING THE LUNAR LANDSCAPE FOR MORE THAN 4.5 BILLION YEARS.

How the Moon's craters formed was not fully understood until the 1960s, when the first robot lunar landers showed that craters of all sizes existed, including tiny ones. This discovery confirmed that the craters must have been caused by impacts from space, rather than by volcanic eruptions.

It is now clear that craters cover all areas of the lunar terrain, although in some places the oldest craters have been covered over by later events, including volcanic eruptions and further impacts. The Moon is not the only heavily cratered body in the solar system, but it is the one we can study in greatest detail.

▽ **How craters form**
The Moon's well-preserved craters have given astronomers a detailed understanding of the crater formation process. The size and shape of a crater are determined mainly by the kinetic energy of the incoming object (a combination of its speed and mass).

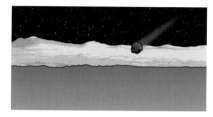

1 Incoming space rock
Meteoroids approach the lunar surface at a variety of speeds, depending on whether they are catching up with the Moon or meeting it head-on.

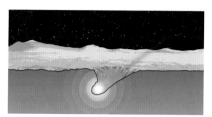

2 Initial impact
The impact creates a shock wave that vaporizes the meteoroid and ripples out into the crust, compressing and heating it along a bowl-shaped shock front.

3 Ejecta blanket
As the shock wave passes, material from the landing site is thrown out, forming a layer of debris on the surrounding landscape known as an ejecta blanket.

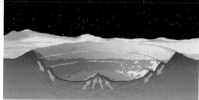

4 Crater
The result is a surface depression. In a large crater, the crust may rebound to form a central peak; the sides may slump under their own weight to form terraces.

◁ **Far side of the Moon**
First revealed by Soviet spacecraft in the late 1950s, the lunar far side looks very different from the side we usually see. It appears more heavily cratered, principally because of the lack of lunar seas, or maria, formed by lava flows. One theory is that the Moon's crust is thicker on the far side, which makes it harder for magma to rise to the surface and create maria. Another possibility is that the far side of the Moon cooled and solidified more quickly than the near side, forming robust rocks that limited the depth of the far side's impact basins.

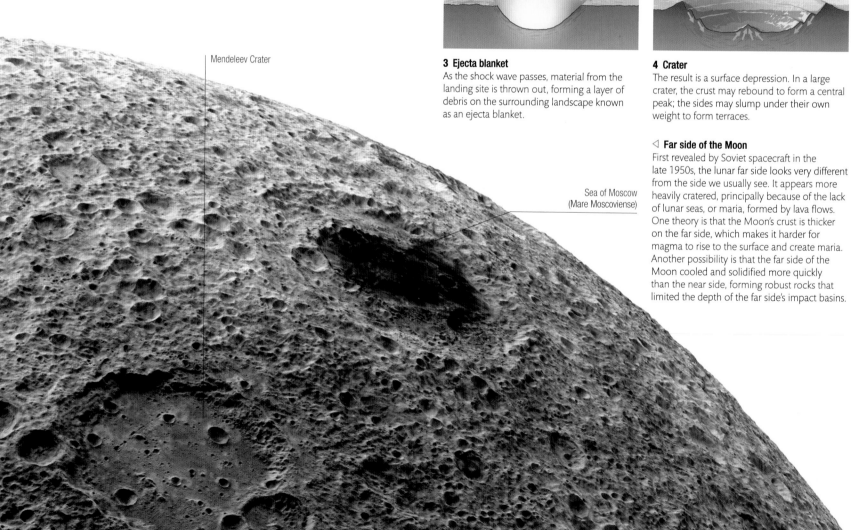

Mendeleev Crater

Sea of Moscow
(Mare Moscoviense)

▷ **Crater map**

This map, created using data from NASA's Lunar Reconnaissance Orbiter, charts the position and size of over 5,000 large impact craters covering parts of the near and far sides of the Moon. The biggest of them have filled in with lava to form basalt plains, producing the lunar maria. Scientists can estimate the age of different parts of the Moon's surface by counting the craters that have accumulated over time.

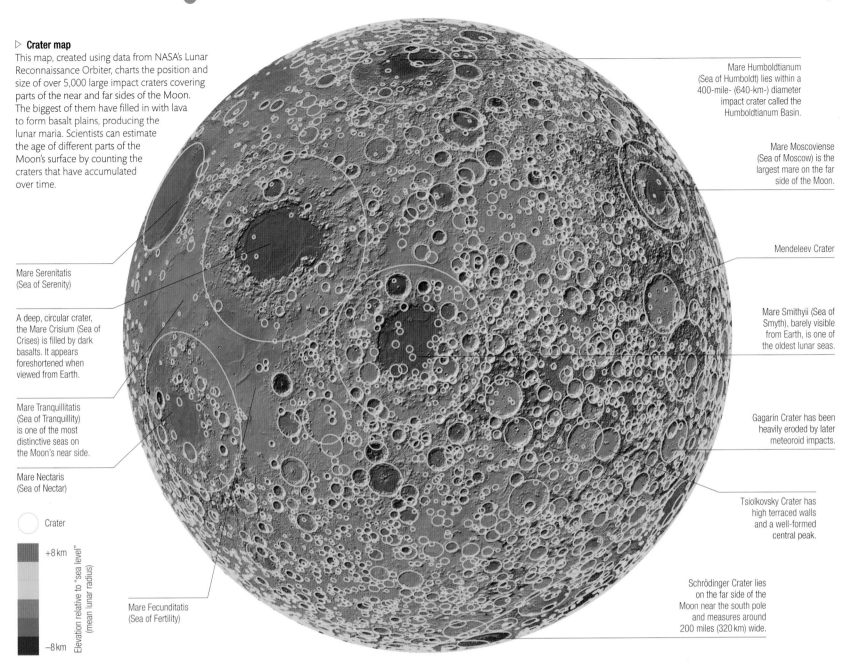

Mare Humboldtianum (Sea of Humboldt) lies within a 400-mile- (640-km-) diameter impact crater called the Humboldtianum Basin.

Mare Moscoviense (Sea of Moscow) is the largest mare on the far side of the Moon.

Mendeleev Crater

Mare Smithyii (Sea of Smyth), barely visible from Earth, is one of the oldest lunar seas.

Gagarin Crater has been heavily eroded by later meteoroid impacts.

Tsiolkovsky Crater has high terraced walls and a well-formed central peak.

Schrödinger Crater lies on the far side of the Moon near the south pole and measures around 200 miles (320 km) wide.

Mare Serenitatis (Sea of Serenity)

A deep, circular crater, the Mare Crisium (Sea of Crises) is filled by dark basalts. It appears foreshortened when viewed from Earth.

Mare Tranquillitatis (Sea of Tranquillity) is one of the most distinctive seas on the Moon's near side.

Mare Nectaris (Sea of Nectar)

Crater

Elevation relative to "sea level" (mean lunar radius)
+8 km
−8 km

Mare Fecunditatis (Sea of Fertility)

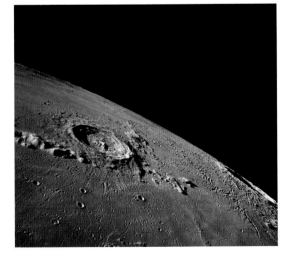

△ **Copernicus Crater**

Easily visible from Earth through binoculars, this young crater, which formed just 800 million years ago, is exceptionally well preserved. Copernicus is surrounded by a ring of enormous, bright rays—debris ejected by the impact—that are most conspicuous at full moon. The crater was the planned landing site for NASA's Apollo 18 manned mission, which was canceled.

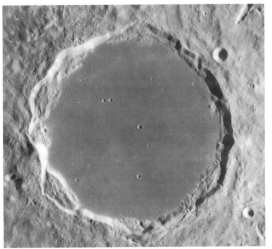

△ **Plato Crater**

This 69-mile- (109-km-) wide crater to the north of the Mare Imbrium has a dark and smooth appearance, thanks to the lavas that flooded its floor. Plato is thought to have formed around 3.84 billion years ago, shortly after the neighboring Mare Imbrium impact basin. The crater's rim has been shaped by a series of landslides.

△ **Plum Crater**

This small crater, measuring only 118 ft (36 m) across, was the first geological stop reached by rover during the Apollo 16 mission of 1972. While searching the crater, 0.9 miles (1.4 km) from their landing site, the astronauts found Big Muley—a 4-billion-year old boulder weighing 26 lb (11.7 kg). Big Muley is the largest rock brought back to Earth by Apollo astronauts.

HIGHLANDS
AND PLAINS

THE LUNAR LANDSCAPE CAN BE BROADLY DIVIDED INTO TWO DISTINCT TYPES OF TERRAIN: BRIGHT, HEAVILY CRATERED HIGHLANDS AND RELATIVELY SMOOTH, DARK PLAINS KNOWN AS LUNAR SEAS, OR MARIA.

The highlands represent the original ancient crust of the Moon, formed as its surface began to solidify from a molten magma ocean 4.5 billion years ago. They are dominated by bright silicate minerals similar to those of Earth's crust, and they feature countless craters laid one on top of another over billions of years. The maria, meanwhile, are flat and sparsely cratered plains consisting of dark basaltic lavas. Studies of the boundaries between the two regions show that the lunar maria are later surfaces that have erased all traces of earlier craters.

Mare Frigoris (Sea of Cold)

△ **Montes Apenninus**
Named after an Italian mountain range, the lunar Apennines form one of the largest and most prominent mountain chains on the Moon. This range of mountains, approximately 370 miles (600 km) long and up to 3 miles (5 km) high, was thrown up during the formation of the Mare Imbrium basin around 3.9 billion years ago.

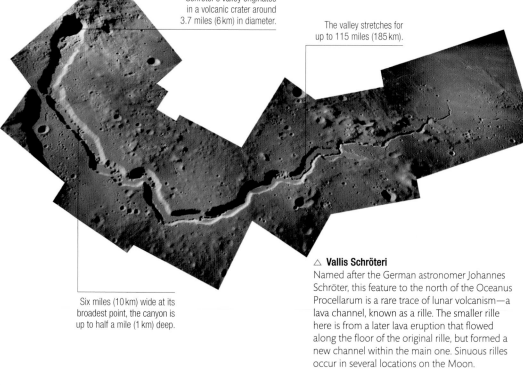

Schröter's Valley originates in a volcanic crater around 3.7 miles (6 km) in diameter.

The valley stretches for up to 115 miles (185 km).

Six miles (10 km) wide at its broadest point, the canyon is up to half a mile (1 km) deep.

△ **Vallis Schröteri**
Named after the German astronomer Johannes Schröter, this feature to the north of the Oceanus Procellarum is a rare trace of lunar volcanism—a lava channel, known as a rille. The smaller rille here is from a later lava eruption that flowed along the floor of the original rille, but formed a new channel within the main one. Sinuous rilles occur in several locations on the Moon.

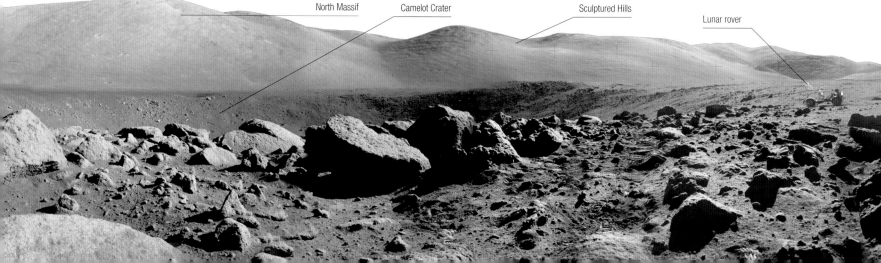

North Massif

Camelot Crater

Sculptured Hills

Lunar rover

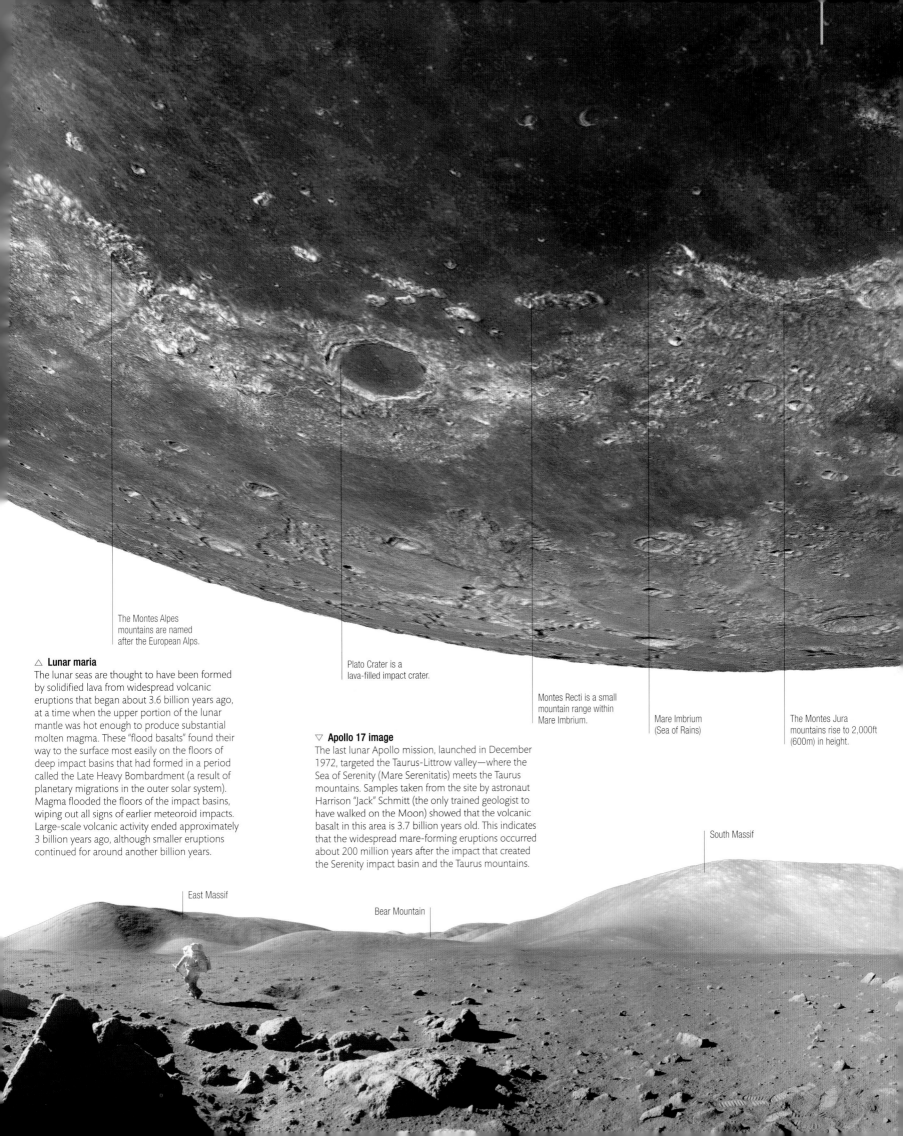

The Montes Alpes mountains are named after the European Alps.

Plato Crater is a lava-filled impact crater.

Montes Recti is a small mountain range within Mare Imbrium.

Mare Imbrium (Sea of Rains)

The Montes Jura mountains rise to 2,000ft (600m) in height.

△ **Lunar maria**
The lunar seas are thought to have been formed by solidified lava from widespread volcanic eruptions that began about 3.6 billion years ago, at a time when the upper portion of the lunar mantle was hot enough to produce substantial molten magma. These "flood basalts" found their way to the surface most easily on the floors of deep impact basins that had formed in a period called the Late Heavy Bombardment (a result of planetary migrations in the outer solar system). Magma flooded the floors of the impact basins, wiping out all signs of earlier meteoroid impacts. Large-scale volcanic activity ended approximately 3 billion years ago, although smaller eruptions continued for around another billion years.

▽ **Apollo 17 image**
The last lunar Apollo mission, launched in December 1972, targeted the Taurus-Littrow valley—where the Sea of Serenity (Mare Serenitatis) meets the Taurus mountains. Samples taken from the site by astronaut Harrison "Jack" Schmitt (the only trained geologist to have walked on the Moon) showed that the volcanic basalt in this area is 3.7 billion years old. This indicates that the widespread mare-forming eruptions occurred about 200 million years after the impact that created the Serenity impact basin and the Taurus mountains.

South Massif

East Massif

Bear Mountain

STORY OF THE **MOON**

THE BIGGEST AND BRIGHTEST OBJECT IN THE NIGHT SKY HAS ALWAYS BEEN AN INVITING SUBJECT TO STUDY, AND PEOPLE HAVE TRACKED ITS MONTHLY CYCLE OF PHASES SINCE PREHISTORIC TIMES.

Moon-watching was important to the first agricultural societies of the Stone Age because the Moon's phases served as a calendar, telling farmers when to sow and harvest crops. By Babylonian times, astronomers not only understood the phases but could predict lunar eclipses, and by Greek times they knew the Moon was spherical and caused tides. Over the following centuries, our understanding of the Moon progressed in small steps as more details came to light: the nature of its rugged surface, its elliptical orbit, and its lack of air. But the giant leap in understanding came in the 20th century when the Moon became the first alien world people have set foot on.

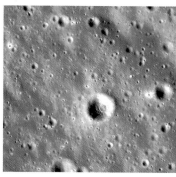

Lunar eclipse

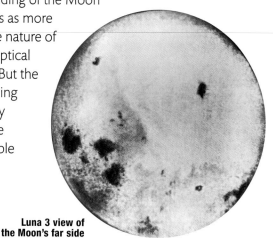

Luna 3 view of the Moon's far side

C. 20,000 BCE
Prehistoric calendar
In the Ishango region of central Africa, people mark a bone with a series of notches that appear to track the monthly cycle and phases of the Moon. Modern researchers believe the Ishango bone is an early lunar calendar.

500 BCE
Predicting eclipses
Babylonian astronomers (based in what is now Iraq) keep detailed records of lunar eclipses. They discover that eclipses occur in a repeating cycle and gain the ability to predict when eclipses will occur.

Impact craters

1959
The far side
The Soviet spacecraft Luna 3 returns the first photographs of the far side of the Moon, which has never been seen before. These images reveal a heavily cratered surface that has fewer dark, flat regions, or maria, than the near side.

1873
Impact theory
English astronomer Richard Proctor proposes that the Moon's craters are caused by meteorite impacts and not, as generally believed, by volcanic activity. Proctor's view is not fully accepted by astronomers until well into the 20th century.

1757
Moon mass measured
The French astronomer Alexis Clairaut, one of the leading mathematicians of the time, makes the first accurate measurement of the mass of the Moon, using the results of his observations to hone Isaac Newton's early calculations.

Luna 9

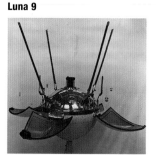

Apollo 17 landing

Lunar formation theory

1966
First soft landing
Another Soviet spacecraft, Luna 9, is the first to make a soft landing on the Moon. It confirms that lunar soil is firm enough to support the weight of a landing craft and that people will be able to walk on the Moon's surface without sinking.

1969–72
Crewed missions
During the Apollo series of lunar missions, US astronauts land on the Moon, place measuring apparatus on its surface, and collect rock samples. Analysis of the samples greatly increases knowledge of the Moon's surface composition, formation, and history.

1980s
Origins understood
There is now agreement among scientists on the origins of the Moon. The favored hypothesis is that the Moon formed from a ring of debris around Earth—the aftermath of a collision between our planet and a planet the size of Mars.

**Hipparchus at the
Alexandria Observatory**

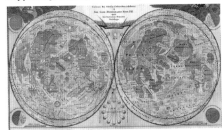

**Galileo's sketch
of the Moon**

▷ c. 450 BCE

Moonshine explained
The Greek scholar Anaxagoras makes the first recorded claim that the Moon shines by reflecting light from the Sun. His theories on the cosmos are advanced for the time. His belief that the Moon and Sun are not deities leads to his prosecution for impiety.

▷ c. 130 BCE

Distance measured
By comparing observations made at the Egyptian cities of Syene (now Aswan) and Alexandria during a total eclipse of the Sun, Greek astronomer Hipparchus calculates the average distance from Earth to the Moon.

▷ 1609 CE

First telescopic study
Italian scientist Galileo Galilei is the first person to scrutinize the Moon through a telescope. He notes that its surface is not smooth, as previously thought, but has mountains, craters, and flat, dark areas that are later called maria (seas).

Newton's cannonball diagram

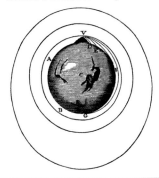

Doppelmayr's comparative map

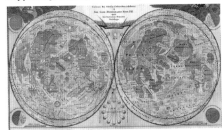

◁ 1753

Thin atmosphere recognized
Croatian astronomer Roger Boscovich argues that the Moon has a negligible atmosphere. This theory is based on his observation that stars disappear instantly as the Moon passes in front of them, rather than fading over a few seconds.

◁ 1680s

Lunar orbit explained
English scientist Isaac Newton develops his theory of gravitation by studying the mathematical properties of elliptical orbits. He uses the analogy of a cannonball to show that the Moon remains in orbit because it is perpetually falling.

◁ 1645–51

First detailed maps
The first detailed Moon maps are made in Germany by Johannes Hevelius and in Italy by Giovanni Riccioli, who includes names still used today. Later (in 1742), German astronomer Johann Doppelmayr makes a comparative map of the two versions.

Excess hydrogen (blue) at south pole

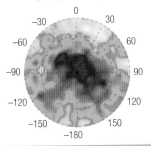

**Lunar
Reconnaissance
Orbiter (USA)**

▷ 1994

Clementine mission
US orbiter Clementine maps the elevation of the lunar surface in detail and returns ultraviolet and infrared images, which enable scientists to map the concentration of different minerals in the Moon's surface.

▷ 1998

Possibility of ice at lunar poles
Another US orbiter, Lunar Prospector, detects excess hydrogen at the Moon's poles. This suggests the presence of water ice in the upper few yards (meters) of the lunar surface, within permanently shadowed craters.

▷ 2004–present

Further missions
The US, Japan, China, India, and the European Space Agency send orbiters to the Moon. These return new data about the Moon's internal structure and the distribution of water and other chemicals in or close to the surface.

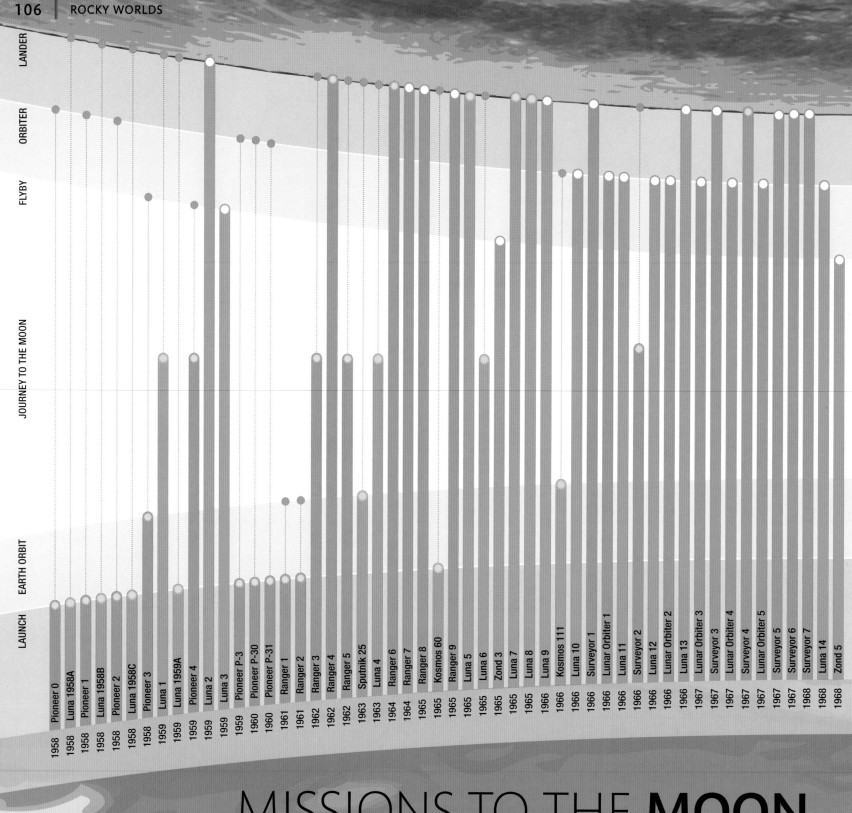

LANDER

ORBITER

FLYBY

JOURNEY TO THE MOON

EARTH ORBIT

LAUNCH

Pioneer 0 — 1958
Luna 1958A — 1958
Pioneer 1 — 1958
Luna 1958B — 1958
Pioneer 2 — 1958
Luna 1958C — 1958
Pioneer 3 — 1958
Luna 1 — 1959
Luna 1959A — 1959
Pioneer 4 — 1959
Luna 2 — 1959
Luna 3 — 1959
Pioneer P-3 — 1959
Pioneer P-30 — 1960
Pioneer P-31 — 1960
Ranger 1 — 1961
Ranger 2 — 1961
Ranger 3 — 1962
Ranger 4 — 1962
Ranger 5 — 1962
Sputnik 25 — 1963
Luna 4 — 1963
Ranger 6 — 1964
Ranger 7 — 1964
Ranger 8 — 1965
Kosmos 60 — 1965
Ranger 9 — 1965
Luna 5 — 1965
Luna 6 — 1965
Zond 3 — 1965
Luna 7 — 1965
Luna 8 — 1965
Luna 9 — 1966
Kosmos 111 — 1966
Luna 10 — 1966
Surveyor 1 — 1966
Lunar Orbiter 1 — 1966
Luna 11 — 1966
Surveyor 2 — 1966
Luna 12 — 1966
Lunar Orbiter 2 — 1966
Luna 13 — 1966
Lunar Orbiter 3 — 1967
Surveyor 3 — 1967
Lunar Orbiter 4 — 1967
Surveyor 4 — 1967
Lunar Orbiter 5 — 1967
Surveyor 5 — 1967
Surveyor 6 — 1967
Surveyor 7 — 1968
Luna 14 — 1968
Zond 5 — 1968

KEY

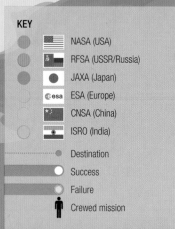

- NASA (USA)
- RFSA (USSR/Russia)
- JAXA (Japan)
- ESA (Europe)
- CNSA (China)
- ISRO (India)
- Destination
- Success
- Failure
- Crewed mission

MISSIONS TO THE **MOON**

OUR NEAREST NEIGHBOR HAS BEEN A TARGET FOR SPACECRAFT FOR OVER 50 YEARS. IT REMAINS THE ONLY DESTINATION BEYOND LOW EARTH ORBIT THAT CREWED CRAFT HAVE VISITED, AND THE ONLY SOLAR SYSTEM BODY BESIDES EARTH THAT PEOPLE HAVE WALKED ON.

After a string of failures in the 1950s, the first craft to reach the Moon's surface was the Soviet probe Luna 2, which deliberately crash-landed in 1959. Three weeks later, Luna 3 returned the first photos of the far side, causing great excitement. Dozens of missions followed as the US and USSR raced to conquer the new frontier of space. More recent missions have aimed to undertake scientific research, but the Moon remains a compelling target for nations eager to demonstrate technological prowess.

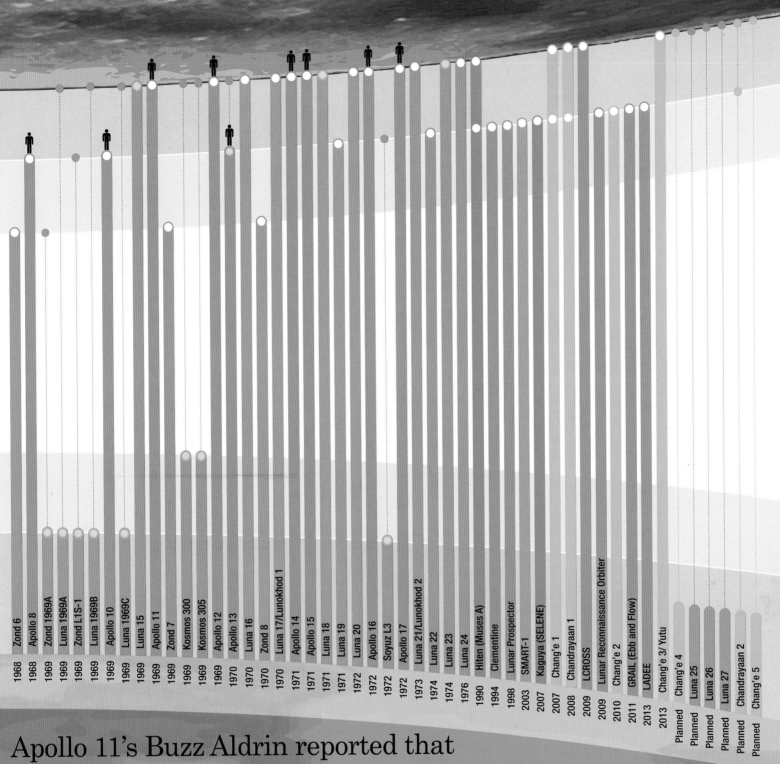

1968	Zond 6
1968	Apollo 8
1969	Zond 1969A
1969	Luna 1969A
1969	Zond L1S-1
1969	Luna 1969B
1969	Apollo 10
1969	Luna 1969C
1969	Luna 15
1969	Apollo 11
1969	Zond 7
1969	Kosmos 300
1969	Kosmos 305
1969	Apollo 12
1970	Apollo 13
1970	Luna 16
1970	Zond 8
1970	Luna 17/Lunokhod 1
1971	Apollo 14
1971	Apollo 15
1971	Luna 18
1971	Luna 19
1972	Luna 20
1972	Apollo 16
1972	Soyuz L3
1972	Apollo 17
1973	Luna 21/Lunokhod 2
1974	Luna 22
1974	Luna 23
1976	Luna 24
1990	Hiten (Muses A)
1994	Clementine
1998	Lunar Prospector
2003	SMART-1
2007	Kaguya (SELENE)
2007	Chang'e 1
2008	Chandrayaan 1
2009	LCROSS
2009	Lunar Reconnaissance Orbiter
2010	Chang'e 2
2011	GRAIL (Ebb and Flow)
2013	LADEE
2013	Chang'e 3/Yutu
Planned	Chang'e 4
Planned	Luna 25
Planned	Luna 26
Planned	Luna 27
Planned	Chandrayaan 2
Planned	Chang'e 5

Apollo 11's Buzz Aldrin reported that moon dust smelled like "spent gunpowder."

Lunar rovers
The vast majority of soft landings on the Moon have involved static spacecraft, but several mobile vehicles have also explored the lunar surface. The first of these was NASA's Lunar Roving Vehicle (LRV)—a "moon buggy" driven by astronauts on the later Apollo missions. The Soviet Union landed two remote-controlled Lunokhod rovers on the Moon in the early 1970s, and in 2013 China landed its Yutu ("Jade Rabbit") vehicle.

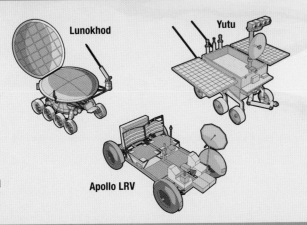
Lunokhod
Yutu
Apollo LRV

Landing sites
The first unpiloted soft landers on the Moon were designed to test surface conditions, amid fears that the soil, having been pulverized by countless impacts, might be too weak to support the weight of a large spacecraft. Later missions, including the Apollo crewed landings, aimed for specific areas and types of lunar landscapes in order to collect data that might shine light on the Moon's formation and early history.

Apollo landing sites

1

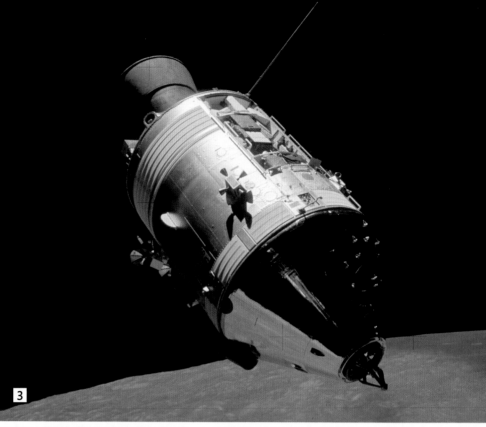

3

2

5

APOLLO PROJECT

1 Test flight
The United States' Apollo project of the 1960s and 70s was the only series of missions ever to put people on another world. The first successful crewed flight was Apollo 7, launched in October 1968 under commander Wally Schirra (pictured). This was a test run for the spacecraft's command and service modules, satisfactorily completed after 163 Earth orbits and nearly 11 days in space.

2 Stepping outside
The task of Apollo 9, launched in March 1969, was crucial to the entire project. On this flight, the lunar landing module was crewed in space for the first time. During their ten-day orbit of Earth, the crew undocked and redocked the lander, tested their equipment and support systems, and performed a spacewalk. Pilot David Scott is seen here emerging into space from the command module.

3 Command and service module
The Apollo spacecraft was made up of three parts: a command module, which served as control center; a service module, which carried a rocket engine, fuel, and oxygen; and a lunar module, which landed on the Moon. Only the cone-shaped command module returned to Earth. Here, the combined command/service module of Apollo 17 is seen in lunar orbit prior to rendezvous with the returning lunar module.

4 Mission accomplished
On July 20, 1969, the Apollo project achieved its aim as Neil Armstrong and Buzz Aldrin of the Apollo 11 mission became the first people on the Moon. Planting his boots on the surface at Tranquility Base (above), Aldrin was intrigued to note that when lunar dust is kicked, "every grain of it lands nearly the same distance away." The astronauts spent 21 hours on the Moon, taking photographs and collecting samples.

5 **Flying the flag**
On every one of the six Apollo landings, it has been a tradition for the astronauts to plant an American flag in lunar soil. Here, near the Moon's Apennine Mountains, Commander David Scott does his duty on the Apollo 15 mission of 1971. Behind him are the landing craft, poised on spiderlike legs, and a small, battery-powered lunar roving vehicle, used for the first time on this mission.

6 **Last lunar excursions**
Apollo 17's lunar module, Challenger, landed astronauts Gene Cernan (right) and Jack Schmitt (reflected in Cernan's helmet) in the Moon's Taurus-Littrow valley. They made long excursions, exploring the terrain and collecting a record number of rock and soil samples. With the cancellation of further planned Apollo missions, nobody has set foot on the Moon since Cernan and Schmitt in 1972.

6

MARS

MARS IS A BITTERLY COLD DESERT WORLD, STAINED A RUSTY RED BY IRON-RICH DUST ON ITS SURFACE. THOUGH HALF THE DIAMETER OF EARTH AND MUCH FARTHER FROM THE SUN'S WARMTH, MARS SHOWS MANY STRIKING SIMILARITIES TO OUR HOME PLANET.

Images of Mars returned by NASA spacecraft reveal a world that looks eerily familiar, with rock-strewn deserts, rolling hills, spectacular canyons, and a hazy sky flecked by occasional white clouds. Mars has a 25-hour day, polar ice caps that wax and wane like Earth's, an axis tilted only two degrees steeper than ours, and dry riverbeds that hint at the past presence of water. Volcanoes and rift valleys suggest tectonic forces were once generated by a hot interior.

Yet despite the many parallels, Mars and Earth are worlds apart. With only a tenth of Earth's mass, Mars lacks the gravity to hold on to a dense atmosphere, and its tenuous air is almost devoid of oxygen. While Earth's large, molten core keeps the planet's fractured crust in motion and generates a protective magnetic shield, Mars's smaller core has cooled and at least partially solidified. Its crust has frozen solid, and its magnetism is too weak to deflect solar radiation.

At one time Mars may have been warm and wet, but today it is an uninhabitable and barren wasteland.

Dust clouds on Mars can reach 3,000 ft (1,000 m) in height and last for **several weeks.**

MARS DATA

Diameter	4,213 miles (6,780 km)
Mass (Earth = 1)	0.11
Gravity at equator (Earth = 1)	0.38
Mean distance from Sun (Earth = 1)	1.5
Axial tilt	25.2°
Rotation period (day)	24.6 hours
Orbital period (year)	687 Earth days
Minimum temperature	−225°F (−143°C)
Maximum temperature	95°F (35°C)
Moons	2

▷ **Northern hemisphere**
A permanent ice cap called the Planum Boreum (Northern Plain) sits on the north pole of Mars. Around 620 miles (1,000 km) across, its perimeter is formed from lobes of ice separated by deep, canyon-like troughs.

▷ **Lava plains**
Enormous lava-covered plains dominate Mars's northern regions. In the south is Hellas Planitia, the largest impact crater on Mars at more than 1,243 miles (2,000 km) wide.

▷ **Southern hemisphere**
At Mars's south pole is the Planum Australe (Southern Plain), an ice cap with an upper layer of carbon-dioxide ice. Beyond it are huge areas of permafrost—water and soil frozen as hard as rock.

The Alba Mons is an enormous flat volcano surrounded by extensive lava fields.

This is the Tharsis region, a huge domed plateau about 2,485 miles (4,000 km) wide and home to giant volcanoes.

Olympus Mons is the largest volcano on Mars.

The southernmost of the three giant Tharsis volcanoes is Arsia Mons.

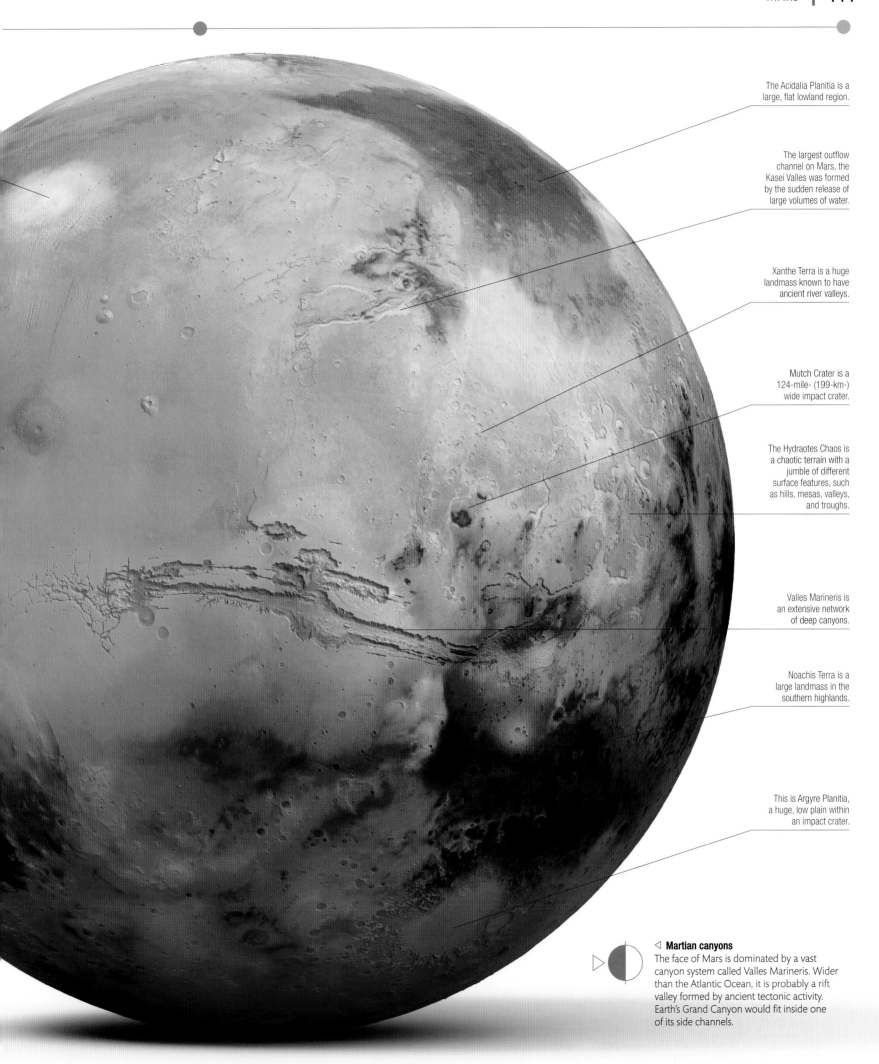

The Acidalia Planitia is a large, flat lowland region.

The largest outflow channel on Mars, the Kasei Valles was formed by the sudden release of large volumes of water.

Xanthe Terra is a huge landmass known to have ancient river valleys.

Mutch Crater is a 124-mile- (199-km-) wide impact crater.

The Hydraotes Chaos is a chaotic terrain with a jumble of different surface features, such as hills, mesas, valleys, and troughs.

Valles Marineris is an extensive network of deep canyons.

Noachis Terra is a large landmass in the southern highlands.

This is Argyre Planitia, a huge, low plain within an impact crater.

◁ **Martian canyons**
The face of Mars is dominated by a vast canyon system called Valles Marineris. Wider than the Atlantic Ocean, it is probably a rift valley formed by ancient tectonic activity. Earth's Grand Canyon would fit inside one of its side channels.

MARS STRUCTURE

NO ONE KNOWS EXACTLY WHAT THE INTERIOR OF MARS IS LIKE, BUT THROUGH VARIOUS STUDIES, INCLUDING UNCREWED SPACECRAFT MISSIONS, SCIENTISTS HAVE BUILT UP A THEORETICAL PICTURE OF THE PLANET'S STRUCTURE.

As a young planet, Mars cooled down more rapidly than Earth, because it is smaller and farther from the Sun, although the outer region of its iron core is thought to still be partially molten. A rocky crust of variable thickness forms the outermost layer of the planet. This is in one solid piece, rather than split into separate moving plates as on Earth. Beneath the crust is a deep mantle of silicate rock, once a fluid layer that was in constant motion. As the mantle shifted, it changed the face of Mars, causing great rifts in the crust and breaking through the surface to form gigantic volcanoes.

Core
Probably partly liquid, the small core of Mars is believed to be composed predominantly of iron. While Mars was still in a molten state, heavy metals sank to the center of the planet and started to solidify as they cooled.

Enormous surface rifts were caused by past movements of the mantle.

▷ **Mars layer by layer**
The outer layer of Mars is a crust of solid rock about 50 miles (80 km) thick in the southern regions and around 22 miles (35 km) thick in the northern hemisphere. Below the crust is a mantle of solid silicate rock. Deeper still is the small core of the planet, which is probably composed of iron as well as lighter materials, including iron sulfide. Layers in this 3D model are not shown to scale: the crust, surface relief, and atmosphere are exaggerated for clarity.

Surface temperatures on Mars can be as low as **−225°F (−143°C)**.

Mantle

Less dense than the core, the mantle is the middle layer of Mars. At the beginning of the planet's life, the mantle was in a liquid state, and its movements and outpourings helped to shape the appearance of the Martian surface. There is now no evidence of activity.

Crust

The outer layer, or crust, of Mars is composed largely of volcanic rock and was formed in one solid piece. Its surface, deeply smothered in soft red dust, bears evidence of a turbulent past marked by volcanic action, flowing water, weathering, and meteoroid impacts.

▽ **Atmosphere**

Mars's atmosphere is 95.3 percent carbon dioxide, with small amounts of other gases, notably nitrogen and argon, and traces of water vapor. Atmospheric pressure varies considerably with the seasons, decreasing in winter as carbon dioxide is locked into ice at the poles and increasing in summer when carbon dioxide returns into the atmosphere as vapor.

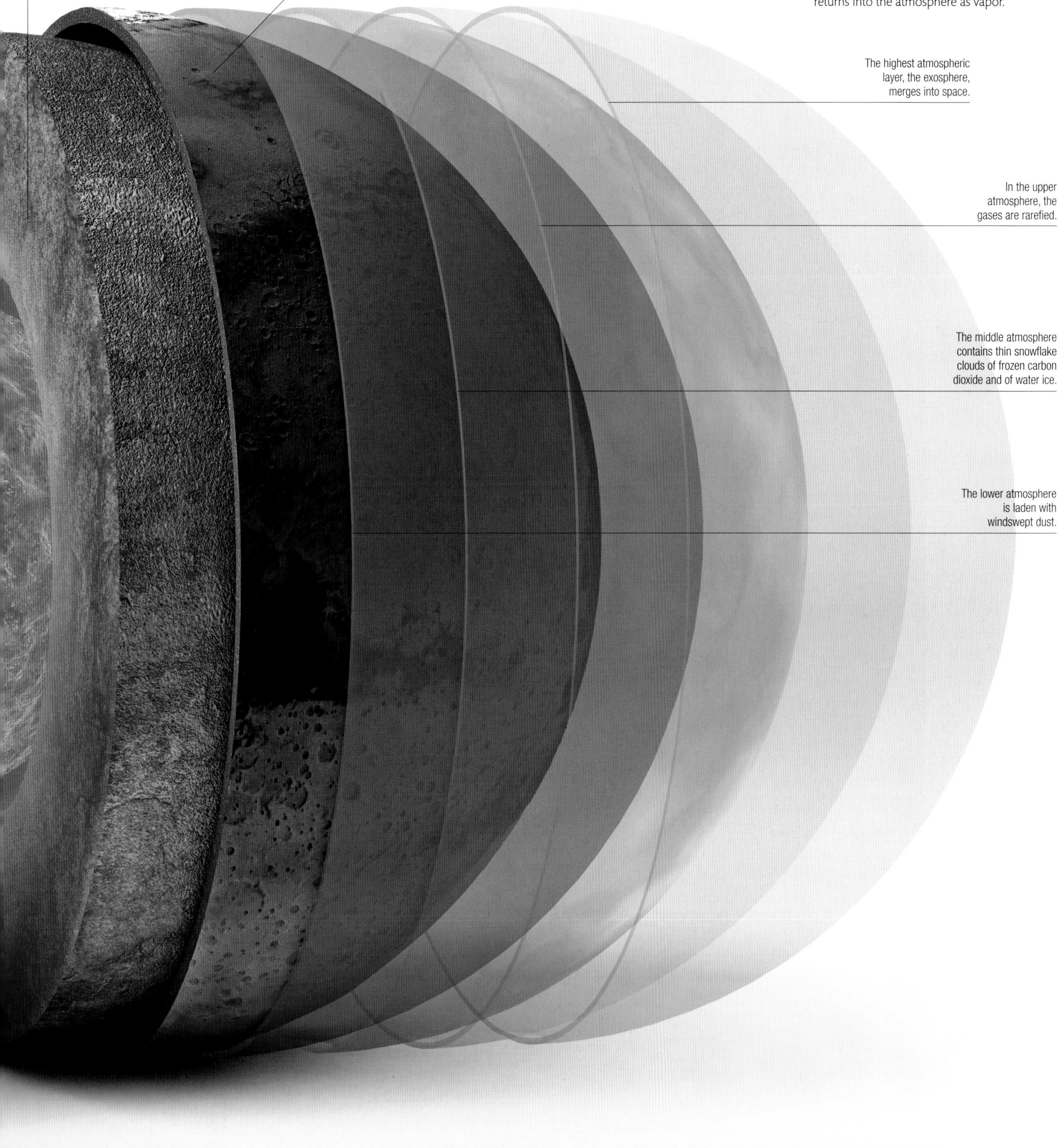

The highest atmospheric layer, the exosphere, merges into space.

In the upper atmosphere, the gases are rarefied.

The middle atmosphere contains thin snowflake clouds of frozen carbon dioxide and of water ice.

The lower atmosphere is laden with windswept dust.

MARS MAPPED

Between the seasonally ice-capped poles, the surface of Mars shows dramatic variation. The northern hemisphere mainly comprises flat lava plains; in the south rise geologically older highlands with vast volcanoes.

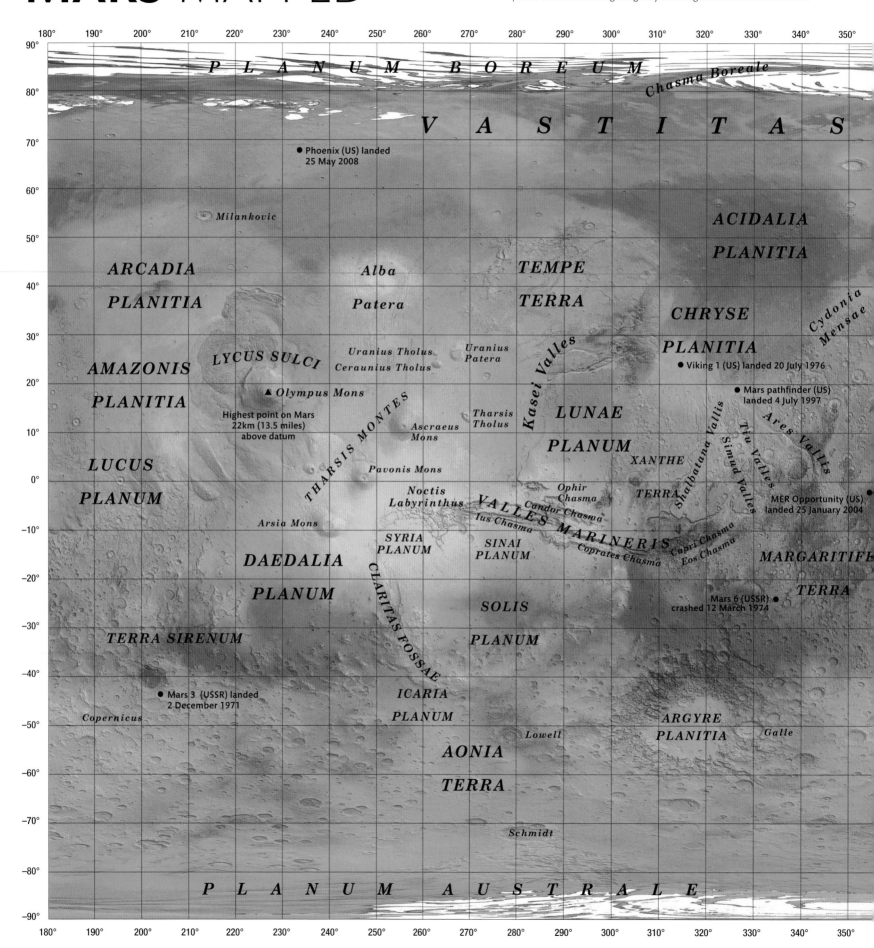

PLANUM BOREUM

Chasma Boreale

VASTITAS

● Phoenix (US) landed
25 May 2008

Milankovic

ACIDALIA

PLANITIA

ARCADIA

Alba

TEMPE

PLANITIA

Patera

TERRA

CHRYSE

Cydonia Mensae

AMAZONIS

LYCUS SULCI

Uranius Tholus
Ceraunius Tholus

Uranius Patera

PLANITIA

● Viking 1 (US) landed 20 July 1976

PLANITIA

▲ Olympus Mons

Tharsis Tholus

Kasei Valles

LUNAE

● Mars pathfinder (US)
landed 4 July 1997

Highest point on Mars
22km (13.5 miles)
above datum

THARSIS MONTES

Ascraeus Mons

PLANUM

XANTHE

Shalbatana Vallis

Simud Vallis

Tiu Valles

Ares Vallis

LUCUS

Pavonis Mons

Noctis
Labyrinthus

Ophir Chasma

Candor Chasma

TERRA

PLANUM

Arsia Mons

VALLES MARINERIS

Ius Chasma

Capri Chasma

Eos Chasma

● MER Opportunity (US)
landed 25 January 2004

DAEDALIA

CLARITAS FOSSAE

SYRIA
PLANUM

SINAI
PLANUM

Coprates Chasma

MARGARITIFE

PLANUM

SOLIS

TERRA

Mars 6 (USSR)
crashed 12 March 1974 ●

TERRA SIRENUM

PLANUM

● Mars 3 (USSR) landed
2 December 1971

ICARIA

ARGYRE

Copernicus

PLANUM

Lowell

PLANITIA

Galle

AONIA

TERRA

Schmidt

PLANUM AUSTRALE

Scale 1:45,884,054

0 250 500 750 1000 km

0 250 500 750 1000 miles

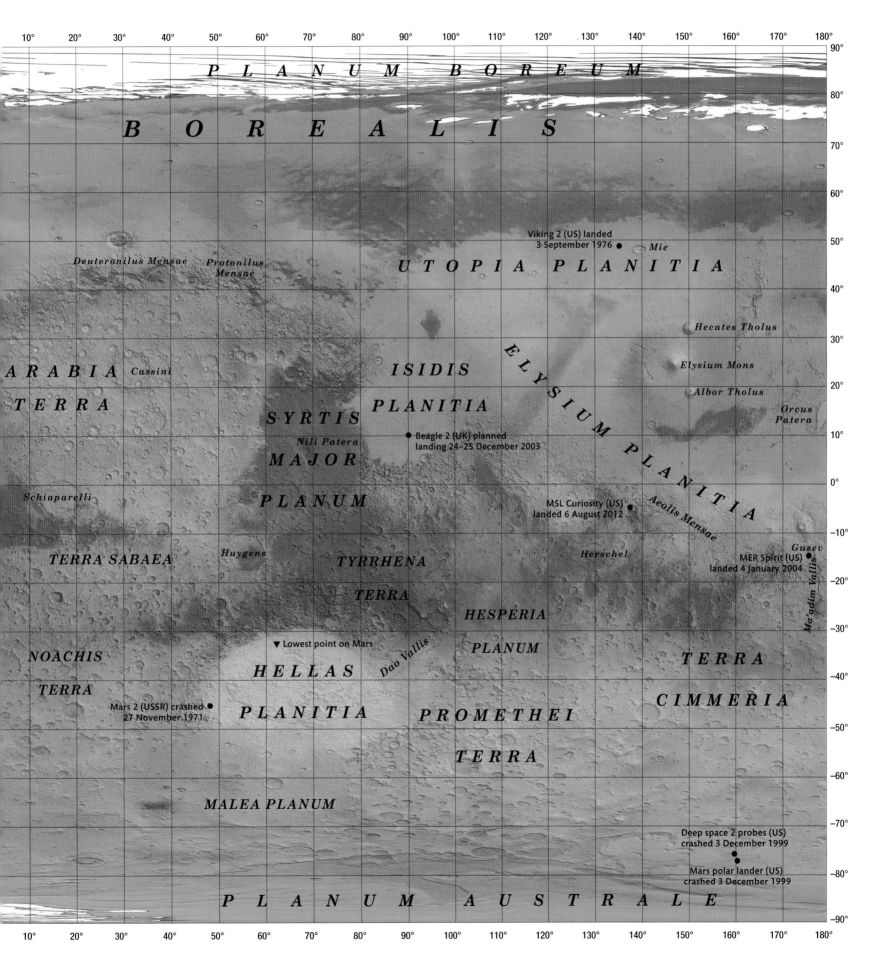

10° 20° 30° 40° 50° 60° 70° 80° 90° 100° 110° 120° 130° 140° 150° 160° 170° 180°

90°
80°

PLANUM BOREUM

B O R E A L I S

70°
60°
50°

Viking 2 (US) landed
3 September 1976 ●

Mie

Deuteronilus Mensae

Protonilus Mensae

U T O P I A   P L A N I T I A

40°

Hecates Tholus

30°

A R A B I A   *Cassini*

I S I D I S

Elysium Mons

20°

T E R R A

P L A N I T I A

Albor Tholus

SYRTIS

Nili Patera

● Beagle 2 (UK) planned
landing 24–25 December 2003

Orcus Patera

10°

MAJOR

ELYSIUM PLANITIA

Schiaparelli

P L A N U M

MSL Curiosity (US)
landed 6 August 2012 ●

Aeolis Mensae

0°
−10°

TERRA SABAEA   *Huygens*

TYRRHENA

Herschel

Gusev

MER Spirit (US)
landed 4 January 2004 ●

−20°

TERRA

Ma'adim Vallis

HESPERIA

−30°

NOACHIS

▼ Lowest point on Mars

Dao Vallis

PLANUM

T E R R A

−40°

H E L L A S

TERRA

Mars 2 (USSR) crashed ●
27 November 1971

P L A N I T I A

PROMETHEI

C I M M E R I A

−50°

TERRA

−60°

MALEA PLANUM

−70°

Deep space 2 probes (US)
crashed 3 December 1999
●

Mars polar lander (US)
crashed 3 December 1999

−80°

P L A N U M   A U S T R A L E

−90°

10° 20° 30° 40° 50° 60° 70° 80° 90° 100° 110° 120° 130° 140° 150° 160° 170° 180°

WATER ON **MARS**

MARS IS A DRY WORLD. IT HAS WATER ABOVE, ON, AND UNDER ITS SURFACE, BUT THE WATER IS IN THE FORM OF VAPOR OR ICE. LIQUID WATER WAS ONCE ABUNDANT ON MARS, AND ITS EFFECT ON THE LANDSCAPE IS STILL EVIDENT.

Today, liquid water cannot exist on the Martian surface because of the low temperature and atmospheric pressure. However, sedimentary rocks built up by water-deposited material, minerals formed by standing water, and landscape features shaped by flowing water all point to the fact that Mars may once have had large volumes of liquid water.

Ancient water

Billions of years ago, when Mars was a warmer planet, riverbeds and channel-like valleys hundreds of miles long formed as fast-flowing water carved through the landscape, and catastrophic floods covered vast areas, leaving floodplains behind. Valleys such as Kasei Valles, the site of two giant waterfalls eight times the height of Earth's Niagara Falls, are now dry. So too are Mars's deltas, lakes, and shallow seas. Increasing our knowledge of the planet's watery past helps in our search for life. Liquid water is essential for life—if it once existed on Mars, then perhaps life did too.

△ **Outflow channel**
The surface of Mars features outflow channels—vast swaths of water-scoured ground. The largest and longest of these is Kasei Valles, at over 1,500 miles (2,400 km) long. It was created by a huge outpouring of fast-flowing water. In this view, the water flowed toward the bottom left, and created an island in the center of the channel.

◁ **Evidence in rocks**
These gray balls, each about 0.2 in (4 mm) wide, lie scattered over a rocky outcrop in Eagle Crater. Analysis by the Opportunity rover in 2004 showed the balls consist of an iron mineral called hematite. Originally embedded in the outcrop, they collect on the ground after the softer rock erodes away. On Earth, hematite typically forms in lakes, so the same could have occurred on Mars. The circular patch is where Opportunity analyzed the underlying rock for comparison.

▽ **Impact meltwater**
Some of the water that flowed on Mars was released by volcanic activity or asteroid impact. This false-color image shows Hephaestus Fossae, a region of impact craters and channels. The impact that created the large crater penetrated the surface and melted underground ice, apparently causing a catastrophic flood.

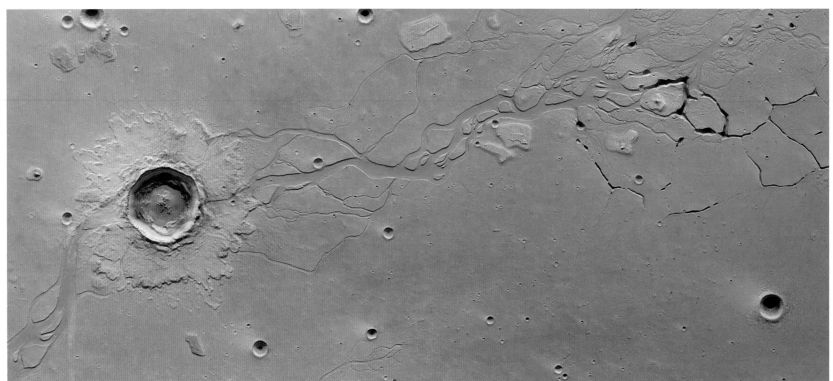

Water today

Most of the water on Mars today is locked within its frozen ice caps or held as vapor in its atmosphere. Orbiting spacecraft have also detected ice below the surface in other locations. Recently formed gullies on crater walls could be evidence of liquid groundwater released onto the surface.

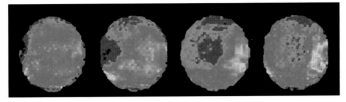

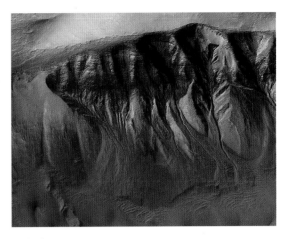

▽ **Water ice**
This huge sheet of water ice is a permanent feature in an unnamed crater near the Martian north pole. The ice is 9 miles (15 km) across and sits on a field of sand dunes. Water ice is also visible on parts of the crater's rim and wall.

△ **Clouds on Mars**
Four Mars Global Surveyor images show the progression of water-ice clouds (in blue) across the planet. These occasional, wispy, cirrus-type clouds occur when atmospheric water vapor forms ice crystals. Water vapor can also form low-lying mist and early morning frost.

◁ **Ice under the surface**
The Phoenix Mars Lander was the first craft to explore Mars's arctic region on the ground. In 2008, it landed near the northern polar cap. Using its robotic arm, it dug into the ground, exposing ice just inches below the surface. Four days later, the ice had vaporized.

△ **Gullies**
Root-shaped gullies on the walls of impact craters may indicate that water still flows. Observations show that the gullies change with the seasons. Mars is too cold for pure water to be liquid, but briny groundwater, which has a lower freezing point, may be released to briefly carry fine-grained sediment down the walls.

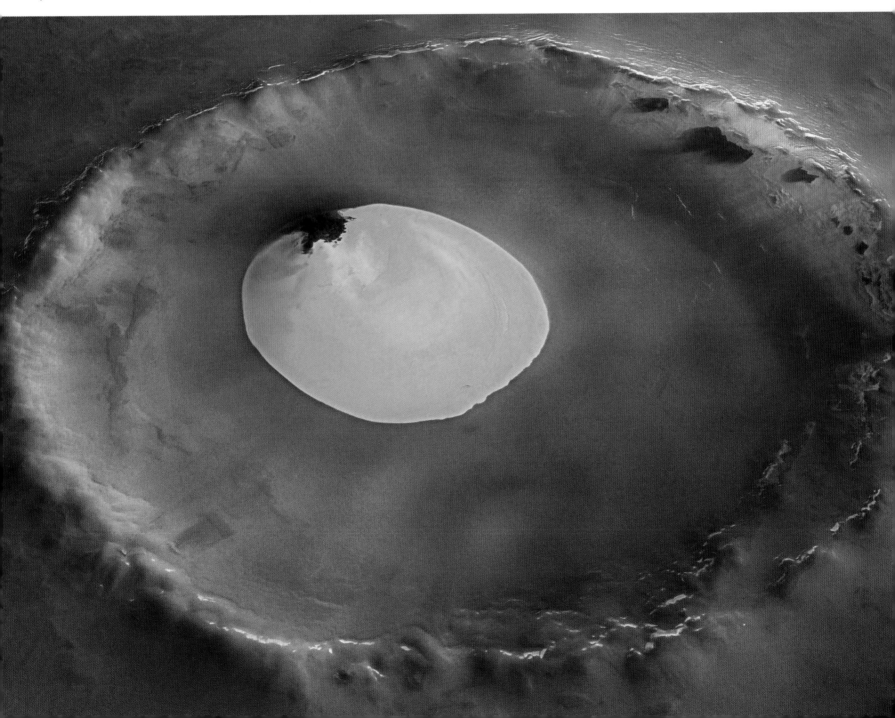

DESTINATION
VALLES MARINERIS

FIVE TIMES DEEPER AND NEARLY TEN TIMES LONGER THAN EARTH'S GRAND CANYON, VALLES MARINERIS STRETCHES ACROSS THE FACE OF MARS LIKE A VAST WOUND.

Named after the Mariner spacecraft that discovered it, Valles Marineris (Mariner Valleys) is a rift valley system that runs nearly a fifth of the way around the Martian equator. While Earth's Great Rift Valley was created by tectonic plate movements, Valles Marineris is thought to have formed as a result of upheaval and collapse of the static Martian crust several billion years ago. Marsquakes, meteorite impacts, and water floods have since triggered numerous landslides in the canyon walls, widening the valley and creating some of the most spectacular terrain in the solar system.

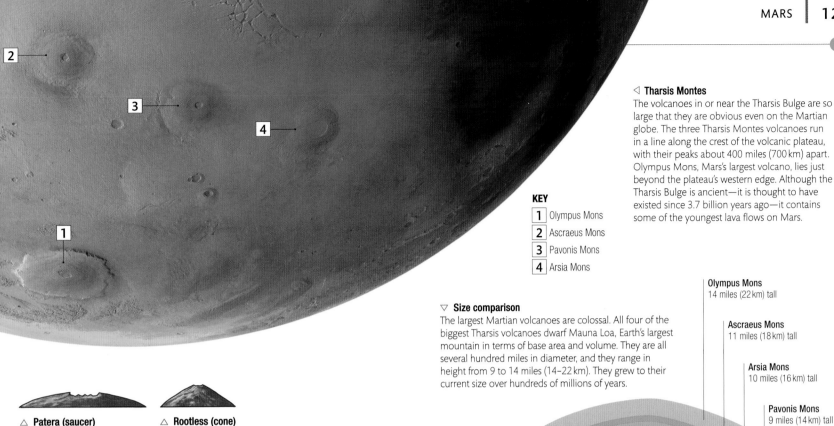

◁ Tharsis Montes

The volcanoes in or near the Tharsis Bulge are so large that they are obvious even on the Martian globe. The three Tharsis Montes volcanoes run in a line along the crest of the volcanic plateau, with their peaks about 400 miles (700 km) apart. Olympus Mons, Mars's largest volcano, lies just beyond the plateau's western edge. Although the Tharsis Bulge is ancient—it is thought to have existed since 3.7 billion years ago—it contains some of the youngest lava flows on Mars.

KEY

1 Olympus Mons
2 Ascraeus Mons
3 Pavonis Mons
4 Arsia Mons

▽ **Size comparison**
The largest Martian volcanoes are colossal. All four of the biggest Tharsis volcanoes dwarf Mauna Loa, Earth's largest mountain in terms of base area and volume. They are all several hundred miles in diameter, and they range in height from 9 to 14 miles (14–22 km). They grew to their current size over hundreds of millions of years.

Olympus Mons
14 miles (22 km) tall

Ascraeus Mons
11 miles (18 km) tall

Arsia Mons
10 miles (16 km) tall

Pavonis Mons
9 miles (14 km) tall

△ **Patera (saucer)**
Paterae are shallow, saucer-shaped bumps in the Martian surface. Like tholi, they may be the tops of buried shield volcanoes, but with larger calderas.

△ **Rootless (cone)**
Small conical structures, less than 800 ft (250 m) wide, are volcanic cones that form on the surface of fresh lava flows. They are rootless as they are not above a magma source.

The flanks of Tharsis Tholus are among the steepest on Mars, with an average slope of 10°.

Impact crater

Lava lands

When lava spills out of Martian volcanoes, it runs down their gentle slopes in sinuous channels before spreading out across the lowlands. Such eruptions leave distinctive formations in the landscape, such as lava tubes and lava plains. Lava tubes form when hot lava continues to flow beneath a solidified crust, like an underground river. When the source is exhausted, an empty tunnel is left behind, the roof of which may later collapse. Lava plains are ancient floods that have cooled and solidified.

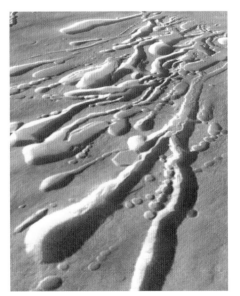

△ **Lava tubes**
Ancient lava tubes have been identified on the slopes of the biggest shield volcanoes. These lava tubes occur on the flanks of Pavonis Mons—the longest of them stretches 40 miles (60 km) from end to end. When the surface of an empty lava tunnel collapses, these long depressions are left behind. Such features indicate that the lava was relatively fluid.

△ **Lava flows**
Hesperia Planum is a 1,000-mile- (1,600-km-) wide lava plain in the southern highlands of Mars. Here, a flood of lava spilled across the land, partially filling a 15-mile- (24-km-) long impact crater. The crater's elliptical shape, formed by a strike at a low angle, is still evident.

Olympus Mons is almost **three times the height** of Earth's **Mount Everest.**

DESTINATION **OLYMPUS MONS**

THE LARGEST VOLCANO IN THE SOLAR SYSTEM, OLYMPUS MONS SITS LIKE A VAST BRUISE ON THE MARTIAN PLAINS. AT 14 MILES (22 KM) IN HEIGHT, IT IS MARS'S TALLEST FEATURE, BUT IT IS SO WIDE THAT A VISITOR LANDING ON THE SUMMIT WOULD SEE NO END TO ITS SHALLOW SLOPES.

Almost as wide as France, Olympus Mons measures 380 miles (610 km) across. It grew to its great size over millions of years as thousands of successive lava flows piled one on top of another. Because Mars's crust is stationary, unlike Earth's, the volcano stays permanently over a hot spot in the planet's mantle. The summit plateau is surrounded by steep, 3-mile- (6-km-) tall cliffs over which lava has cascaded like a waterfall to spill onto the surrounding plain. Olympus Mons is dormant at present but could easily erupt again. Originally discovered by astronomers in the 19th century, this Martian giant was not recognized as a volcano until the Mariner 9 spacecraft went into orbit around Mars in 1971.

3D reconstruction from MOLA (Mars Orbiter Laser Altimetre) elevation data with accurate vertical relief.

LOCATION

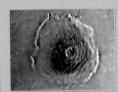

Latitude 19°N; **longitude** 226°E

LAND PROFILE

Olympus Mons dwarfs Earth's tallest volcano, Mauna Kea in Hawaii. Both are asymmetrically shaped shield volcanoes with an average hill slope of about 5 degrees.

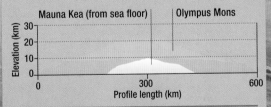

Mauna Kea (from sea floor) | Olympus Mons
Elevation (km): 30, 20, 10, 0
Profile length (km): 0, 300, 600

CALDERA

The summit caldera is about 37 miles (60 km) across. It contains at least six individual craters that formed after lava flow ceased and the magma chambers below collapsed. The figures are the craters' approximate ages in millions of years.

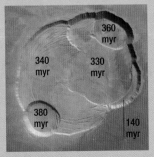

360 myr
340 myr
330 myr
380 myr
140 myr

1

DUNES OF **MARS**

1 Noachis Terra
Orbiting cameras have captured many of Mars's stunningly beautiful dune fields, which are created by wind-blown surface materials forming rippling patterns. This false-color image shows sand dunes trapped inside an impact crater in Noachis Terra, a region in the southern hemisphere of the planet.

2 North polar erg
Fantastically sculptured dunes created from grains of basalt and gypsum decorate an icy plain in the high northern region known as the north polar erg. This area, which encircles the north polar ice cap, contains immense dune fields. The crescent formations seen here occur when the sand cover is relatively thin.

3 Seasonal changes
The "tree plantation" on this dune field in the northern polar region is an illusion. It is caused by streaks of dark, basaltic sand emerging from gaps in the carbon dioxide frost that covers the dunes in winter. Blown by wind, the sand spreads over the ice. The phenomenon occurs in spring as the ice layer thins.

4 Dunes on the move
Like dunes on Earth, those on Mars show significant movement, reflecting the effect of local winds. In this image, the dunes are migrating from left to right. The dark arcs in the lower right are barchans—wind-sculpted, crescent-shaped dunes that also form in sandy deserts on Earth.

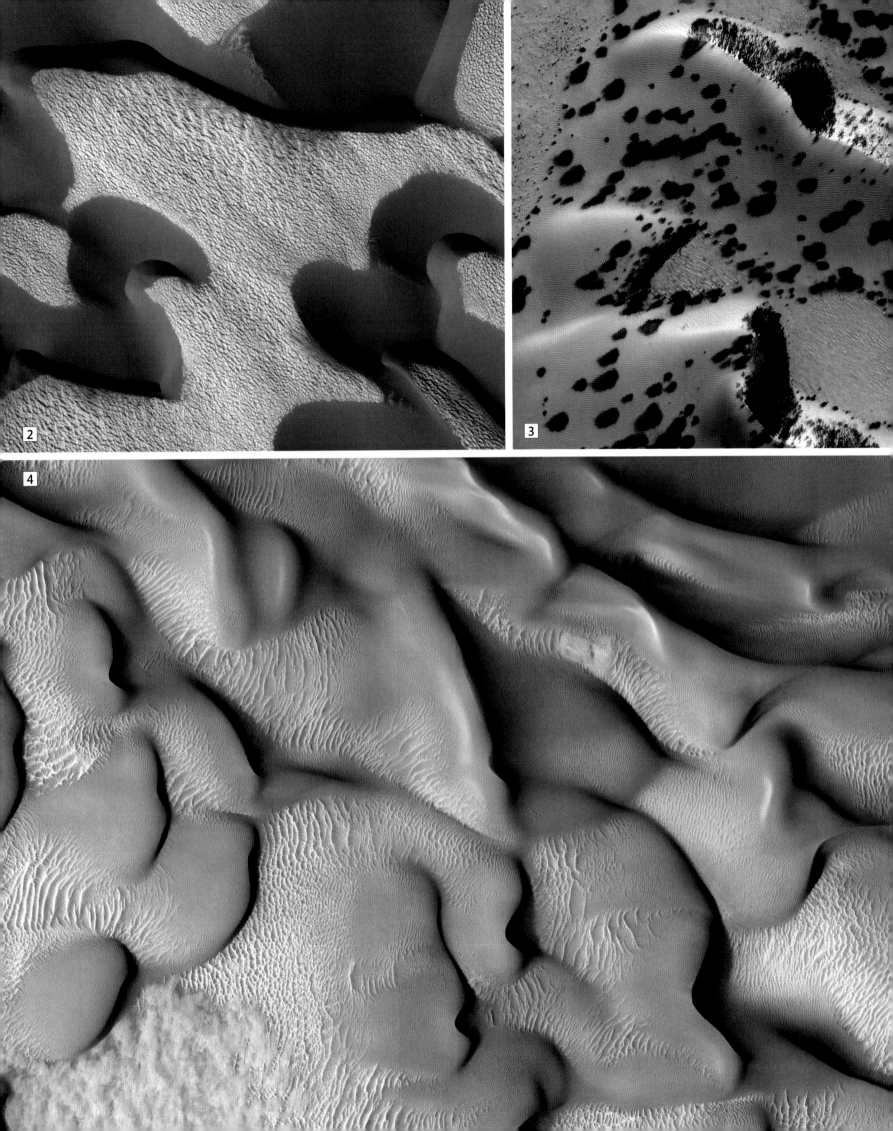

POLAR **CAPS**

A WHITE CAP MADE PREDOMINANTLY OF FROZEN WATER SITS ON EACH OF MARS'S POLES. ALTHOUGH THESE ALMOST CIRCULAR CAPS ARE A PERMANENT FEATURE OF THE MARTIAN LANDSCAPE, BOTH CHANGE WITH THE SEASONS.

The polar caps are huge mounds of ice that stand above the land that surrounds them. Cliffs at their edges reveal that the caps are made of layer upon layer of ice, sand, and dust laid down over millions of years. In winter, the caps extend as they are covered with new deposits of carbon dioxide snow and ice. With rising temperatures in summer, the carbon dioxide returns to the atmosphere as gas and the caps shrink.

North cap

The northern cap, Planum Boreum (Northern Plain), is the larger of the two—about 620 miles (1,000 km) across and 1.2 miles (2 km) thick—and is 90 percent water ice. Data on the thickness and composition of the cap's layers, collected by NASA's Mars Reconnaissance Orbiter, is being used to study the planet's history of climate change.

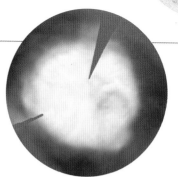

▷ **Seasonal change**
These two images of the north cap from the Hubble Space Telescope show the change from winter to spring. By late winter, the ice extends southward to almost 60°N latitude— nearly its maximum extent. Three months later, it is warmer and the carbon dioxide ice and frost south of 70°N have evaporated. By early summer, only the remnant core of water ice will remain.

Late winter

▽ **Chasma Boreale**
This 3-D reconstruction looks into Chasma Boreale, the northern cap's largest canyon. It is about 350 miles (570 km) long—a little longer than Earth's Grand Canyon—and up to 0.87 miles (1.4 km) deep. At its mouth it is 75 miles (120 km) wide, tapering as it runs into the cap. The walls are stacked layers of ice, and the dark terrain is frozen sand.

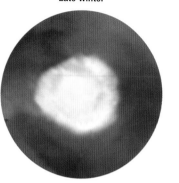

Mid-spring

▽ **Spiral pattern**
The distinctive spirals of dark troughs at the north cap were caused by strong polar winds over millions of years. The troughs probably began as slight depressions that gradually deepened into valleys. The vast dark sea of dunes extending from the cap formed when Mars was warmer and still ice-free.

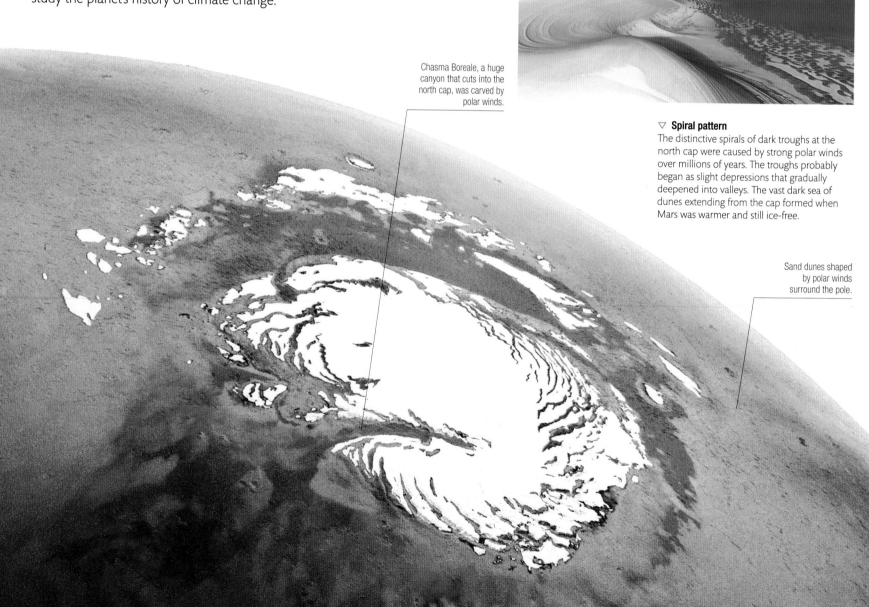

Chasma Boreale, a huge canyon that cuts into the north cap, was carved by polar winds.

Sand dunes shaped by polar winds surround the pole.

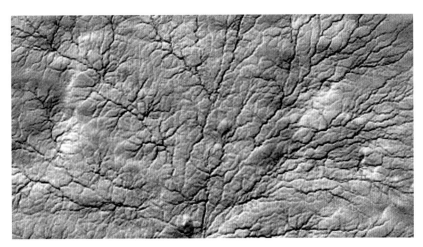

South cap

The south cap, Planum Australe (Southern Plain), has a thick base of water ice topped with a 25-ft (8-m) layer of carbon dioxide ice. At its minimum size in summer, it measures about 260 miles (420 km) across. During the southern winter, the cap is in permanent darkness, the temperature drops, and carbon dioxide both freezes as frost and falls as snow.

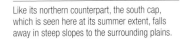

Like its northern counterpart, the south cap, which is seen here at its summer extent, falls away in steep slopes to the surrounding plains.

▷ Starburst

In springtime, as carbon dioxide gas beneath the seasonal ice makes its way to the surface, it carves out troughs in the ground. These troughs form branchlike patterns often referred to as starbursts or spiders. In some locations, dust carried by the gas falls to the ice surface in fan-shaped deposits.

△ Frozen solid

The south cap is the only place on Mars where carbon dioxide, which freezes at around −193°F (−125°C), persists as ice on the surface year-round. As in the north, the south polar region is encircled by a vast area of permafrost (water ice mixed with soil and frozen to the hardness of solid rock).

▽ Icy pits

This view shows an effect created in the late summer in the southern polar region. The carbon dioxide ice here is about 10 ft (3 m) thick and penetrated by flat-floored pits. For most of the year, the pit walls are covered by bright frost. But an upper layer of the ice has turned to gas, revealing the edges of the pits. The smallest of the pits are roughly the size of a sports stadium—about 195 ft (60 m) across.

THE MOONS OF **MARS**

MARS HAS TWO MOONS, PHOBOS AND DEIMOS. THEY ARE IRREGULARLY SHAPED, ROCKY LUMPS WITH CRATERED SURFACES. TINY COMPARED TO EARTH'S MOON, THEY HURTLE AROUND MARS IN LESS THAN A DAY AND A HALF.

On **Phobos**, the temperature in the shade is –170°F (–112°C).

The pair were discovered within days of each other by American astronomer Asaph Hall, in 1877. Their names are from Greek mythology—Phobos and Deimos were the twin sons of Ares, the god of war. Phobos was the god of fear and Deimos of terror, and the brothers accompanied their father Ares into battle. In the solar system, the two moons accompany Mars, the Roman equivalent of Ares. The moons have been seen in detail only in relatively recent times. Phobos has been studied most closely; it was the subject of a series of flybys by Mars Express in 2010. The origin of the moons is uncertain; some astronomers think the pair are asteroids captured by Mars's gravity, others that Phobos formed from debris left over from the formation of Mars.

Earth's Moon (diameter)
2,160 miles
(3,476 km)

Phobos (average width)
13.8 miles
(22.2 km)

Deimos (average width)
7.7 miles
(12.4 km)

△ **Comparing the moons of Mars and Earth**
Earth's Moon is around 155 times wider than Phobos and 280 times larger than Deimos. But Mars's two moons are much closer to their parent planet than the Moon is to Earth. An observer on the Martian surface would see Phobos at just over a third of the size that the Moon appears in Earth's sky.

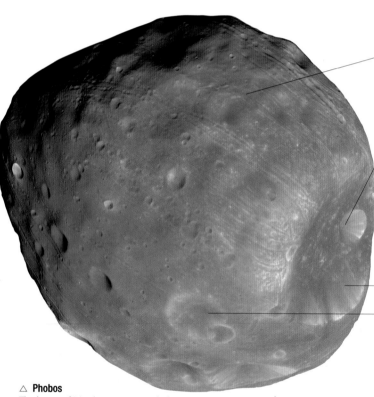

Phobos looks solid, but it is largely a pile of rubble held together by gravity.

Limtoc Crater is 1.2 miles (2 km) wide, and named after a character from Lilliput in Jonathan Swift's *Gulliver's Travels*.

The lines on the inner walls of Stickney Crater are landslides of rock and dust.

Reldresal Crater is 1.8 miles (2.9 km) across; like Limtoc, it gets its name from a character in *Gulliver's Travels*.

△ **Phobos**
The larger of Mars's two moons, Phobos is a cavity-ridden rocky body about 17 miles (27 km) long. Its heavily cratered, barren surface is covered in a thick, loose layer of fine dust. Almost all of its 20 named surface features are craters. The largest, Stickney, is about 5.5 miles (9 km) across. The grooves and rows of craters surrounding it could have been formed by the impact that created Stickney or by debris ejected from Mars when meteoroids hit the planet's surface.

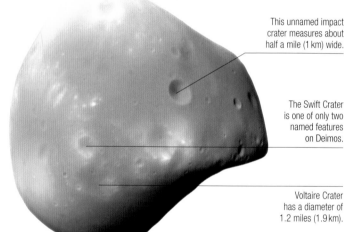

This unnamed impact crater measures about half a mile (1 km) wide.

The Swift Crater is one of only two named features on Deimos.

Voltaire Crater has a diameter of 1.2 miles (1.9 km).

△ **Deimos**
At 9 miles (15 km) long, Deimos is about half the size of Phobos. Like its bigger companion, it is a rock body blanketed in a reddish soil of rock fragments and dust. It has fewer craters; all but the most recent contain soil, giving Deimos a smoother surface. The surface color varies: it is least red around the freshest craters, where the soil has slipped down the slopes and exposed the bedrock.

Mars rotates on its axis in 24 hours 37 minutes.

Deimos orbits Mars in 30 hours 18 minutes.

One orbit of Mars by Phobos takes 7 hours 39 minutes.

◁ **Moon orbit and spin**
Phobos and Deimos follow near-circular orbits above Mars's equator. Phobos is closest, at 5,826 miles (9,376 km) from Mars, and it is getting closer by an inch or so (a few centimeters) each year. In 50 million years, it will have either crashed into Mars or, more likely, disintegrated from the stress of being pulled by Mars's gravity. Deimos is 14,576 miles (23,458 km) away, over twice the distance of Phobos. Both moons are in synchronous orbits, keeping the same face pointed toward Mars at all times.

△ **Phobos over Mars**
When the Viking 1 orbiter was imaging the surface of Mars as it flew around the planet in September 1977, it took a snapshot of the largest Martian moon. Seen here as an almost black ball, Phobos was between Viking 1 and the surface when the picture was taken. Phobos orbits the planet faster than Mars spins on its axis. Anyone on the surface would see Phobos rise in the west, move rapidly across the sky, and then set in the east—a feat it performs twice each Martian day.

THE **RED** PLANET

MARS HAS BEEN KNOWN SINCE ANCIENT TIMES. EARLY ASTRONOMERS NOTED ITS COLOR AND MOVEMENT ACROSS THE SKY, AND TELESCOPES LATER CAPTURED SURFACE DETAIL.

The color of Mars led the ancient Greeks and Romans to associate the planet with blood and war. It wasn't until much later that telescopes revealed more than just a reddish point of light. The mistaken sighting of channels led to the idea that Mars might be home to an advanced civilization, but when spacecraft visited, they found a dry, lifeless desert. Nevertheless, evidence suggests that Mars had a watery past. Several generations of rovers have now explored the surface, and Mars is the planet most likely to be visited by people in the future.

Mars, Roman god of war

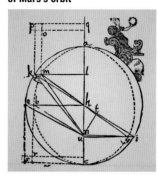

Johannes Kepler's illustration of Mars's orbit

500 BCE

The red planet
Mars, the red planet, is named after the Roman god of war. Astrologically, the planet becomes associated with passion, fighting, and lust. Apparent variations in its movements and brightness baffle astronomers until the 17th century.

1609 CE

Orbit calculation
German astronomer Johannes Kepler works out the shape of Mars's orbit. Kepler realizes that planets have elliptical rather than circular orbits, and he derives three laws of planetary motion. These will later inspire Isaac Newton's revolutionary work on gravity.

Mariner 4 image of cratered surface

Orson Welles in the CBS radio studio

1965

First spacecraft and surface photos
NASA's Mariner 4 spacecraft performs the first successful flyby of Mars, passing within 6,118 miles (9,846 km) of the surface. It takes 21 images of the southern hemisphere. The area imaged is billions of years old and cratered much like Earth's moon.

1947

Atmosphere
US astronomer Gerard Kuiper, working at the Yerkes Observatory in Wisconsin, finds that the thin atmosphere of Mars consists mainly of carbon dioxide. The discovery helps overturn the widespread belief that Mars is like Earth.

1938

Mars and science fiction
The idea that Mars is inhabited is popular in science fiction. On October 30, Orson Welles makes a radio broadcast of H. G. Wells' *War of the Worlds*. Presented in the style of a news bulletin, it convinces some listeners that Martian invaders are taking over Earth.

Summit of Olympus Mons volcano

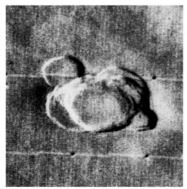

Viking 2 lander on Utopia Planitia

1971

First orbiter
Mariner 9 is the first spacecraft to orbit a planet other than Earth. It finds huge, dormant volcanoes, a giant system of canyons, and signs of erosion by fluids. The southern hemisphere is more cratered than the younger northern hemisphere.

1975

Landers on Mars
Two identical Viking craft leave Earth for Mars. Each consists of an orbiter and a lander. The Viking 1 lander is the first to the surface, and within five minutes of touchdown it returns the first images from the ground. Both landing craft search for evidence of life, past and present. The orbiters see what appear to be dried-up, branching river beds.

Herschel's 1784 drawings of Mars show
ice caps and surface features

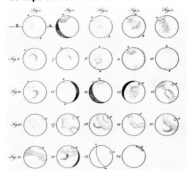

The two hemispheres of Mars by Schiaparelli

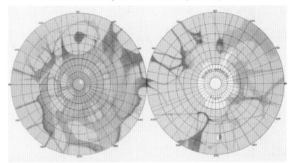

1659

First surface observations
Dutch scientist Christiaan Huygens looks at
Mars through a telescope and sees markings
on the surface. By watching them disappear
and reappear he finds that Mars spins on its
axis every 24 hours 40 minutes. In 1672,
Huygens discovers Mars's polar caps.

1784

Seasons on Mars
English astronomer William Herschel
improves the measurement of Mars's
rotation period and finds that its axis is tilted
by 25.2°. As a result, Mars has seasons.
Herschel notes that the size of Mars's ice
caps changes with the seasons.

1863

First maps
Italian astronomer Angelo Secchi produces
the first color map of Mars. Then, in 1879,
fellow Italian Giovanni Schiaparelli produces
more detailed maps that include fine lines
labeled *canali*—Italian for "channels." English
versions mistranslate the word as "canals."

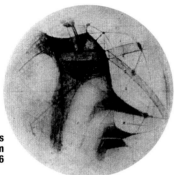

One of Percival Lowell's
drawings of Martian
canals, 1896

US Naval Observatory
26-in (66-cm)
refracting telescope

1924

Temperature
Using the Hooker telescope on Mount
Wilson, California, US astronomers Edison
Pettit and Seth Nicholson measure Mars's
surface temperature. It is 45°F (7°C) at the
equator and –90°F (–68°C) at the pole. The
wind and temperature vary seasonally.

1896

Intelligent life on Mars
Using the 24-in (60-cm) refractor at his
private observatory in Arizona, astronomer
Percival Lowell maps Mars. Inspired by
Schiaparelli's "canals," he argues in his book
Mars as the Abode for Life that the planet is
inhabited by intelligent beings.

1877

Discovery of moons
With Mars in a favorable position, US
astronomer Asaph Hall discovers its two
moons, Phobos and Deimos. He uses the
largest telescope in the world at the time,
a 26-in (66-cm) refractor at the US Naval
Observatory, Washington, DC.

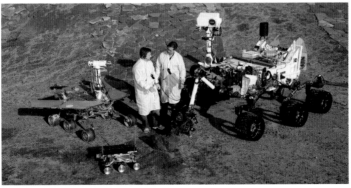

Three generations of rovers:
Sojourner (front), Opportunity
(left), and Curiosity (right)

1984

Martian meteorite
Meteorite ALH84001 is found on Earth,
in the Allan Hills region of Antarctica. It
was ejected from Mars 16 million years
ago, reaching Earth 13,000 years ago. It
contains structures that look like
fossilized microbes.

2012

Rovers on Mars
Curiosity, the latest and largest of the four rovers to
roam on Mars, arrives in Gale Crater. Sojourner was the
first and explored the floodplain Chryse Planitia in 1996,
staying close to its mother craft. The twin rovers Spirit
and Opportunity arrived in 2004 and covered many
miles as they explored the planet.

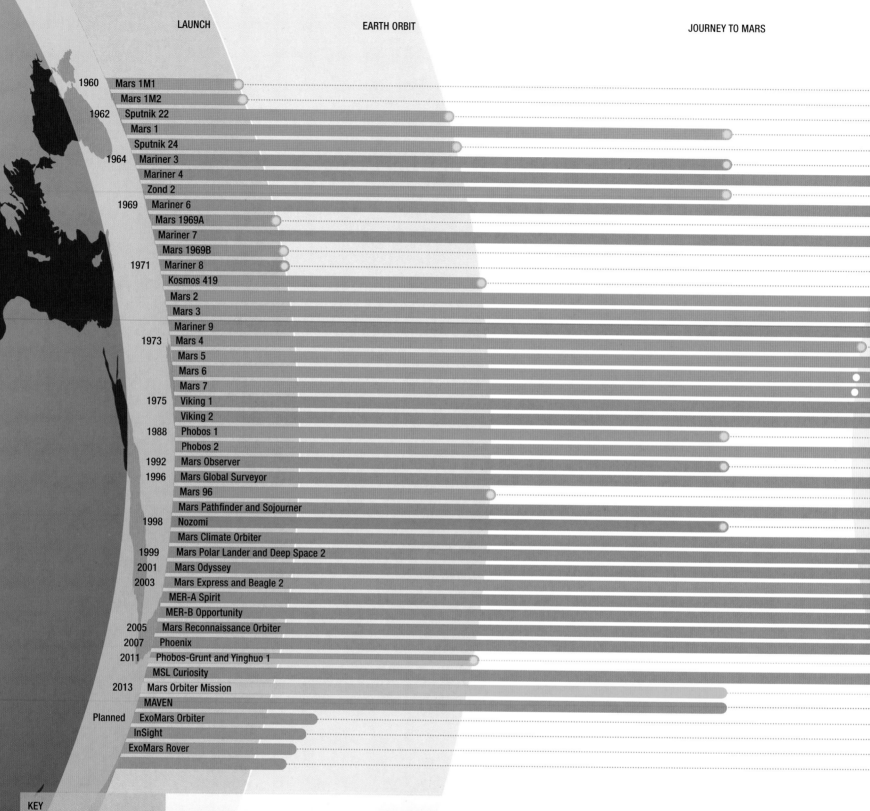

LAUNCH   EARTH ORBIT   JOURNEY TO MARS

1960	Mars 1M1
	Mars 1M2
1962	Sputnik 22
	Mars 1
	Sputnik 24
1964	Mariner 3
	Mariner 4
	Zond 2
1969	Mariner 6
	Mars 1969A
	Mariner 7
	Mars 1969B
1971	Mariner 8
	Kosmos 419
	Mars 2
	Mars 3
	Mariner 9
1973	Mars 4
	Mars 5
	Mars 6
	Mars 7
1975	Viking 1
	Viking 2
1988	Phobos 1
	Phobos 2
1992	Mars Observer
1996	Mars Global Surveyor
	Mars 96
	Mars Pathfinder and Sojourner
1998	Nozomi
	Mars Climate Orbiter
1999	Mars Polar Lander and Deep Space 2
2001	Mars Odyssey
2003	Mars Express and Beagle 2
	MER-A Spirit
	MER-B Opportunity
2005	Mars Reconnaissance Orbiter
2007	Phoenix
2011	Phobos-Grunt and Yinghuo 1
	MSL Curiosity
2013	Mars Orbiter Mission
	MAVEN
Planned	ExoMars Orbiter
	InSight
	ExoMars Rover

KEY

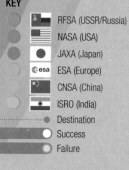

	RFSA (USSR/Russia)
	NASA (USA)
	JAXA (Japan)
	ESA (Europe)
	CNSA (China)
	ISRO (India)

- - - - Destination
○ Success
◐ Failure

▷ **Landing sites**
Seven craft have touched down successfully on Mars. Three stayed where they landed and investigated their immediate surroundings. These were Vikings 1 and 2, which arrived in 1976, and Phoenix in 2008. The other four craft, two of which are still working, were rovers designed to drive over the Martian landscape, stopping now and then to investigate.

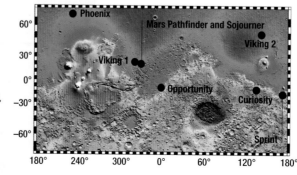

▷ **First surface image**
The American craft Viking 1 was the first to return images from Mars's surface. Although the earlier Soviet craft Mars 3 had a TV camera on board, it stopped transmitting seconds after landing and nothing was seen of its surroundings. Viking 1 took its first image (right) just after it arrived on July 20, 1976; one of the craft's footpads is seen in the photograph.

FLYBY   ORBITER   LANDER

ROVER

MISSIONS TO **MARS**

IN THE PAST 60 YEARS, MORE THAN 40 MISSIONS HAVE BLASTED OFF FROM EARTH FOR MARS. THE PLANET HAS BEEN FLOWN BY, ORBITED, LANDED ON, AND ROVED OVER, AND WAS THE FIRST PLANET EVER SEEN IN CLOSE-UP.

Missions sent to Mars in the 21st century have been extraordinarily successful, sometimes far exceeding expectations. But success has been built on earlier disappointments, with more than half of all Mars missions either failing to get away from Earth or losing contact with their controllers as they closed in on their target. The first attempts at Martian exploration were undertaken by the USA and the then Soviet Union in the 1960s and 70s, after which there was little interest in Mars until the mid-1990s. Now, six countries have sent craft to Mars, more missions are planned, and a privately funded project is underway to develop a spaceflight system capable of taking a human crew to Mars.

▷ **Landmark missions**
The American Mariner series provided the first successful missions to Mars. Mariner 4 was the first craft to fly by the planet and to take close-up images. Mariner 9 was the first craft to orbit Mars. The first soft landing on Mars was made by the Soviet Mars 3, but no data were returned.

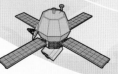

Mariner 9
The first craft to orbit any planet, Mariner 9 arrived in 1971 and provided the first global map of Mars.

Sojourner
The size of a microwave oven, Mars's first rover worked for almost three months from July 1997.

Vikings 1 and 2
Twin craft, each consisting of an orbiter and a lander, treached Mars in 1976 and made soil tests.

Mars Express
The orbiter, Europe's first planetary mission, has been mapping Mars since December 2003.

ROVING ON **MARS**

MARS IS THE ONLY PLANET THAT ROBOTIC ROVERS HAVE EXPLORED. FOUR ROVERS HAVE SUCCESSFULLY VISITED MARS—SOJOURNER, SPIRIT, OPPORTUNITY, AND CURIOSITY. WE NOW UNDERSTAND MORE ABOUT THE SURFACE OF MARS THAN ANY OTHER PLANET EXCEPT EARTH.

Designed to drive across alien terrain, robotic rovers are mobile science labs that hunt out interesting sites and conduct on-the-spot investigations. With their own power supply, they are operated by onboard computers and armed with scientific instruments, including cameras and rock analysis tools. Back on Earth, ground controllers decide where the rovers should go and what they should do. Directions take a few minutes to get through. Collected data is relayed directly to Earth or via orbiters like Mars Reconnaissance Orbiter, a spacecraft circling the planet.

▷ **Curiosity self-portrait**
Curiosity is investigating the floor of Gale Crater, a 96-mile- (154-km-) wide impact crater formed more than 3 billion years ago. This self-portrait shows the rover in the Yellowknife Bay area of the crater, where sedimentary rocks called mudstones indicate an ancient lake bed. The image is a mosaic of dozens of individual views taken in February 2013 using MAHLI (Mars Hand Lens Imager), one of Curiosity's 17 cameras.

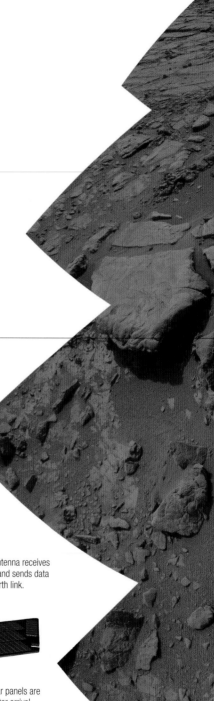

This patch of flat outcrop is named John Klein and was the site of Curiosity's first rock drilling.

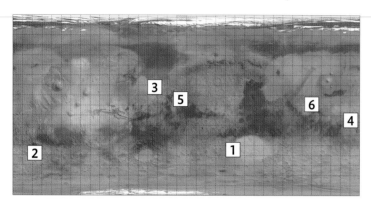

KEY
1 Mars 2 (1971)
2 Mars 3 (1971)
3 Sojourner (1997)
4 Spirit (2004)
5 Opportunity (2004)
6 Curiosity (2012)

△ **Rover sites**
The first two attempts to put rovers on Mars ended in failure. The Soviet Mars 2 lander, carrying a tethered rover equipped with skis, crash-landed. Its twin Mars 3 failed seconds after touchdown. Since then, four rovers have made successful landings. They have explored a variety of terrains, all low-lying for ease of landing and smooth enough to drive over.

▽ **Opportunity**
Opportunity touched down in Meridiani Planum in 2004, and has investigated sites including four impact craters—Endurance, Erebus, Victoria, and Endeavour. It travels at around 0.5 in (1 cm) per second, sending back images of the terrain and results of its rock analysis. Designed to operate for about three months, it is now in its eleventh year of work.

▽ **Martian rovers**
The first rover, Sojourner, was the size of a microwave oven. It stayed close to its landing site and worked for about three months. The twin craft Spirit and Opportunity arrived on opposite sides of Mars in 2004. Spirit no longer works, but Opportunity continues to explore. Curiosity is the size of a small car and has a laser tool to gauge the composition of a rock in seconds.

Pancam consists of two digital cameras, which take 360° views.

Low-gain antenna sends images to orbiters for relay to Earth.

High-gain antenna receives commands and sends data via direct Earth link.

Hinged solar panels are unfolded after arrival.

Rock analysis tools at end of jointed arm

Rocker-bogie suspension keeps wheels in contact with the ground.

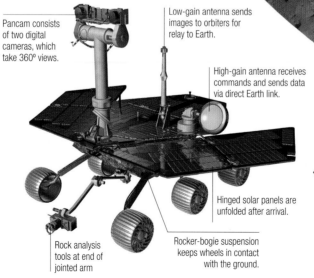

Sojourner July–September 1997
Distance traveled: 330 ft (100 m)

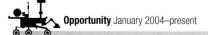
Spirit January 2004–March 2010
4.75 miles (7.7 km)

Opportunity January 2004–present
24 miles (38.7 km)

Curiosity 2012–present
3 miles (4.89 km)

The rover's ChemCam tool fires a laser at target rock or soil. The flash of reflected light is analyzed to identify elements in the target.

A plutonium power source provides electricity.

▽ **Landing on Mars**
Curiosity arrived at Mars in a shell-shaped capsule. Once this and a parachute were jettisoned, Curiosity used a sky crane system to touch down. About 65 ft (20 m) above the ground, three tethers and a cable providing power and communication linked Curiosity to the descent stage. Once touchdown was detected, the links were cut and the descent stage flew clear of the landing site.

Curiosity drives across the rocky surface at 1.5 in (3.8 cm) per second.

EXPLORING **MARS**

1 **Endurance Crater**
This view of wind-whipped sand dunes inside Endurance Crater is one of many incredible views of Mars returned by Opportunity—the longest-running rover on Mars. Opportunity was unable to ride directly over the sand because of the risk of becoming stuck—a fate that befell its twin, the Spirit rover, in 2009.

2 **Santa Maria Crater**
This montage of images from Opportunity reveals the view east across the 295-ft- (90-m-) wide Santa Maria Crater, with the rim of Endurance Crater visible in the far distance. Camera filters were used to highlight different rocks and soils in false color; to human eyes, the scene would appear reddish brown.

3 **Gale Crater**
The rolling hills on the horizon in this image from NASA's Curiosity rover are part of the rim of Gale Crater. Curiosity touched down in this ancient, 96-mile- (154-km-) wide meteor crater in 2012. The site was chosen because it may once have contained running water and—possibly in the distant past—microbial life.

4 **Home Plate**
The Spirit rover visited this rust-red rocky plateau, named for its similarity to a baseball home plate, in 2006. The plateau is thought to have formed in an ancient volcanic explosion, perhaps when lava came into contact with water. One of Spirit's radio antennae, used to send signals to Earth, is visible on the right.

5 Payson Outcrop

Captured by the Opportunity rover's panoramic camera, this image shows Payson Outcrop, the crumbling, eroded wall of Erebus Crater. False colors have been used to enhance subtle differences in layers of rock and soil. The outcrop is about 3 ft (1 m) deep and 82 ft (25 m) long.

ASTEROIDS

ASTEROIDS ARE ROCKY BODIES THAT VARY IN SIZE FROM A FRACTION OF AN INCH TO HUNDREDS OF MILES WIDE. THEY EXIST THROUGHOUT THE SOLAR SYSTEM, BUT MOST ARE FOUND IN THE ASTEROID BELT BETWEEN MARS AND JUPITER.

Sometimes called minor planets, asteroids orbit the Sun in the same direction as planets, but only the very largest have sufficient mass to pull themselves into a regular, rounded shape.

Asteroids were much more numerous in the solar system's early years. As they orbited the Sun, they collided and sometimes joined through gravity, accumulating to form larger bodies. Some of these embryonic worlds were destined to become today's terrestrial planets, but those near Jupiter's orbit were disturbed by the giant planet's powerful gravity, which caused them to crash violently and fragment. As a result, a ring of rocky debris has remained between the orbits of Mars and Jupiter ever since, forming the asteroid belt.

Today, the asteroid belt is sparsely populated; the total mass of the main belt is equal to only 4 percent of the Moon's mass. Collisions dominate this part of the solar system, and most asteroids are fragments of larger bodies that were destroyed.

253 Mathilde
(NEAR Shoemaker image)

△ **Carbonaceous asteroids (C-type)**
Asteroids can be classified by the materials they are made up of. About 75 percent of all known asteroids are carbonaceous. These carbon-rich asteroids have very dark surfaces, typically reflecting only 3–10 percent of the light that falls on them. Carbonaceous asteroids are found in the outer regions of the main asteroid belt.

A series of concentric troughs circle Vesta's equator. These are fractures produced when the largest craters formed.

▷ **Many sizes**
The largest body in the asteroid belt is Ceres, which is 592 miles (952 km) wide and classed as a dwarf planet because of its spherical shape. While there are few very large asteroids in the belt, there are an estimated 200 million asteroids larger than 0.6 miles (1 km) in diameter, and millions of smaller ones. They are irregular in shape and bear the scars of repeated impacts and collisions. The smallest asteroids are just a fraction of an inch across; smaller still are countless specks of asteroid dust.

Largest asteroids by diameter

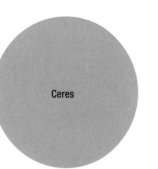
Ceres

Pallas

Vesta

Hygeia

Interamnia

Europa

The Moon

433 Eros
(NEAR Shoemaker image)

216 Kleopatra
(Arecibo radio telescope image)

▽ Asteroid evolution

If an asteroid grows sufficiently large, heat released inside it by decay of radioactive elements can cause it to melt. The molten materials then separate out due to gravity, heavy elements such as iron sinking to form a core, and more lightweight rocky minerals settling on top as mantle and crust. An asteroid continues to evolve as a result of impacts with other asteroids. Small impacts merely break off fragments, which become new asteroids. A large impact can smash an asteroid, scattering the fragments, but the parts may slowly reaggregate under gravitational attraction to form a loose mass of rubble.

△ Gray silicaceous asteroids (S-type)

These rocky bodies consist mainly of iron and magnesium silicates—the same materials that makes up Earth's mantle. Their surfaces reflect 10–22 percent of the light that falls on them, and they make up about 17 percent of the asteroid belt. Eros, the asteroid visited and orbited in 2000 by the NEAR Shoemaker spacecraft, is an S-Type.

△ Metallic asteroids (M-type)

These bodies appear to be a mixture of iron and nickel, similar in composition to Earth's core. This material has been molten and well mixed in the past, and then slowly cooled. The 0.75-mile- (1.2-km-) diameter Barringer Crater in Arizona was formed 50,000 years ago when a 164-ft- (50-m-) wide M-type asteroid hit Earth at about 30,000 mph (50,000 km/h).

◁ Vesta

The second most massive member of the asteroid belt, Vesta rotates once every 5.3 hours and is 326 miles (525 km) wide. Its surface is extensively cratered, and ejecta from these impacts has subsequently fallen to Earth, creating around 1,200 meteorites. Vesta is large enough to have melted completely as a result of radioactive heating and to have separated into a rocky mantle and metallic core. It was visited by NASA's Dawn spacecraft between July 2011 and September 2012.

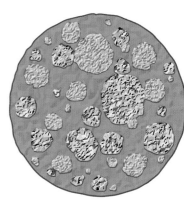

Accretion of smaller bodies

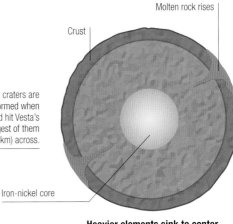

Molten rock rises

Crust

The Snowman craters are thought to have formed when another asteroid hit Vesta's surface. The largest of them is 43 miles (70 km) across.

Iron-nickel core

Heavier elements sink to center

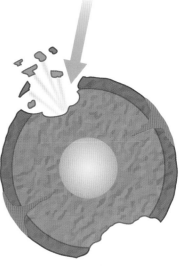

Impacts break off fragments

THE **ASTEROID BELT**

THERE ARE MILLIONS OF ASTEROIDS. MOST ARE IN THE ASTEROID BELT—A DOUGHNUT-SHAPED RING BETWEEN MARS AND JUPITER. EACH ONE FOLLOWS ITS OWN PATH AROUND THE SUN, BUT THEY ALL SHARE A COMMON ORIGIN.

The asteroid belt, or main belt, stretches between 195 million and 300 million miles (315–480 million km) from the Sun. Frequent collisions send asteroids hurtling out of the belt—its overall mass has decreased with time. Today, the combined mass of its asteroids is 4 percent of the mass of the Moon. The largest asteroid of all, Ceres, lies within the belt and makes up 30 percent of the belt's mass. It is one of only eight asteroids that are more than 186 miles (300 km) across and are spherical. The rest are irregular and much smaller.

Itokawa
A near-Earth asteroid that orbits outside the belt. It is 3.4 miles (5.4 km) in length and orbits in 1.52 years.

A near-Earth asteroid, Eros is 21 miles (34 km) long and takes 1.76 years to orbit the Sun.

The main belt is about 2.8 times farther from the Sun than Earth is.

Gaspra
Measuring 11.2 miles (18 km) long, Gaspra orbits every 3.29 years, near the inner edge of the belt. Its surface is pitted with craters from collisions with other asteroids.

Toutatis is a 2.7-mile- (4.3-km-) long near-Earth asteroid. Its orbit is outside the belt and takes 4.03 years to complete.

Belt profile

Around 200,000 belt asteroids are bigger than 6 miles (10 km) across, 200 million are over 0.6 miles (1 km) wide, and billions more are smaller still. They orbit the Sun in the same direction as the planets—that is, counterclockwise if they were seen from above. Their individual orbits are non-circular and slightly inclined to the plane of the planetary orbits, making the asteroid belt not flat but doughnut-shaped. One orbit of the Sun typically takes four to five years to complete. Jupiter's gravitational pull can change asteroid orbits, pushing or pulling them out of the belt. Asteroids outside the belt include near-Earth asteroids, and several thousand Trojans—two swarms of asteroids with orbits similar to that of Jupiter.

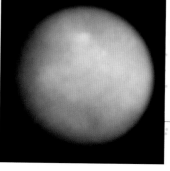

Ceres
The largest asteroid, Ceres, is classed as a dwarf planet. Its orbit takes 4.6 years to complete and is inclined by 10.6°.

These Trojans move 60° behind Jupiter.

Origins and collisions

Astronomers think that the belt asteroids are the remains of a planet that almost formed between Mars and Jupiter when the solar system was young. At that time, the gap between Mars and Jupiter contained about four times the material that makes up Earth. This rocky and metallic material started to clump together into ever-larger bodies. However, the gravity of the young Jupiter disrupted this process by changing the orbital paths of the bodies, causing them to collide and break up.

The main belt originally contained 1,000 times as much material as it does now. Since then, the number of asteroids has decreased. Collisions forced some asteroids out of the belt; most of these were destroyed long ago when they struck planets and moons. Collisions still occur in the belt today. They result in impact craters and, less frequently, the internal fracturing of asteroids; rarely, an asteroid may shatter into pieces that disperse. Most collisions occur at thousands of miles per hour. A collision's outcome depends mainly on the sizes of the bodies involved.

Both sets of Trojans orbit in roughly the same time as Jupiter—11.8 years. This group travels 60° in front of Jupiter.

Ida, which is 37 miles (60 km) long, orbits the Sun in 4.84 years.

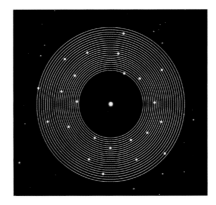

△ **Orbits before Jupiter formed**
The chunks of material between Mars and Jupiter initially followed nearly circular orbits. Collisions between them were at relatively low speeds, so material stuck together until some bodies grew as big as Mars.

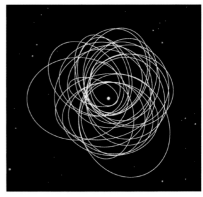

△ **Orbits after Jupiter formed**
Jupiter's gravity pulled at the bodies and changed their orbits into ellipses. This caused collisions to occur at much higher velocities. As a result, the impacting bodies smashed into pieces, producing a belt of asteroids.

Small impactor strikes | Crater forms on large asteroid

△ **Cratering**
Most collisions involve a small asteroid striking a larger one. The small asteroid is destroyed, leaving a crater in the large asteroid's surface that measures about ten times the size of the impactor. Most of the material blasted from the crater moves into its own orbit around the Sun.

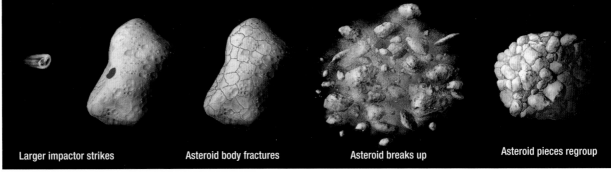

Larger impactor strikes | Asteroid body fractures | Asteroid breaks up | Asteroid pieces regroup

△ **Rubble pile**
When the impacting asteroid is bigger—about one fifty-thousandth the size of the large asteroid—it strikes with greater force, and the body of the large asteroid breaks up. The combined gravitational pull of the pieces soon pulls them back together. The result is an asteroid that is not one solid body but a ball of rubble.

▽ **Asteroid family**
An even bigger impactor—over one fifty-thousandth the size of the large asteroid—is more devastating. The large asteroid shatters, but the combined gravitational pull of the fragments cannot pull them back together. Instead, they form a family of asteroids that spread out around the orbit of the original large asteroid.

Very large impactor strikes | Large asteroid shatters | Family of asteroids forms

NEAR-EARTH **ASTEROIDS**

THOUSANDS OF ASTEROIDS PASS RELATIVELY CLOSE TO EARTH ON THEIR ORBITS, AND WITH SOME THERE IS A GENUINE RISK OF A COLLISION WITH OUR PLANET; HUGE CRATERS ON EARTH'S SURFACE ARE THE SCARS OF PAST ENCOUNTERS.

Near-Earth asteroids (NEAs) started life in the asteroid belt. At some point, Jupiter's gravitational pull or collisions with other asteroids set them in new orbits that now bring them within 121 million miles (194.5 million km) of the Sun, which is classed as being "near" Earth. Asteroids closer to Earth than 4.7 million miles (7.5 million km)—less than 20 times the average Earth–Moon distance—and at least 500 ft (150 m) across are called potentially hazardous asteroids (PHAs). Anything this size or larger would have a devastating impact on Earth, producing a huge tsunami if it landed in the ocean or vaporizing an area the size of Manhattan if it struck land.

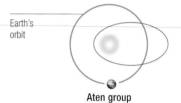

Eros

△ **Close enough**
Eros is an NEA and a member of the Amor group (see right). In January 2012, it passed within 16.6 million miles (26.7 million km) of Earth. In the same month, a 26-ft- (8-m-) wide Aten asteroid, 2012 BX34, made one of the closest recorded flybys, at a distance of 40,400 miles (65,000 km)—one-sixth of the Earth-Moon distance.

▷ **Mapping asteroids**
In this edge-on view of the solar system, tiny dots are NEAs that scientists believe exist. The data is from the NEOWISE survey, which was carried out by the Wide-Field Infrared Survey Explorer telescope between 2010 and 2011. Astronomers have discovered 10,000 NEAs at least half a mile (1 km) across—perhaps 90 percent of the total number. There are thought to be about 5,000 PHAs. In early 2014, around 1,500 PHAs were being monitored.

KEY
— Earth's orbit
• Potentially hazardous asteroids
● Near-Earth asteroids

▽ **Orbit types**
Near-Earth asteroids are classified by their orbital paths. The estimated 5,200 Apollo asteroids follow paths that cross Earth's orbit. The 750 or so members of the Aten group have orbits that stay mainly inside Earth's orbit. The Atiras, a small subgroup of the Atens, travel entirely within Earth's orbit. The orbits of the Amor group lie mostly between Earth and Mars.

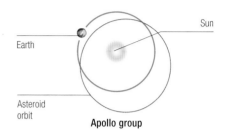

Sun
Earth
Asteroid orbit
Apollo group

Earth's orbit
Aten group

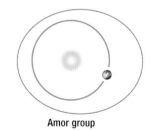

Atira subgroup

Amor group

◁ **Chelyabinsk meteorite**
In February 2013, a brilliant fireball blazed across the morning sky over the city of Chelyabinsk, Russia. It was a previously undetected asteroid, 60 ft (18 m) across and with a mass of 11,000 tons, speeding through Earth's atmosphere. The asteroid exploded at an altitude of 14 miles (23 km), producing a shower of pieces that fell to the ground as meteorites. It was the largest object to enter Earth's atmosphere since a similar event occurred over Tunguska, Siberia, in 1908.

▽ **Detection and monitoring**
Astronomers use optical telescopes to detect and track NEAs and PHAs, and radio telescopes to image any PHAs that get close enough. Once detected, an object is verified and cataloged by the Minor Planet Center in Cambridge, MA. Its orbital path is updated, and improved predictions are made about the asteroid's future close approaches to Earth.

Impact on Earth

Thousands of tons of asteroid material enter Earth's atmosphere each year. Most are small pieces that burn up before reaching the ground. Pieces big enough to survive the journey are known as meteorites. Earth was heavily bombarded by asteroids when it was young. The rate of impact has decreased, but it hasn't stopped; an asteroid at least 450 ft (150 m) wide strikes roughly every 10,000 years, and one more than half a mile (1 km) wide hits Earth every 750,000 years.

> Earth is **twice as likely** to be hit by something we **don't know** as by something we do.

The telescope sends out radio waves, and the dish collects the "echoes" that bounce back off objects, such as asteroids, in space.

▽ **Earth's impact craters**
Measuring 0.7 miles (1.2 km) across, Barringer Crater in Arizona (below) was created by a 165-ft- (50-m-) wide meteorite. The largest of Earth's 180 known impact craters is Vredefort, South Africa, which was 185 miles (300 km) across when it formed over 2 billion years ago. Many other craters have been wiped out by volcanic or tectonic resurfacing, and by erosion.

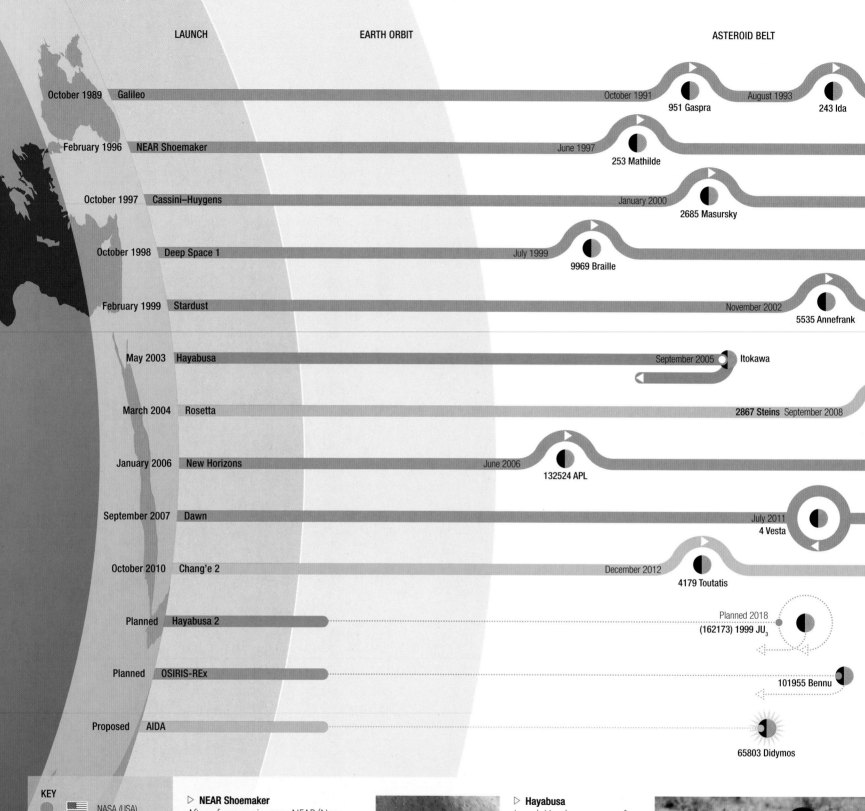

LAUNCH | EARTH ORBIT | ASTEROID BELT

October 1989 — Galileo — October 1991 · 951 Gaspra — August 1993 · 243 Ida

February 1996 — NEAR Shoemaker — June 1997 · 253 Mathilde

October 1997 — Cassini–Huygens — January 2000 · 2685 Masursky

October 1998 — Deep Space 1 — July 1999 · 9969 Braille

February 1999 — Stardust — November 2002 · 5535 Annefrank

May 2003 — Hayabusa — September 2005 · Itokawa

March 2004 — Rosetta — 2867 Steins · September 2008

January 2006 — New Horizons — June 2006 · 132524 APL

September 2007 — Dawn — July 2011 · 4 Vesta

October 2010 — Chang'e 2 — December 2012 · 4179 Toutatis

Planned — Hayabusa 2 — Planned 2018 · (162173) 1999 JU$_3$

Planned — OSIRIS-REx — 101955 Bennu

Proposed — AIDA — 65803 Didymos

KEY

- NASA (USA)
- JAXA (Japan)
- ESA (Europe)
- CNSA (China)
- Destination
- Flyby
- Orbit
- Sample return
- Lander
- Collision

▷ **NEAR Shoemaker**

After a four-year journey, NEAR (Near-Earth Asteroid Rendezvous) Shoemaker arrived at Eros in February 2001 and moved into orbit around the asteroid. Over the next 12 months, its orbit took it ever closer to the surface of Eros, enabling the craft to capture increasingly detailed images. Planned and built as an orbiter, NEAR Shoemaker's mission was later changed and it made a soft landing on Eros—the first landing on an asteroid.

Eros surface from NEAR Shoemaker

▷ **Hayabusa**

Japan's Hayabusa spacecraft reached Itokawa in September 2005. It surveyed the asteroid from a few miles away and then touched down to collect a surface specimen. The craft broke up on reentry into Earth's atmosphere, but the previously ejected sample capsule made a parachute landing in the South Australian outback on June 13, 2010.

Retrieving the sample capsule

MISSIONS TO **ASTEROIDS**

MORE THAN A DOZEN ASTEROIDS HAVE BEEN VISITED BY SPACECRAFT, BUT ONLY FOUR MISSIONS HAVE BEEN DEDICATED TO STUDYING THESE ROCKY BODIES. THE MOST RECENT SUCCEEDED IN RETURNING A SAMPLE TO EARTH.

The first close-up image of an asteroid came in 1991 when the Galileo spacecraft sent back remarkable images of Gaspra—an 11.2-mile- (18-km-) long, crater-covered boulder—during the spacecraft's journey to Jupiter. The first dedicated asteroid mission was NEAR Shoemaker, which landed on Eros in 2001. Nearly five years later, the Japanese spacecraft Hayabusa touched down on the half-mile- (1-km-) wide asteroid Itokawa, collected a sample, and brought it back. After orbiting Vesta, Dawn is on target for a 2015 encounter with the largest and first discovered asteroid, Ceres. Even more ambitious projects are under discussion, including a NASA mission to capture an asteroid and tow it into lunar orbit, where astronauts can visit it.

433 Eros
February 2001

July 2010
21 Lutetia

Planned 2015
Pluto

Planned 2015
Ceres

Hayabusa brought back around **1,500** particles of asteroid dust.

△ **Itokawa dust sample**
Analysis of the asteroid dust returned by Hayabusa's sample capsule showed that it had lain on Itokawa's surface for about 8 million years. It also revealed that Itokawa probably formed from the fragments of a larger asteroid that broke up in a collision.

The small, transparent container held dust grains a tiny fraction of an inch (less than 0.1 mm) wide.

▽ **Dawn**
Dawn's mission is to orbit the two most massive asteroids, Vesta and Ceres. The craft entered orbit around Vesta in July 2011 after a voyage that took it past Mars. It returned thousands of images that have enabled scientists to study the geology of Vesta's surface in detail. Dawn left for Ceres in September 2012 and will image the dwarf planet's entire surface.

With solar arrays extended, Dawn is 65 ft (20 m) wide.

◁ **Asteroid capture**
NASA is considering a mission to capture a near-Earth asteroid about 500 tons in weight and 25 ft (8 m) wide. The asteroid would be towed into a lunar orbit, where a crewed Orion capsule could dock with the capture craft, allowing astronauts to study the rock. Lunar orbit would be safer than Earth orbit because the risk of accidental collision with Earth would be lower.

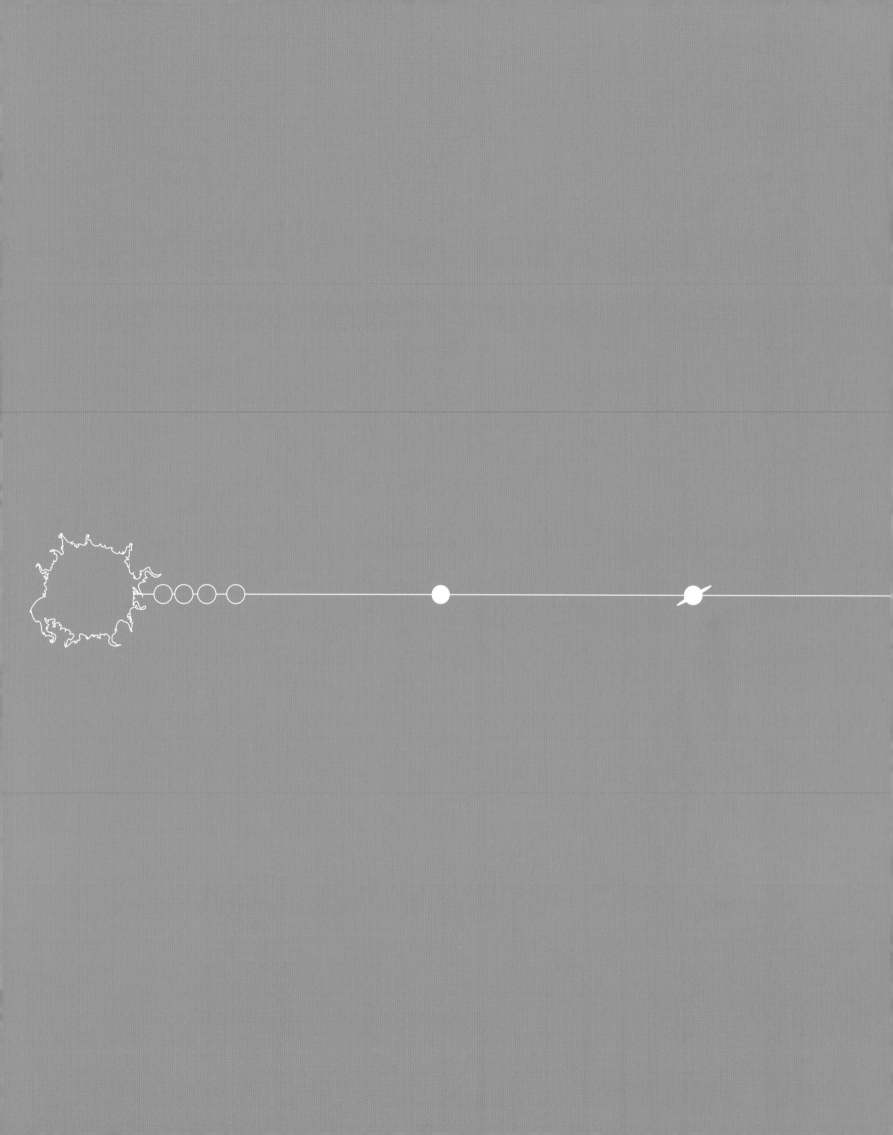

GAS **GIANTS**

The planets of the cold outer reaches of the solar system are not worlds on which a spacecraft could ever land. Jupiter, Saturn, Uranus, and Neptune, known collectively as the gas giants, are colossal globes of hydrogen and helium that are solid only at their cores. They formed toward the far edge of the spinning nebula of dust from which our Sun was born. At first mere clumps of rock and ice, they

REALM OF **GIANTS** ———————————————○

grew big enough to exert gravitational pull, ballooning into huge planets as they attracted layer after layer of gases. While the Sun contains 98 percent of all matter in the solar system, vast Jupiter, biggest of the four giants, comprises nearly all the rest. The outer planets take their time to circle our star: Jupiter's year is nearly 12 Earth years, Neptune's almost 165. These are incredibly active worlds whose hot interiors generate phenomenal cosmic weather. Jupiter's much-photographed Great Red Spot is a gigantic storm system three times the size of Earth. On Neptune, the fastest recorded winds in the solar system rage at over 1,200 mph (2,000 km/h). All of the gas giants are surrounded by rings of debris, the most famous being the rings of Saturn. These form a gleaming disk visible through binoculars—if placed around Earth, they would stretch nearly all the way to the Moon. And each of the outer planets is attended by an orbiting retinue of moons of diverse shapes and sizes.

◁ **Far-away worlds**
In the outer solar system there are small worlds as well as large ones. In this dramatic photograph taken by NASA's Cassini spacecraft, Io, the innermost of Jupiter's major moons, appears as an insignificant dot against the swirling cloud bands of its giant parent.

JUPITER

THE LARGEST OBJECT IN THE SOLAR SYSTEM AFTER THE SUN, JUPITER IS A BLOATED BALL OF GAS STREAKED WITH MULTICOLORED CLOUDS. THIS RAPIDLY SPINNING PLANET IS CIRCLED CEASELESSLY BY WINDS AND STORMS.

The first of the giant planets beyond the asteroid belt, Jupiter is nearly five times farther away from the Sun than Earth is. Composed of gas at increasingly high pressure, almost like a miniature star, it has a gravitational pull strong enough to have captured a large family of orbiting moons. Even with the naked eye, Jupiter is easily identifiable as one of the brightest objects in the night sky.

Rotating on its axis in just under ten hours, Jupiter has the shortest day of all the planets in the solar system. The planet spins so fast that its equator is forced outward in a noticeable bulge. The zones of high and low atmospheric pressure wrapped around the planet, identifiable by the different colors of clouds found within them, are stretched out by the rapid rotation. Nonstop winds race in both directions, stirring up giant storms large enough to engulf Earth. The Great Red Spot, Jupiter's most prominent feature, is a storm that has been raging for more than 300 years.

Winds in Jupiter's equatorial region can reach speeds in excess of **250 mph (400 km/h).**

JUPITER DATA

Equatorial diameter	88,846 miles (142,984 km)
Mass (Earth = 1)	318
Gravity at equator (Earth = 1)	2.36
Mean distance from Sun (Earth = 1)	5.20
Axial tilt	3.1°
Rotation period (day)	9.93 hours
Orbital period (year)	11.86 Earth years
Cloud-top temperature	−162°F (−108°C)
Moons	67+

▷ **Northern hemisphere**
Until 2003, Jupiter's north polar regions hid a secret—a dark spot twice the size of the planet's best-known feature, the Great Red Spot. The dark spot, which is visible only intermittently, appears to be in the highest layers of Jupiter's atmosphere.

▷ **Tilt**
Jupiter orbits with almost no tilt in its axis, so it has no seasons, and the equator always receives much more heat from solar radiation than the poles. This may contribute to the planet's remarkably stable large-scale weather systems.

Jupiter has a thin, barely discernible ring system with four distinct regions.

▷ **Southern hemisphere**
Both of Jupiter's poles are partly obscured by a haze, caused by radiation making chemical changes in atmospheric gases. Enormous electrical energy at the poles creates aurorae thousands of times more extensive than those seen in polar latitudes on Earth.

The North Temperate Belt has a strong jet stream blowing in the same direction as Jupiter's rotation.

The Great Red Spot is a giant storm that sits between the South Equatorial Belt and the South Tropical Zone.

The South Tropical Zone is Jupiter's most active weather region, with a strong jet stream moving in the opposite direction of the planet's rotation.

Complex, ribbonlike features called festoons form in the turbulent boundaries between belts and zones.

The North Equatorial Belt marks a clearing of the atmosphere, where darker clouds are seen deeper down.

The Equatorial Zone is a belt of bright, high-altitude clouds.

The South Equatorial Belt is usually the broadest and darkest cloud band on the planet.

◁ **Stormy face**
Jupiter's turbulent cloud belts and zones are very long-lived features, but their intensity varies according to weather conditions and the changing combinations of chemicals dredged up from the interior.

Atmosphere
Jupiter's atmosphere, mostly hydrogen gas with some helium, extends upward for more than 3,100 miles (5,000 km) to merge with interplanetary space.

JUPITER STRUCTURE

GIGANTIC THOUGH JUPITER IS, THE MATERIALS THAT FORM THE PLANET ARE COMPARATIVELY LIGHT. DESPITE THIS, FORCES OF GRAVITATIONAL CONTRACTION DEEP INSIDE JUPITER TURN THE PLANET'S INTERIOR INTO A POWERHOUSE OF ENERGY.

While Jupiter's interior is almost entirely pure hydrogen, the planet's upper layers are enriched with more complex gases that form the well-defined striped atmosphere. Around 600 miles (1,000 km) below this apparent "surface," pressures are high enough to transform hydrogen gas into liquid. Some 12,500 miles (20,000 km) farther inward, pressure is so intense—many millions of times the atmospheric pressure on Earth—that it tears the hydrogen atoms apart, freeing their hold over electrons and causing the hydrogen to behave like liquid metal.

Within the planet, denser materials sink downward, while the lighter materials rise up. The power this generates allows Jupiter to pump out more energy than it receives from the Sun, mostly in the form of heat and radio waves. Huge electrical currents in the metallic hydrogen layer create the most powerful magnetic field of any planet in the solar system.

Core
The existence of a solid core at Jupiter's heart is unproven but likely. It could be the original seed around which the planet coalesced, or possibly a growing nucleus formed by Jupiter's ongoing contraction.

Liquid metallic hydrogen layer
Liquid hydrogen atoms break down under heat and pressure to create a layer of liquid metallic hydrogen. This fluid, produced under extreme conditions, never occurs naturally on Earth.

Liquid layer
Below Jupiter's cloud layer, increasing pressure gradually causes the planet's hydrogen to act like a liquid rather than a gas.

Jupiter has **2.5 times the mass** of all the other planets **put together.**

Temperatures at the center may be higher than 36,000ºF (20,000ºC), which is hotter than the surface of the Sun.

Swirling currents within the liquid metallic hydrogen layer generate a gigantic magnetic field around Jupiter.

Jupiter's upper layers contain a chemical cocktail that includes ammonia, methane, water, and hydrogen sulfide.

◁ **Jupiter's layers**
This model shows Jupiter's internal structure divided into sharply defined layers. However, the transformation of hydrogen from gas to liquid in the depths of the planet is gradual and no obvious meeting point marks the boundary between the phases.

JUPITER UP CLOSE

ALTHOUGH JUPITER HAS NO SOLID SURFACE, THE TURBULENT CLOUDS THAT COVER ITS FACE ARE PACKED WITH DETAIL, AND INDIVIDUAL WEATHER SYSTEMS CAN PERSIST FOR YEARS OR EVEN CENTURIES IN THE SWIRLING ATMOSPHERE.

Jupiter's most conspicuous features are the bands of cloud that encircle the planet parallel to its equator. Astronomers classify them as either light-colored zones or dark-colored belts. Zones are high-pressure areas where clouds pile up at high altitude, while belts are low-pressure clearings in which sinking, cloud-free air provides a window through to darker clouds below. Storms, such as the Great Red Spot, are areas of high pressure where the clouds tower high above everything else.

The giant planet's weather is created by the interaction of a number of different factors, including heat rising from Jupiter's deep interior, the differential rotation that causes equatorial regions to move faster than polar latitudes, and convection in the upper atmosphere, which redistributes heat between Jupiter's warm equator and its colder poles.

The complex boundaries between belts and zones are shaped by powerful jet-stream winds that blow in opposite directions around the planet. These winds cause the zones to flow in an eastward direction (with the planet's rotation) and the belts, in contrast, to flow in a westward (or retrograde) direction. The general system of belts and zones appears to be stable over long periods of time, although the width of specific bands can vary significantly, as can the hue and intensity of the clouds in the belts.

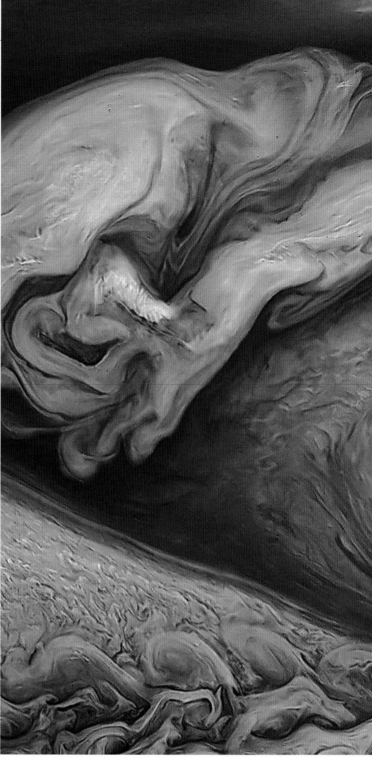

▽ **Banded planet**
Individual belts and zones are named according to their geographical location, such as the North Temperate Belt and the Equatorial Zone. Jupiter's rotation can be measured approximately by monitoring the movement of the dark cloud belts, but scientists can obtain more accurate results by measuring the rotation of the planet's magnetosphere.

▷ **Great Red Spot**
Jupiter's most spectacular feature is the Great Red Spot, observed since at least 1830 and possibly since the 17th century. It is an anticyclonic (counter-clockwise-rotating), hurricane-like weather system, twice the size of Earth and with a high-pressure center. The origins of its color are uncertain, but its intensity can vary substantially and seems to be linked to the appearance of the neighboring South Equatorial Belt.

North Polar Region

North North Temperate Belt

North Temperate Belt

North Equatorial Belt

North Temperate Zone

North Tropical Zone

Direction of movement

Equatorial Zone

South Equatorial Belt

South Tropical Zone

South Temperate Belt

South Temperate Zone

South South Temperate Belt

South South Temperate Zone

South Polar Region

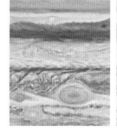

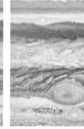

△ **Baby spots**
This sequence of images from the Hubble Space Telescope shows a succession of close encounters between Jupiter's Great Red Spot and two smaller storms in summer 2008. Red Spot Junior (bottom) survived unscathed after passing the Great Red Spot several times, but the Baby Red Spot (smallest spot) was captured and destroyed by the giant storm.

◁ Cloud temperature and height

This infrared image from the Gemini Observatory shows temperature differences in color. Zones appear blue because they are higher and colder than the belts, which are reddish. The cloud tops in the Great Red Spot and other high-altitude storms appear white because they are even higher and colder than the zones.

High, cold, light-colored clouds form in the zones.

Jet-stream winds blow in opposite directions.

Upwelling warm gases from Jupiter's interior

Low, warmer, dark-colored clouds form in the belts.

▷ Convection cycle

Convection of gases maintains the structure of zones and belts. Zones occur where clouds well up and cool; belts occur where they descend and warm up. Bright ammonia-ice clouds at the top of the zones hide the underlying clouds. Deeper in the atmosphere, the clouds are made of ammonium hydrosulfide and water.

Gases cool and then sink back down.

THE **JUPITER** SYSTEM

FITTINGLY FOR THE LARGEST PLANET IN THE SOLAR SYSTEM, JUPITER ALSO HAS THE BIGGEST FAMILY OF SATELLITES—AT LEAST 67 ARE KNOWN AT PRESENT. HOWEVER, JUST FOUR OF THESE ARE PLANET-SIZED AND DOMINATE THE JUPITER SYSTEM.

Jupiter's satellites are divided into three major groups: four small inner satellites, sometimes called the Amalthea group; the four huge Galilean moons (discovered in 1610 by Italian astronomer Galileo Galilei); and 59 or more small outer moons, most of which are just a few miles across, though some are much larger. The Amalthea group and the Galilean moons are together referred to as regular satellites, which means they orbit in the same direction as Jupiter's rotation and are all on roughly the same plane. The outer, irregular satellites are small bodies captured by Jupiter's gravity throughout its life.

Moons to scale
The four Galilean moons account for most of the material in Jupiter's satellite system. The other, irregular satellites include captured asteroids, centaurs, and comets. Most are lumps of ice or rock, but a few are dozens of miles across or even bigger.

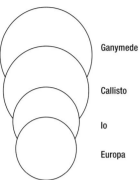

Ganymede

Callisto

Io

Europa

Himalia
Amalthea
Thebe
Elara
Pasiphaë
Carme
Metis
Sinope
Lysithea
Ananke
Adrastea
Leda
Callirrhoe
Themisto
Praxidike
Iocaste
Taygete
Kalyke
Megaclite
S/2000 J11
Helike
Harpalyke
Hermippe
Thyone
Chaldene
Aoede
Eukelade
Isonoe
S/2003 J5
Autonoe
Carpo
Euanthe
Aitne
Erinome
Eurydome
Hegemone
Arche
Euporie
S/2003 J3
S/2003 J18
Thelxinoe
Orthosie
S/2003 J16
Mneme
Herse
Kale
S/2003 J19
S/2003 J15
S/2003 J10
S/2003 J23
Kallichore
Pasithee
S/2010 J1
Kore
Cyllene
S/2003 J4
Sponde
S/2003 J2
S/2003 J12
S/2001 J1
S/2010 J2
S/2011 J2
S/2003 J9

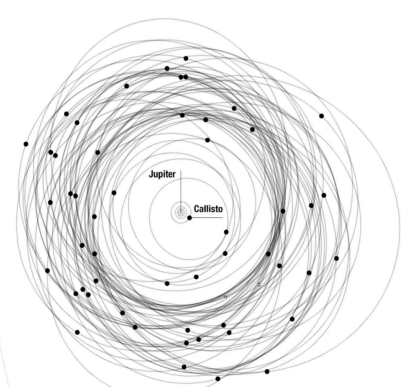

Jupiter

Callisto

Ganymede
Ganymede, the solar system's largest moon, is locked in orbital resonance with other moons, orbiting Jupiter in 7 days 3.7 hours—twice the period of Europa and four times that of Io. Ganymede doesn't experience significant tidal heating, but its surface shows signs that it did so in the past.

△ Outer moons
The orbits of the irregular satellites trace a chaotic cloud around Jupiter, some following Jupiter's rotation and others orbiting the opposite way. There are distinct groups among the outer moons. The Himalia, Carme, Ananke, and Pasiphaë groups all consist of a single large moon and a number of smaller ones in related orbits. Each group probably formed from the breakup of a larger object.

Callisto
The outermost Galilean moon orbits Jupiter in 16 days 16.5 hours, at a distance of almost 1.2 million miles (1.9 million km). Its heavily cratered surface shows it never suffered the extreme tidal heating that probably caused the widespread resurfacing evident on the other Galileans.

Io
The innermost Galilean moon orbits at 262,000 miles (421,700 km) from the center of Jupiter, just slightly more than the Moon–Earth distance. As a result, it is subject to enormous tidal stresses that heat its interior and drive constant volcanic activity on its surface.

Europa
Jupiter's smallest Galilean moon is locked in orbital resonance with Io, so that its orbital period is exactly twice as long as that of Io's. Like Io, it takes a pummeling from Jupiter's powerful tidal forces, and beneath its icy crust, volcanic activity is thought to warm a hidden ocean.

The Thebe gossamer ring is fed by dust spiraling inward from the surface of Thebe.

Adrastea
The smallest of Jupiter's regular satellites, misshapen Adrastea has an average diameter of 10 miles (16 km). It orbits at the outer edge of Jupiter's main ring. Like the other inner moons, it is thought to contribute dust to the rings as a result of micrometeorites impacting its surface.

Amalthea
The largest inner moon, and the first to be discovered, Amalthea is an ovoid roughly 155 miles (250 km) long with a remarkably red surface. It may have originated much farther from Jupiter than its current orbit, which is slightly eccentric (non-circular) as a result of Io's gravitational influence.

Jupiter

Thebe
This misshapen satellite is the outermost and second-largest of Jupiter's inner moons. Like Amalthea, it has a distinctly reddish surface, and is probably made from either a porous, loose collection of rubble, or water ice and other chemicals.

Metis
Jupiter's innermost known moon was discovered in 1979 during the Voyager 1 flyby. It orbits in just 7 hours 4 minutes (less than a Jovian day), in a distinct gap within Jupiter's main ring. Roughly oval in shape, Metis makes a significant contribution to the inner ring's dusty material.

The main ring is relatively narrow and centered at about 1.8 Jupiter radii.

The Amalthea gossamer ring is a broad disk fed by dust from Amalthea.

Ganymede is larger than Mercury and almost the size of Mars. It **would be classified as a planet** if it were orbiting the Sun rather than Jupiter.

△ **Inner moons**
Jupiter's inner moons include four small satellites associated with the tenuous ring system, and the four giant Galilean moons—Ganymede, Callisto, Io, and Europa. The large size of the Galilean moons makes them vulnerable to tidal forces. Io, Europa, and Ganymede have settled into resonant orbits, in which their mutual gravitational tugs help to keep the orbits of Io and Europa stable.

IO

THE INNERMOST OF JUPITER'S LARGE GALILEAN MOONS IS A HELLISH WORLD TORTURED BY TIDAL FORCES AND WRACKED BY POWERFUL VOLCANIC ERUPTIONS.

Io is the third-largest of Jupiter's moons. Its location near the center of the Jovian system puts it in the middle of a gravitational tug-of-war between Jupiter and the large moons Europa and Ganymede orbiting farther out. Powerful tides squeeze the moon in various directions, flexing its surface by as much as 330 ft (100 m). In comparison, the most dramatic tidal range of the sea on Earth is just 60 ft (18 m). Io's tidal activity heats the moon's interior, which consists of sulfurous rock with a much lower melting point than the silicate rocks of Earth. As a result, Io is the most volcanically active world in the solar system, with numerous volcanoes pouring sulfur-rich magma onto the surface or launching geyser-like plumes of sulfurous chemicals as high as 190 miles (300 km) into the sky. Io owes its colorful, pizza-like appearance to the unique properties of the element sulfur, which can take several different forms (allotropes) with different physical properties. The moon has little atmosphere—only a thin layer of gases, mostly sulfur dioxide, surrounds it.

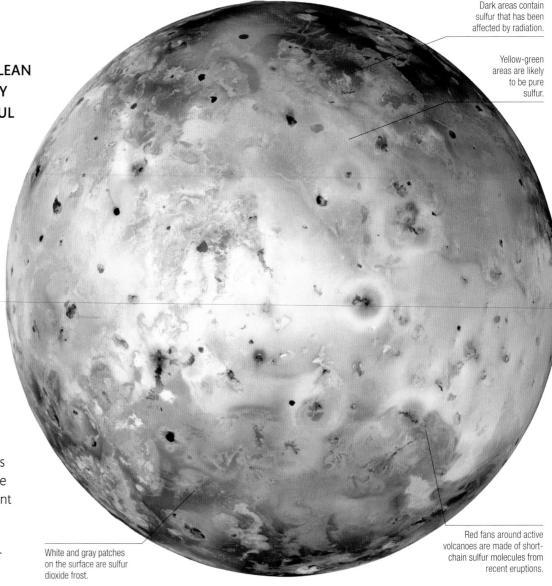

Dark areas contain sulfur that has been affected by radiation.

Yellow-green areas are likely to be pure sulfur.

White and gray patches on the surface are sulfur dioxide frost.

Red fans around active volcanoes are made of short-chain sulfur molecules from recent eruptions.

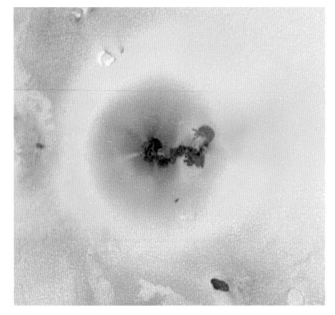

△ **Volcanic plume**
A volcanic feature known as Prometheus has been nicknamed Old Faithful on account of its reliable outbursts. Prometheus sends geyser-like plumes of molten sulfur up into the sky, and the fallout creates an ever-changing halo of color around the vent.

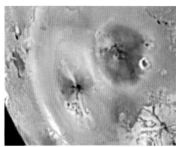

△ **A new eruption**
Io's landscape is dynamic. These images from the Galileo orbiter, taken five months apart, show the growth of a 250-mile (400-km) dark spot on Pillan Patera volcano.

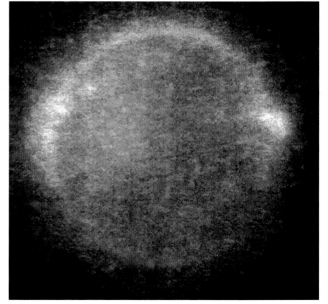

△ **Aurorae on Io**
Io's location within Jupiter's magnetic field means that it is constantly bombarded by high-energy particles trapped in the planet's radiation belts. When the particles collide with gases in Io's thin atmosphere, they produce glowing aurorae with vivid red and green colors.

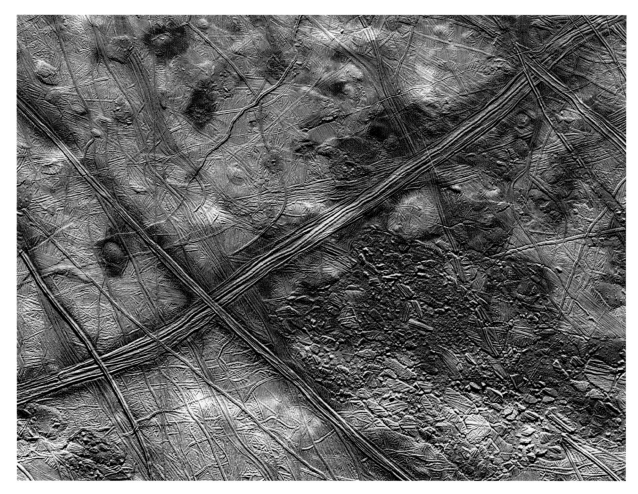

◁ Cracking ice
The way features on opposite sides of lineae match reveals that Europa's icy crust is being pulled apart as the moon flexes under changing tidal forces. Warmer ice, stained red by salts and sulfur, wells up from beneath and heals the cracks, but liquid water sometimes also erupts violently, gushing out to form huge plumes more than 125 miles (200 km) high.

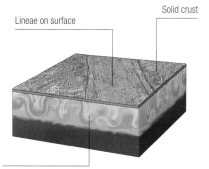

Lineae on surface

Solid crust

Liquid water or a slushy layer of convective ice lies below the surface.

△ Water world
The ocean beneath Europa's crust is thought to be around 60 miles (100 km) deep, but it is covered by a crust of solid ice that may be dozens of miles thick. The recent discovery of liquid water eruptions suggests the crust could be thinner in places, though it is not clear whether the eruptions are fed by the ocean or isolated pockets of water in the crust.

EUROPA

THE SMALLEST OF JUPITER'S GALILEAN MOONS, EUROPA HIDES ITS INTERNAL RESEMBLANCE TO ITS VOLCANIC NEIGHBOR IO BENEATH A SUPERFICIALLY PLACID ICY SURFACE.

Europa's icy crust gives it the smoothest surface of any large world in the solar system. Any significant features—formed, for example, by impact craters—gradually slump back to the average surface level. Enhanced-color images, however, reveal networks of discolored lines called lineae crisscrossing the ground, showing that Europa is far from dormant. Like Io, this moon is squeezed and stretched by competing tidal forces—from Io and Jupiter on the one side and planet-sized Ganymede on the other. The volcanic activity this generates is thought to warm a vast ocean of liquid water trapped beneath the frozen crust. This hidden sea may be one of the few places in the solar system that is hospitable to life.

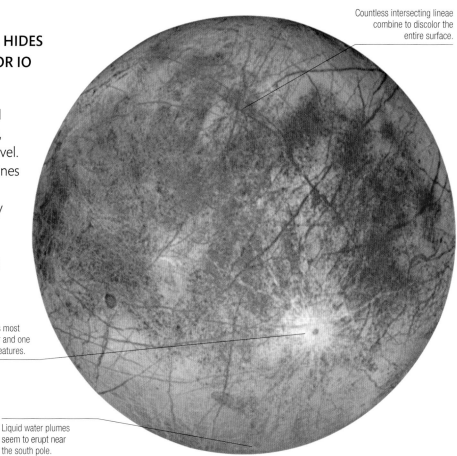

Countless intersecting lineae combine to discolor the entire surface.

Pwyll is Europa's most prominent crater and one of its youngest features.

Liquid water plumes seem to erupt near the south pole.

Europa is one of the **main targets** in the **search for life** beyond Earth.

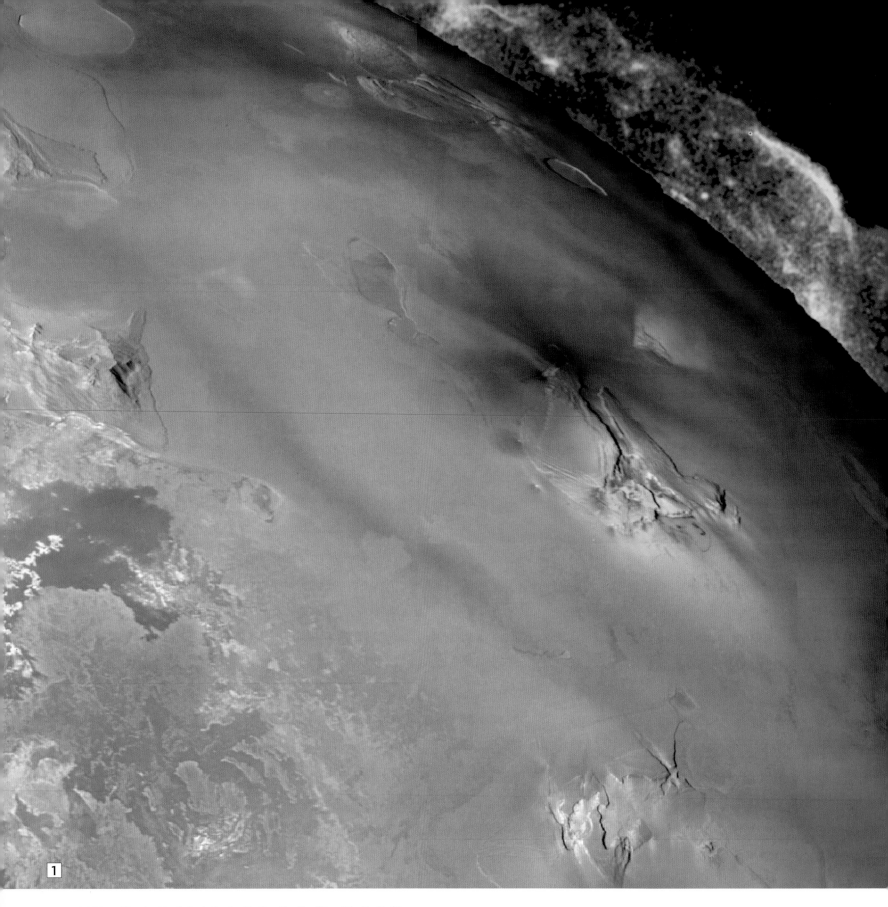

1

THE **GALILEAN MOONS**

1 Pele erupts on Io
This Voyager 1 image shows the eruption of Io's huge volcano, Pele. A plume of gas and dust rises from the volcano's vent to a height of 185 miles (300 km). The plume, invisible against the surface, can be seen as a bright umbrella shape against the dark sky. Fallout from the plume covers a heart-shaped area the size of Alaska around Pele.

2 Europa
When Galileo imaged Europa, it revealed vast plains of bright ice crisscrossed by cracks snaking away to the horizon, and dark patches that probably contain both ice and dirt. There are few highlands or big impact craters. Plumes of water vapor 125 miles (200 km) tall have been seen spurting from Europa's surface, and oceans of briny liquid water—and perhaps even life—may exist below the icy exterior.

3 Ganymede
In this Voyager 2 view from a range of 185,000 miles (300,000 km), the ancient dark area of Galileo Regio is at the upper right. In the lower center is a relatively young impact crater, surrounded by white rays of water-ice debris. The lighter regions—the younger parts of the surface—have grooves and ridges caused by tectonic activity. Like Europa and Callisto, Ganymede probably has subsurface saltwater.

4 Callisto
Though Callisto formed at the same time as Io, the two moons are very different. While Io's surface is young, constantly renewed by volcanism, Callisto's surface is old, scarred by the highest density of impact craters in the solar system and devoid of volcanoes and large mountains. In fact, it is one vast ice field, laced with cracks and craters from billions of years of collisions with interplanetary debris.

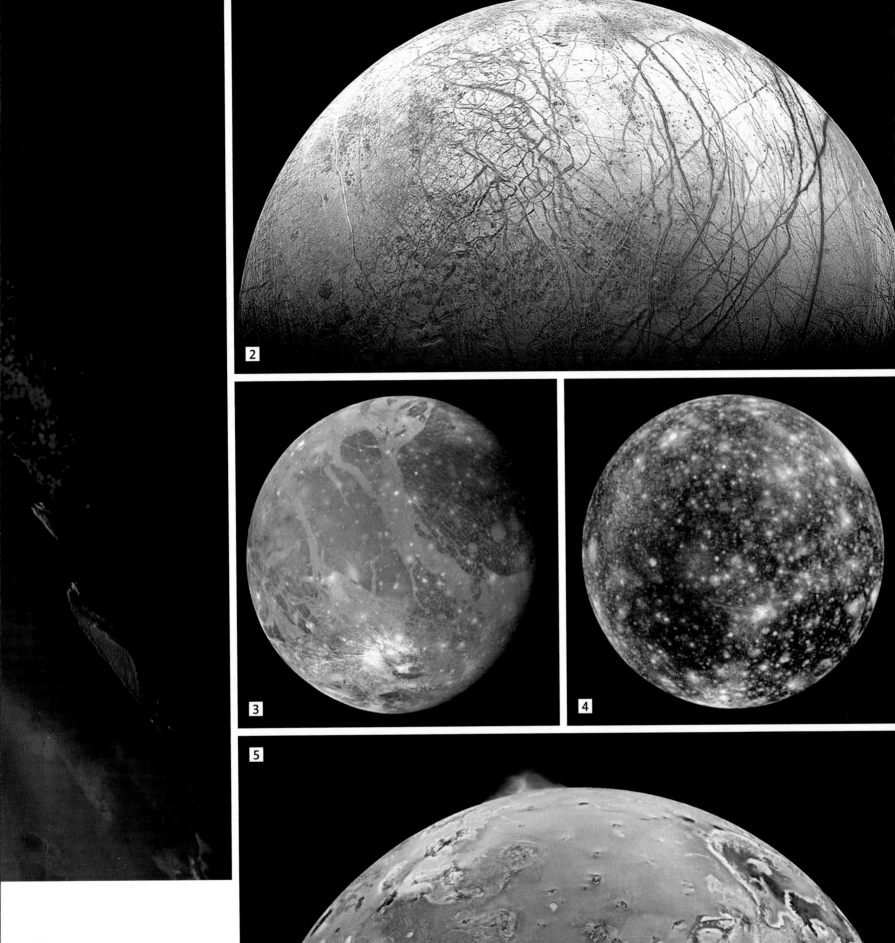

5 | **Volcanic plumes on Io**
Io's surface has virtually no impact craters and is continually being repaved by lava from its many volcanoes. Two sulfurous eruptions are visible in this Galileo image. At the top, on Io's limb, a bluish plume towers about 97 miles (140 km) above Pillan Patera, a volcanic caldera. At bottom center, a ring-shaped plume can be seen over the vent of the volcano Prometheus, rising about 47 miles (75 km) into space.

GANYMEDE

THE LARGEST MOON IN THE SOLAR SYSTEM, GANYMEDE IS BIGGER THAN THE PLANET MERCURY. WHILE IT SHOWS FEW SIGNS OF ACTIVITY TODAY, ITS SURFACE—A JIGSAW PUZZLE OF DIVERSE TERRAIN—BEARS THE SCARS OF A COMPLEX HISTORY.

With a diameter of 3,272 miles (5,268 km), Ganymede is 8 percent wider than Mercury and 25 percent larger in terms of volume. However, its much lower density suggests that it consists of a rock-ice mix similar to its inner neighbor, the moon Europa. Ganymede has a thin atmosphere dominated by oxygen, and its landscape is a jumble of bright and dark regions, with far fewer impact craters in the bright regions. This suggests that the dark areas have been exposed to bombardment from space for significantly longer than the lighter patches, which have been resurfaced. The lighter areas are scarred by parallel grooves and ridges—evidence of tectonic activity.

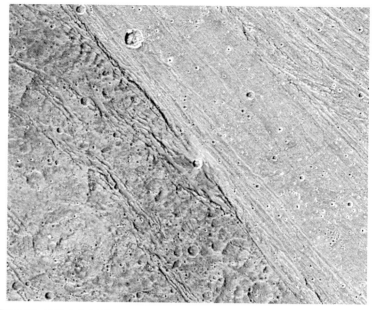

△ **Icy tectonics**
Ganymede's surface probably solidified early in its history, though Jupiter's tidal pull caused the moon's interior to remain molten. The crust split into tectonic plates like those on Earth, and a slushy mix of rock and ice welled up to bridge the gap between moving plates, forming grooved terrain similar to younger parts of Earth's crust.

Ganymede's magnetic field interacts with that of Jupiter's.

△ **Magnetic moon**
Ganymede is the only moon in the solar system known to have a significant magnetic field—evidence of an interior with distinct layers and a core that likely contains liquid iron. In 2002, scientists detected features in the magnetic field that suggest the presence of an ocean layer roughly 125 miles (200 km) below the surface, between layers of ice.

Lighter patches form where the dark plates have drifted apart.

Relatively young craters expose fresh ice at the surface.

Dark cratered areas are Ganymede's oldest terrain.

Water ice accounts for some **90 percent** of Ganymede's surface terrain.

Bright, fresh ice exposed
by recent impact

CALLISTO

THE OUTERMOST GALILEAN MOON IS A
DARK, HEAVILY CRATERED BALL OF ICE AND
ROCK, CONTRASTING IN STRUCTURE AND
APPEARANCE WITH ITS INNER NEIGHBORS.

Callisto seems to have changed relatively little since its formation,
and spacecraft images show a surface covered in craters accumulated
over 4.5 billion years of solar system history. The largest features are
enormous, ringed impact basins such as those named Asgard and
Valhalla. Solar radiation has caused the surface of the moon to darken
gradually over time, making the youngest impact craters look like
bright starbursts on the otherwise dull landscape.

Callisto is the least dense of the Galilean moons, indicating that it
contains more ice and less rock than its neighbors. It is thought to
be a relatively homogenous blend of rock and ice throughout, a
structure that may have been common to all the Galilean moons
before tidal heating took hold and caused the interiors of the other
moons to melt and separate into layers.

Major impact basins
are surrounded by
concentric rings

Valhalla is Callisto's
largest impact crater.

Fresh ice has welled
up to fill the center
of the basin.

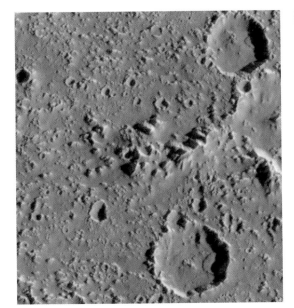

△ **Craters**
Callisto's location close to Jupiter puts it directly in the firing
line for comets and asteroids pulled to their doom by the
giant planet's gravity. As a result, Callisto is often described
as the most heavily cratered world in the solar system.

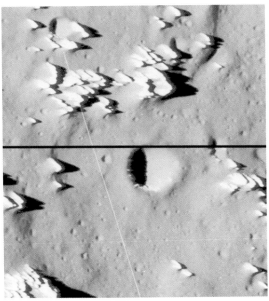

△ **Jagged hills**
Erosion by solar radiation has caused much of the ice in raised
crater rims to evaporate from the rock–ice mix, degrading their
structure and leaving chains of jagged, knoblike peaks across the
land. Landslips in the remaining material are common.

△ **Scarps**
Callisto's largest impact basins contain long scarps—cliffs
separating areas of different elevation. The scarps mark the top
of deep faults where the crust has fractured after impact and
blocks of terrain have shifted vertically in relation to each other.

Jupiter's gravity pulls comets and asteroids **to their doom**, shredding them into fragments that **collide with its moons.**

Artist's impression based on NASA images

LOCATION

Latitude 39°N; longitude 14°W

FORMATION

The comet or asteroid that formed Enki Catena was probably drawn into orbit around Jupiter before its breakup and impact.
1. Object strays too close to Jupiter.
2. Object breaks up, and fragments spread out along orbit.
3. Collision with Ganymede.

Jupiter

Ganymede

1.

3.

2.

CRATER CHAIN

Icy debris flung out during the impact surrounds one end of Enki Catena, where the underlying surface is younger. Darker material in older terrain may disguise the ejecta.

No ejecta on older, darker landscape

Bright ejecta on young, light terrain

DESTINATION **ENKI CATENA**

A SPECTACULAR CHAIN OF CRATERS MARCHES IN A STRAIGHT LINE ACROSS 100 MILES (160 KM) OF GANYMEDE'S SURFACE. THIS REMARKABLE FEATURE IS THE RESULT OF A SERIES OF IMPACTS IN THE MOON'S RELATIVELY RECENT PAST.

Enki Catena consists of at least 13 overlapping craters, each around 6 miles (10 km) or more in diameter, running diagonally across a boundary between darker and lighter areas of Ganymede's terrain (see page 162). The chain is the most prominent of several such features identified on Ganymede and Callisto. It almost certainly formed from the near-simultaneous impact of fragments of a comet or asteroid, broken apart under the force of Jupiter's gravitational pull in the same way as Comet Shoemaker-Levy 9, which struck the giant planet itself in 1994.

KING OF THE PLANETS

JUPITER'S BRIGHTNESS AND STATELY MOVEMENT THROUGH EARTH'S SKIES LED EARLY STARGAZERS TO GIVE IT A PROMINENT PLACE IN THEIR MYTHOLOGY. FROM THE DAWN OF SCIENTIFIC ASTRONOMY, IT HAS PLAYED A PIVOTAL ROLE IN MANY DISCOVERIES.

Because of its great size, Jupiter is visible as a disk rather than a point through even a basic telescope, and its four largest moons are easy to observe. However, the planet's shifting surface markings mystified early astronomers, and it was not until the 20th century that Jupiter's gaseous nature was widely accepted. Since the 1970s, spacecraft such as Galileo have revealed many more of the Jupiter system's secrets.

Greek bust of Zeus

Galileo's record of Jovian moons

C. 500 BCE

Ruling planet
The ancient Greeks and Romans associate the planet with the king of the gods, known as Zeus to the Greeks and Jupiter or Jove to the Romans. Long before, astronomers in Babylon associated Jupiter with Marduk, the ruling god in the Babylonian pantheon.

1610 CE

Galilean moons
Italian scientist Galileo Galilei studies Jupiter with his telescope and sees four faint "stars" nearby, which prove to be satellites. The existence of moons around other worlds contradicts the prevailing idea that everything in the universe circles Earth.

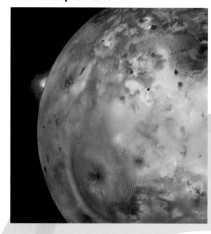

Volcanic eruption on Io

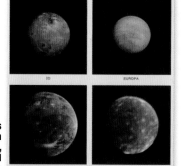

Voyager images of the Galilean moons Io, Europa, Ganymede, and Callisto

Stamp commemorating Pioneer 10

1979

Volcanoes over Io
Voyager 2 captures an image of a huge plume of material arching high above the surface of Io. This moon is the most volcanically active body in the solar system, with sulfurous eruptions driven by heat generated by Jupiter's tidal forces.

1979

Voyagers
The Voyager 1 and 2 spacecraft provide the first detailed views of Jupiter's Galilean moons, revealing four complex worlds, each the size of a small planet. Voyager 1 also discovers a tenuous ring system, composed of sparse particles, encircling Jupiter.

1973

Jupiter flyby
Launched in 1972, Pioneer 10 flies close to Jupiter the following year and returns the first close-up images of the planet. It suffers radiation damage while passing through Jupiter's magnetic equator, confirming the great strength of Jupiter's magnetic field.

Aftermath of Shoemaker–Levy impact

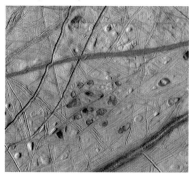

Close-up of Europa's surface

1994

Comet impact
Fragments of the comet Shoemaker-Levy 9 strike Jupiter, creating fireballs larger than Earth and stirring up material from deep inside the planet. The resulting "bruises" on Jupiter's face provide an insight into the planet's internal chemistry.

1995

Probing the atmosphere
NASA's Galileo spacecraft releases a probe that plunges into Jupiter's clouds. The probe sends back data about weather conditions and atmospheric chemistry as it descends 97 miles (156 km) through the upper atmosphere, until contact is lost.

1995–2003

Orbiting Jupiter
The Galileo orbiter studies the Jovian system for over eight years, investigating the planet and its major moons in detail and making countless discoveries. It finds evidence of a liquid-water ocean deep under the icy surface of the Galilean moon Europa.

Cassini's sketches of Jupiter

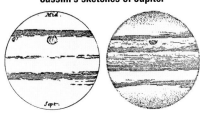

Ole Rømer observing Jupiter

1665–90

Jovian weather

Italian-French astronomer Giovanni Cassini makes sketches of Jupiter's atmosphere and identifies cloud bands and spots, which he uses to measure the planet's rotation. By 1690 he has concluded that different parts of Jupiter rotate at different rates.

1676

Measuring the speed of light

Danish astronomer Ole Rømer notices that eclipses and transits of Jupiter's moons don't always occur at predicted times, because of variations in the time it takes for light to reach Earth. This allows him to make the first estimate of the speed of light.

1733

Calculating Jupiter's diameter

English astronomer James Bradley measures the size of Jupiter's disk through a telescope and uses his result to calculate the planet's immense diameter. Bradley also tracks the movements of Jupiter's moons and studies their shadows and eclipses.

A 19th-century map of Jupiter

1955

Jupiter's magnetic field

In the US, Kenneth Franklin and Bernard Burke detect bursts of radio waves, known as synchrotron radiation, coming from Jupiter. This shows that Jupiter has a magnetosphere, since this type of radiation is emitted when high-speed electrons spin in a magnetic field.

1903

Jupiter is a gas giant

American astronomer George W. Hough states that Jupiter is dominated by a deep envelope of gases, transforming into liquid at great depth and high pressure—the first suggestion that Jupiter is a gas giant and not a solid body with a thin atmosphere.

1830

Great Red Spot

Giovanni Cassini and English scientist Robert Hooke may have seen the giant storm called the Great Red Spot in the 1660s, but the first confirmed sighting is made by German astronomer Heinrich Schwabe in 1830. It has been regularly viewed ever since.

Io and Jupiter from Cassini

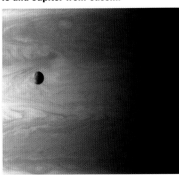

Three red spots (Junior at lower left)

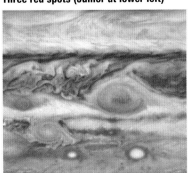

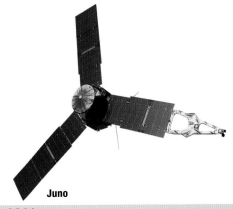

Juno

2000

Cassini flyby

The Cassini spacecraft takes 26,000 images of Jupiter from a distance of 6.2 million miles (10 million km) during a flyby en route to Saturn. Together with Galileo's close-ups, the Cassini images lead to new findings about the giant planet's weather systems.

2006

Red Spot Junior

Astronomers notice that a large storm, formed by the merging of three smaller white storms in 1998–2000, is turning red. Over the next few years, "Red Spot Junior" grows to more than half the size of the more famous Great Red Spot.

2011

Launch of Juno

Upon arrival in 2016, NASA'S Juno will map Jupiter's magnetic field, measure atmospheric levels of water and ammonia, observe Jupiter's aurorae, and investigate whether Jupiter has a solid core. It is hoped that Juno's findings will reveal much about how giant planets form.

		LAUNCH	EARTH ORBIT	JOURNEY TO JUPITER
1972	Pioneer 10			
1973	Pioneer 11			
1977	Voyager 1			
1977	Voyager 2			
1989	Galileo			
1997	Cassini			
2006	New Horizons			
2011	Juno			
Planned	JUICE			

MISSIONS TO JUPITER

MOST OF THE SPACECRAFT THAT HAVE VISITED JUPITER HAVE MADE ONLY A BRIEF FLYBY DURING A GRAVITY-ASSIST MANEUVER ON THE WAY TO ANOTHER PLANET. ONLY ONE SPACECRAFT HAS GONE INTO ORBIT AROUND JUPITER ITSELF.

The first spacecraft to journey beyond the inner solar system were Pioneers 10 and 11. After proving that the asteroid belt could be safely crossed, they sent back the first close-ups of Jupiter in 1973 and 1974. They were followed by the more sophisticated Voyagers 1 and 2, which returned breathtaking images of Jupiter's moons. The Galileo spacecraft entered orbit in 1995 and spent eight years surveying the Jupiter system in great detail, before self-destructing. Cassini and New Horizons, bound for Saturn and Pluto, respectively, followed later.

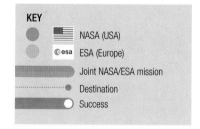

KEY
- NASA (USA)
- ESA (Europe)
- Joint NASA/ESA mission
- Destination
- Success

Jupiter's rings from Voyager 2

△ Voyager 1 and 2
The two Voyager flybys of Jupiter provided the first detailed views of the giant planet and its major moons. These missions confirmed the existence of a tenuous ring system around Jupiter (above) and discovered three new inner moons orbiting among the rings. The Voyager spacecraft returned beautiful images of Io's active volcanoes and Europa's fractured, icy crust, and time-lapse movies of Jupiter brought the planet's swirling cloud bands and rotating Great Red Spot to life.

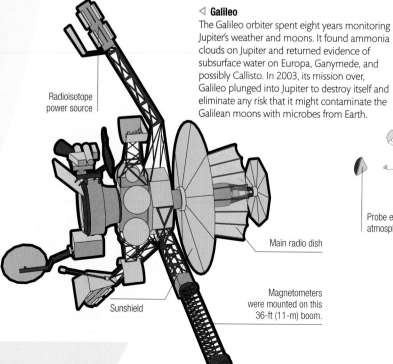

Radioisotope power source

Main radio dish

Magnetometers were mounted on this 36-ft (11-m) boom.

Sunshield

◁ Galileo
The Galileo orbiter spent eight years monitoring Jupiter's weather and moons. It found ammonia clouds on Jupiter and returned evidence of subsurface water on Europa, Ganymede, and possibly Callisto. In 2003, its mission over, Galileo plunged into Jupiter to destroy itself and eliminate any risk that it might contaminate the Galilean moons with microbes from Earth.

▽ Atmosphere probe
Shortly after arriving in orbit, Galileo released a probe that parachuted into Jupiter's atmosphere, descending 95 miles (150 km) through the upper cloud layers. The heat and pressure soon became too intense for the probe, but for 78 minutes it succeeded in collecting data about temperatures, winds, lightning, and the clouds and gases through which it passed.

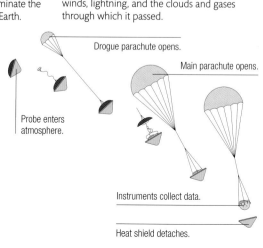

Drogue parachute opens.

Main parachute opens.

Probe enters atmosphere.

Instruments collect data.

Heat shield detaches.

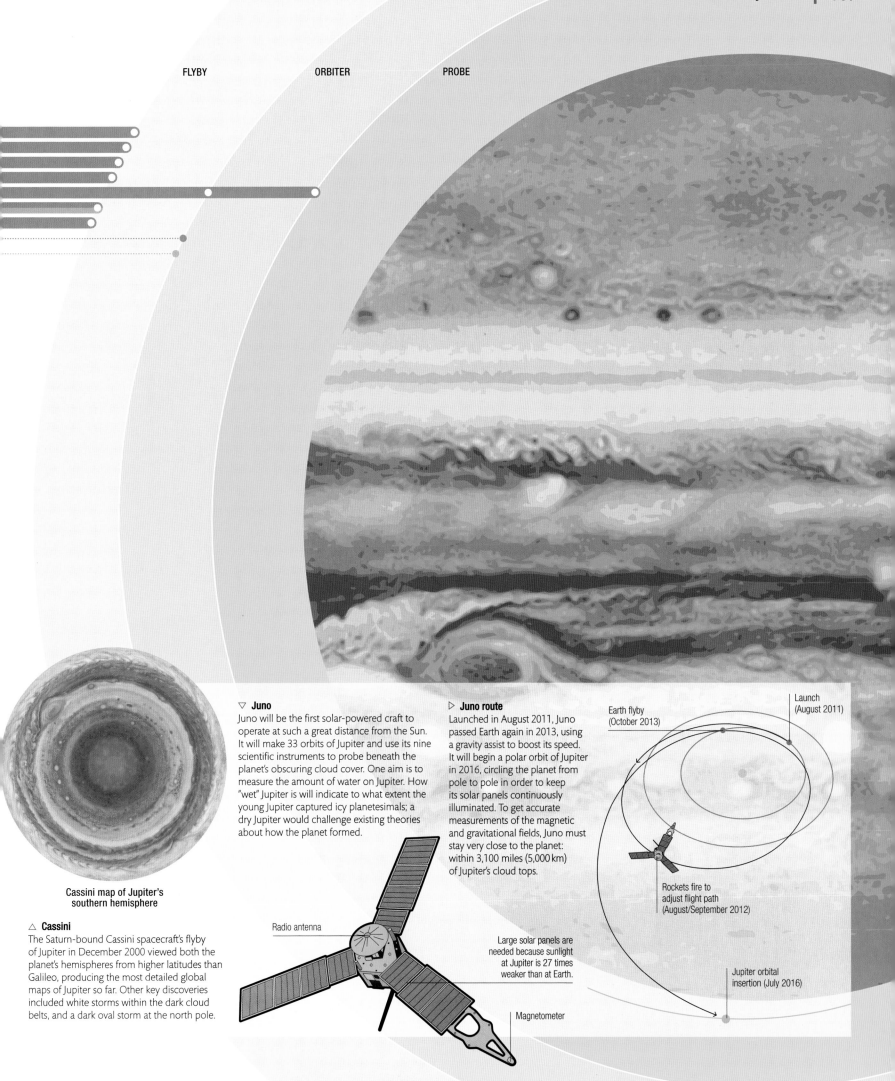

FLYBY   ORBITER   PROBE

▽ Juno

Juno will be the first solar-powered craft to operate at such a great distance from the Sun. It will make 33 orbits of Jupiter and use its nine scientific instruments to probe beneath the planet's obscuring cloud cover. One aim is to measure the amount of water on Jupiter. How "wet" Jupiter is will indicate to what extent the young Jupiter captured icy planetesimals; a dry Jupiter would challenge existing theories about how the planet formed.

▷ Juno route

Launched in August 2011, Juno passed Earth again in 2013, using a gravity assist to boost its speed. It will begin a polar orbit of Jupiter in 2016, circling the planet from pole to pole in order to keep its solar panels continuously illuminated. To get accurate measurements of the magnetic and gravitational fields, Juno must stay very close to the planet: within 3,100 miles (5,000 km) of Jupiter's cloud tops.

Launch (August 2011)

Earth flyby (October 2013)

Rockets fire to adjust flight path (August/September 2012)

Jupiter orbital insertion (July 2016)

Cassini map of Jupiter's southern hemisphere

△ Cassini

The Saturn-bound Cassini spacecraft's flyby of Jupiter in December 2000 viewed both the planet's hemispheres from higher latitudes than Galileo, producing the most detailed global maps of Jupiter so far. Other key discoveries included white storms within the dark cloud belts, and a dark oval storm at the north pole.

Radio antenna

Large solar panels are needed because sunlight at Jupiter is 27 times weaker than at Earth.

Magnetometer

SATURN

SATURN IS A LONG WAY FROM THE SUN. VIEWED FROM EARTH, IT IS EASILY OUTSHONE IN THE NIGHT SKY BY SUCH LUMINARIES AS JUPITER AND VENUS. BUT SEEN FROM SPACE, IT IS ARGUABLY THE MOST BEAUTIFUL OF ALL THE PLANETS.

Data

Equatorial diameter	74,898 miles (120,536 km)
Mass (Earth = 1)	95.2
Gravity at equator (Earth = 1)	1.02
Mean distance from Sun (Earth = 1)	9.58
Axial tilt	26.7°
Rotation period (day)	10.66 hours
Orbital period (year)	29.46 Earth years
Cloud-top temperature	−220°F (−140°C)
Moons	62+

Clothed in a creamy white blanket of high-altitude ammonia clouds, with softly muted color bands just visible through the hazy cover, Saturn looks deceptively placid. But beneath this disguise is a turbulent atmosphere. Saturn spins fast, generating high winds that race nonstop around the planet. Colossal electrical storms occur frequently and can last for months, hurling down bolts of lightning thousands of times more powerful than those on Earth.

All the giant planets have ring systems, but Saturn's is the glory of the solar system. These concentric disk-like platters are composed of countless ringlets, each of which consists of millions of orbiting ice fragments of varying size and composition.

> Saturn's **density** is less than water—placed in a large enough ocean, the planet **would float.**

Saturn's rapid rotation forces its gases outward, causing a distinct bulge at the equator.

▽ **Northern hemisphere**
Saturn's north polar region is remarkable for a long-lived hexagonal cloud structure, more than 17,000 miles (27,000 km) across, with a huge storm at its center. This weather system, which is different from any other so far seen in the solar system, is thought to be caused by a circumpolar jet stream.

▽ **Tilted axis**
Saturn's axis of rotation is tilted at an angle of 26.7°, so we view the planet and its rings at different angles throughout its 29.5-Earth-year orbit, as either the north or south pole tips toward the Sun. When the rings lie edge-on, they are invisible to observers on Earth.

▽ **Southern hemisphere**
The south polar region of Saturn is dominated by a hurricane-like storm almost the diameter of Earth and rotating about 340 mph (550 km/h) faster than the planet itself. The eye of the storm is ringed by clouds up to 45 miles (75 km) high.

The polar regions acquire a blue tinge in winter.

Saturn is encircled by broad cloud bands running parallel to the equator.

Though vast in terms of diameter, the main rings are just dozens of yards thick.

◁ **Rings and bands**
Like Jupiter, Saturn has a distinctive banded appearance, although with much paler colors. The enormous ring system extends far beyond the planet; its main elements have a total diameter of over 170,000 miles (270,000 km).

Upper atmosphere gas forms bands that encircle the planet. Clouds and storms are generated within them.

High winds reach speeds of up to 1,120 mph (1,800 km/h).

SATURN STRUCTURE

SATURN IS SIMILAR IN COMPOSITION AND STRUCTURE TO JUPITER, BUT IT IS CONSIDERABLY LESS MASSIVE THAN ITS NEIGHBOR. ITS WEAKER GRAVITY ALLOWS ITS LAYERS TO EXPAND OUTWARD, LOWERING ITS OVERALL DENSITY.

Saturn's low density and its greater distance from the Sun combine to make its outer layers significantly cooler than those of Jupiter—a feature that is most evident in the formation of ammonia-ice clouds across the entire upper atmosphere. These yellowish-white clouds give Saturn its color.

Beneath the visible cloud layers, Saturn is roughly 96 percent hydrogen, 3 percent helium, and 1 percent other, heavier elements that concentrate at the center. As with Jupiter, the gradual sorting of elements by density drives a "heat engine" that allows Saturn to pump out 2.5 times more energy than it receives from the Sun.

Descending into the planet, the interior can be broadly divided into layers, dominated by gaseous hydrogen, liquid molecular hydrogen, and liquid metallic hydrogen, around a solid core.

Saturn's lightning has **10,000** times the power of lightning on Earth.

Core
Saturn's core has a diameter of around 15,500 miles (25,000 km). Heated to more than 22,000°F (11,700°C), it may be a molten mix of rock and metal rather than a solid body, and may have 9–22 times the mass of Earth.

Liquid metallic hydrogen
At a depth of around 9,300 miles (15,000 km), hydrogen molecules begin to break down into individual atoms, creating a sea of electrically conducting liquid metal with currents that generate Saturn's powerful magnetic field.

Liquid hydrogen
Molecular hydrogen (H_2) condenses into liquid form gradually with increasing depth. Liquid hydrogen becomes dominant below about 600 miles (1,000 km).

Atmosphere
Saturn's outermost layer is about 600 miles (1,000 km) deep and is dominated by pure hydrogen gas. Clouds in this region form from the condensation of different chemical compounds, including ammonia and water.

Hydrogen molecules break down into metallic form under pressures equal to a million Earth atmospheres.

Temperatures at the base of the liquid hydrogen layer reach 10,800°F (6,000°C).

The ring system consists of many individual rings and a number of gaps between them.

◁ **Complex atmosphere**
Saturn's placid appearance belies its dynamic interior and stormy atmosphere. Enhanced-color images from spacecraft have revealed the presence of turbulent cloud layers beneath the outer ammonia haze. These cloud layers are dominated by ammonium hydrogen sulfide at high altitudes and by water ice at lower levels.

SATURN'S RINGS

SATURN IS ENCIRCLED BY THE MOST SPECTACULAR RING SYSTEM IN THE SOLAR SYSTEM. THE BRIGHT PLATTERS VISIBLE FROM EARTH CONSIST ALMOST ENTIRELY OF ICE FRAGMENTS THAT WHIRL AROUND THE PLANET IN CONCENTRIC RINGLETS.

Saturn's rings contain billions of pieces of ice, varying from house-sized boulders to minute crystals. Jostling together, these particles are constrained by the planet's gravity to orbit in a flat plane above Saturn's equator. The system is complex, with each large ring being made up of many narrow ringlets. Several distinct gaps between the rings are created by the gravitational pull of Saturn's more distant moons and the clumping together of material within the rings themselves. The particles consist predominantly of water ice, which makes them naturally reflective. Although their surfaces become dust-coated over time, constant collisions within the rings cause them to fracture, exposing bright new facets.

The origin of the rings is something of a mystery. They may be the remains of a small, icy moon that was either torn apart by Saturn's powerful gravity or destroyed in a collision with another body.

← Colombo gap

D ring → ← C ring
46,300 miles (74,700 km) from Saturn's center

In places, Saturn's **main rings** are a mere **10 yards (10 m)** thick.

▷ **Rings within rings**
Astronomers have identified at least nine major rings. The A and B rings are the brightest and contain the largest ice particles; white and purple denote particles larger than 2 in (5 cm) in this false-color image. A wide gap called the Cassini division separates the A and B rings. The paler C and D rings extend inward from the B ring and contain particles less than 2 in (5 cm) in size (here, colored green and blue).

▷ **Ringside view**
Spacecraft can see far more detail in the rings than could ever be observed from Earth, though even the best images cannot resolve individual ring particles. In this ultraviolet image of Saturn's outer C (left) and inner B (right) rings seen from the Cassini spacecraft, chemical and physical properties are highlighted in specific colors. Dust-covered ice particles appear red, while purer water ice is turquoise. The more densely packed B ring appears cleaner and purer, indicating that collisions between ice particles are more frequent here, repeatedly opening up fresh new surfaces where the ice has fractured.

▽ **Shepherd moons**
Small satellites orbiting within the rings, or very close to them, are known as shepherd moons. The gravitational influence of these bodies can create complex structures within the ring plane, including fine, braided ringlets, narrow gaps, and even vertical bumps. At Saturn's equinox, these inner satellites can cast long shadows across the rings, as shown below in an image of Daphnis, a small moon that maintains the Keeler gap within the A ring.

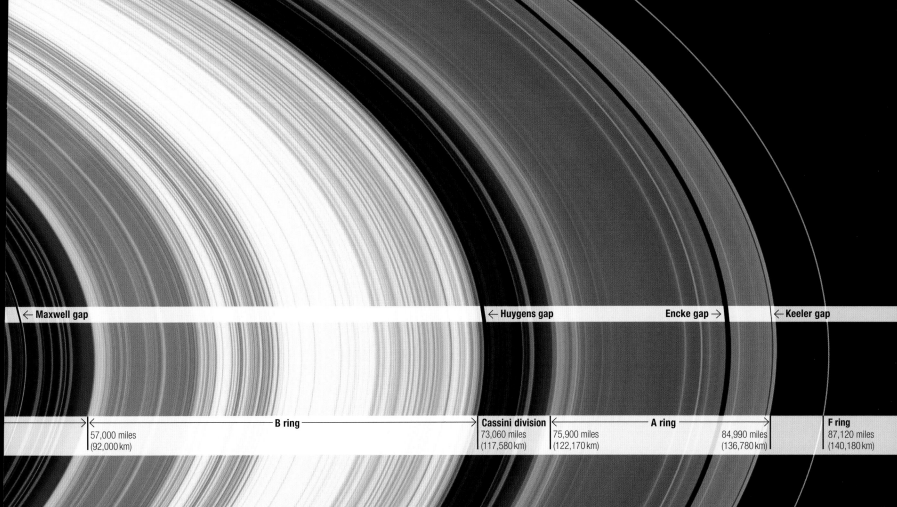

← Maxwell gap   ← Huygens gap   Encke gap →   ← Keeler gap

	B ring	Cassini division	A ring	F ring
57,000 miles (92,000 km)		73,060 miles (117,580 km)	75,900 miles (122,170 km)	84,990 miles (136,780 km)	87,120 miles (140,180 km)

▽ Outer rings

Beyond Saturn's familiar main rings are several hazier, darker, and much less sharply defined outer rings. These tenuous haloes of dust and ice become visible only with the use of special imaging techniques. Below, a backlit view of Saturn, with the Sun obscured by the planet's disk, reveals the faint E ring. This cloud of microscopic particles is fed by the plumes of ice that erupt from the surface of Enceladus, one of Saturn's most interesting moons. Unlike the slim main rings, the E ring is more than 1,250 miles (2,000 km) thick.

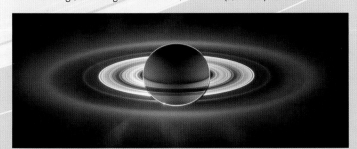

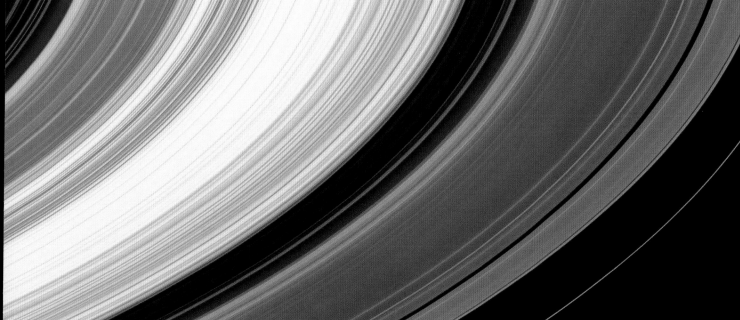

◁ The Phoebe ring

In 2009, astronomers using NASA's infrared Spitzer Space Telescope discovered a vast ring of dust thought to be produced by meteor impacts on one of Saturn's outer moons, Phoebe. Tilted at 27 degrees to the other rings, the Phoebe ring begins at around 2.5 million miles (4 million km) from Saturn and extends outward for more than three times that distance.

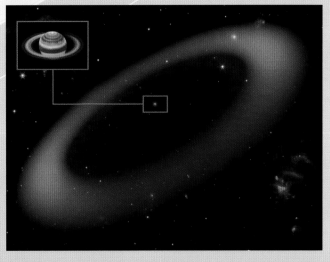

DESTINATION
SATURN'S RINGS

THE B RING IS THE LARGEST, BRIGHTEST, AND MOST DENSELY PACKED OF SATURN'S RINGS. HERE, GIANT BOULDERS OF SPARKLING ICE FLOAT ALONGSIDE ONE ANOTHER IN A SEEMINGLY IMPOSSIBLE ORBITAL BALLET. THE DENSE DISK OF DEBRIS SPELLS DOOM FOR ANYTHING THAT ATTEMPTS TO CROSS ITS PATH.

While the plane of particles orbiting Saturn extends to many times the planet's own diameter, and contains trillions of objects, the particles' individual paths are remarkably uniform—each follows a near-perfect circular orbit in a plane directly above Saturn's equator. Objects straying into more elliptical orbits or attempting to cross the plane soon collide with their neighbors and are nudged back into more orderly paths. Fragments produced by recent collisions are everywhere, gleaming brightly in the sunlight as they slowly attempt to reassemble under their own gravitational attraction.

Artist's impression of B ring

The main rings lie within Saturn's **Roche lobe**—a region where the planet's gravity prevents them from coalescing into a **single moon.**

LOCATION

B ring, 31,000 miles (50,000 km) above Saturn's cloud tops

SPIRAL WAVES

Material in the B ring is not uniform—it is spread out unevenly due to density waves. These are caused by changes in Saturn's gravity, when the planet is shaken by internal tremors.

30 **MILLION BILLION TONS—** **THE TOTAL MASS OF SATURN'S RINGS**

CLUMPING

This computer simulation, based on observations by Cassini, shows how ring particles gradually coalesce. They slowly clump together to form more substantial moonlets that are eventually shattered in collisions, thus causing the cycle to repeat.

SATURN UP CLOSE

BENEATH AN OUTER HAZE OF BRIGHT AMMONIA CLOUDS THAT GIVE THE ENTIRE PLANET A SEPIA TINT, SATURN'S DEEP, GASEOUS ATMOSPHERE IS JUST AS ACTIVE AND TURBULENT AS THAT OF ITS INNER NEIGHBOR, JUPITER.

Orbiting almost twice as far from the Sun as Jupiter, Saturn receives only a quarter as much solar heat, and so the upper layers of its atmosphere are considerably colder, averaging about −220°F (−140°C). At such low temperatures, atmospheric ammonia freezes into ice crystals, cloaking the planet in a layer of thin, hazy clouds. Beneath this outer cloud layer, however, Saturn is wracked by storms, high winds, and lightning, driven not only by heat from the Sun but also by Saturn's own internal energy.

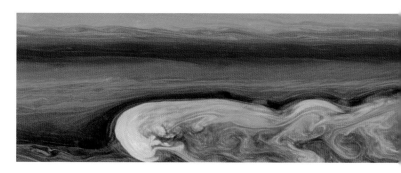

△ **Colorful stream**
This infrared image from the Cassini spacecraft unwraps the full extent of a great white spot that appeared in 2010 and grew rapidly through 2011. High clouds at the head of the storm system (left) suggest that the original spot formed from an upwelling of warm material from inside the planet, perhaps linked to seasonal changes.

▽ **White storms**
Saturn's most prominent weather features are large white spots that periodically erupt in its northern hemisphere. The spots recur roughly every 29 years and usually coincide with the onset of the northern summer, suggesting they are triggered by an increase in heat from the Sun. As they develop, the spots can wrap themselves around the planet to form pale, turbulent bands.

Stormy skies

Saturn's atmosphere has a banded appearance with some resemblance to that of Jupiter, albeit with broader bands and less contrast between light and dark regions. Hidden within these bands, long-lived storms crackle with powerful lightning. They can be detected from the radio signals they emit, but occasionally the storms also erupt into visibility on Saturn's surface as seasonal "great white spots."

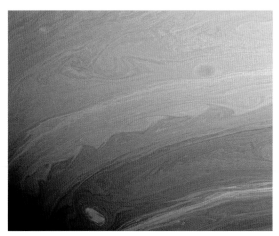

△ **Cloud bands**
Saturn's bluish-colored clouds tend to be made up of water vapor, while the higher red-orange ones are largely formed from ammonium hydrosulfide. Color and temperature variations are exaggerated in this Cassini infrared image. Dark and light bands seem to move in opposite directions, but this is an illusion caused by their rotation at different rates.

Saturn's winds are the second-fastest in the solar system.

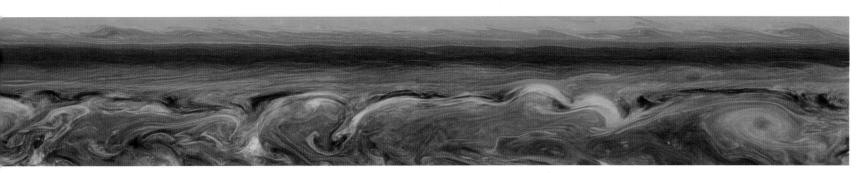

Polar regions

Saturn's axis of rotation is tilted at an angle similar to that of Earth, giving it a cycle of seasons like our own planet's, with each pole spending roughly half of Saturn's long year in permanent darkness. This gives polar regions very different weather from the rest of the planet. Each pole is dominated by a swirling, hurricane-like vortex with a cloudless "eye" at its center.

▷ **Southern lights**

Saturn's powerful magnetic field draws in charged particles from the solar wind and channels them into the upper atmosphere around the poles. There they collide with gas molecules, causing the molecules to emit light and produce beautiful aurorae, as seen in these images from the Hubble Space Telescope.

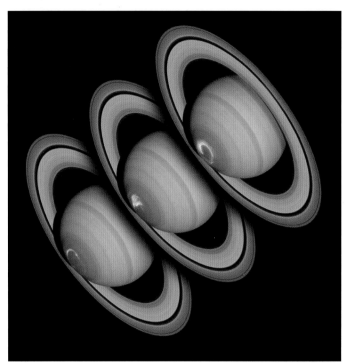

▷ **Southern hot spot**

This infrared image from Cassini reveals heat emanating from deep inside Saturn, with colder, overlying cloud bands revealed in silhouette. Powered by internal contraction, Saturn radiates 2.5 times more energy than it receives from the Sun, but astronomers are not sure why so much of it escapes around the south pole.

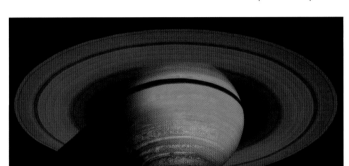

▽ **Hexagonal hurricane**

Saturn's north polar vortex is surrounded by a remarkable hexagonal cloud structure that has persisted at least since the Voyager flybys of the early 1980s. The geometric pattern is thought to arise at the boundary between different atmospheric zones that are moving at contrasting speeds. Each side of the hexagon is longer than the diameter of Earth.

△ **Northern rose**

This Cassini close-up focuses on the eye of Saturn's north polar vortex, revealing the sharp division between high surrounding clouds (colored green in the image) and the much deeper clouds within the eye (colored red). The eye is an impressive 1,250 miles (2,000 km) across, and wind speeds around it reach 330 mph (530 km/h).

SATURN IN THE SPOTLIGHT

This remarkable, natural-color view of Saturn and its rings is a mosaic of more than 120 photographs. The images were captured by the NASA spacecraft Cassini in 2004, a few months after it arrived at Saturn to begin an initial four-year study of the gas giant and its system of rings and moons. Still in orbit today, Cassini is the first and only spacecraft to orbit Saturn and has returned images of unprecedented detail and clarity. At the top of this image, shadows cast by the rings can be seen sharply etched across the north polar region. When these photographs were taken, Saturn had just passed its northern winter solstice, and the pole had assumed the azure-blue tint characteristic of Saturnian winters. The pale blue oval spots just discernible in a band around the southern hemisphere are storms in Saturn's atmosphere.

Titan

Moons to scale
Saturn's satellites are dominated by the huge bulk of Titan. All the other moons are far smaller, and scientists speculate that Titan's formation stunted the growth of its neighbors.

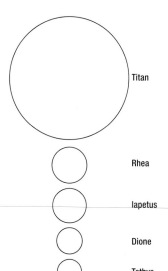

Titan

Rhea

Iapetus

Dione

Tethys

Enceladus

Mimas
Hyperion
Phoebe
Janus
Epimetheus
Prometheus
Pandora
Siarnaq
Helene
Albiorix
Atlas
Pan
Telesto
Paaliaq
Calypso
Ymir
Kiviuq
Tarvos
Ijiraq
Erriapus
Skathi
Hyrrokkin
Daphnis
Tarqeq
Mundilfari
Narvi
Suttungr
Thrymr
Bestla
Kari
S/2007 S 2
Bebhionn
Skoll
S/2004 S 13
Greip
Jarnsaxa
S/2006 S 1
Bergelmir
Hati
Aegir
S/2004 S 7
S/2006 S 3
Surtur
Loge
Fornjot
Pallene
S/2004 S 12
Farbauti
S/2007 S 3
S/2004 S 17
Fenrir
Methone
Polydeuces
Anthe
Aegaeon
S/2009 S 1

THE **SATURN** SYSTEM

A HUGE FAMILY OF MOONS SURROUND SATURN. THEY RANGE FROM PLANET-SIZED WORLDS WITH COMPLEX ATMOSPHERES AND ACTIVE SURFACES TO SMALL LUMPS OF ROCK AND ICE TRAPPED IN ORBIT BY THE PLANET'S GRAVITY.

Saturn has 62 officially recognized satellites, 53 of which have been named. The dividing line between moons, "moonlets," and large particles of ring material is not clear-cut, so a precise count of Saturn's moons may never be agreed upon. The innermost moons orbit within the planet's ring system, sitting in small gaps in the rings that are kept clear by the moons' gravity; these are known as shepherd moons. The outermost moons follow wildly eccentric orbits that can take them tens of millions of miles away from Saturn. In contrast, Saturn's largest moons mostly follow circular orbits relatively close to the planet, but outside the main rings.

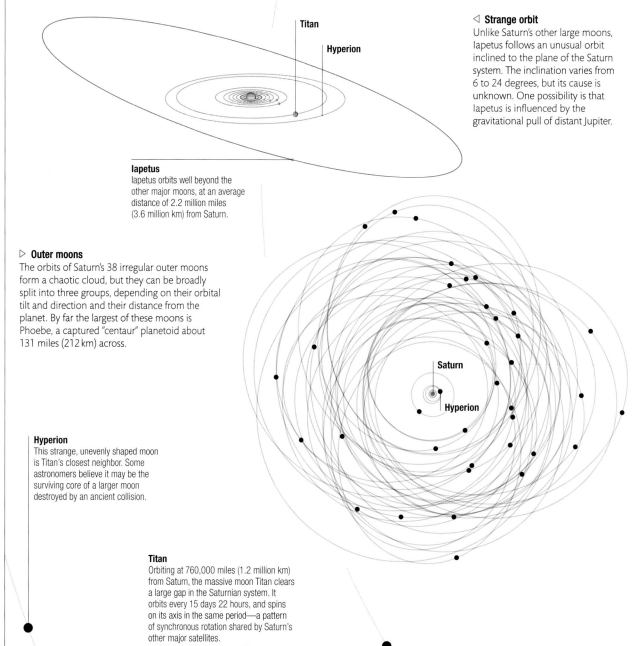

Titan

Hyperion

Iapetus
Iapetus orbits well beyond the other major moons, at an average distance of 2.2 million miles (3.6 million km) from Saturn.

◁ **Strange orbit**
Unlike Saturn's other large moons, Iapetus follows an unusual orbit inclined to the plane of the Saturn system. The inclination varies from 6 to 24 degrees, but its cause is unknown. One possibility is that Iapetus is influenced by the gravitational pull of distant Jupiter.

▷ **Outer moons**
The orbits of Saturn's 38 irregular outer moons form a chaotic cloud, but they can be broadly split into three groups, depending on their orbital tilt and direction and their distance from the planet. By far the largest of these moons is Phoebe, a captured "centaur" planetoid about 131 miles (212 km) across.

Saturn

Hyperion

Hyperion
This strange, unevenly shaped moon is Titan's closest neighbor. Some astronomers believe it may be the surviving core of a larger moon destroyed by an ancient collision.

Titan
Orbiting at 760,000 miles (1.2 million km) from Saturn, the massive moon Titan clears a large gap in the Saturnian system. It orbits every 15 days 22 hours, and spins on its axis in the same period—a pattern of synchronous rotation shared by Saturn's other major satellites.

More than **150 moonlets** have been detected within Saturn's rings.

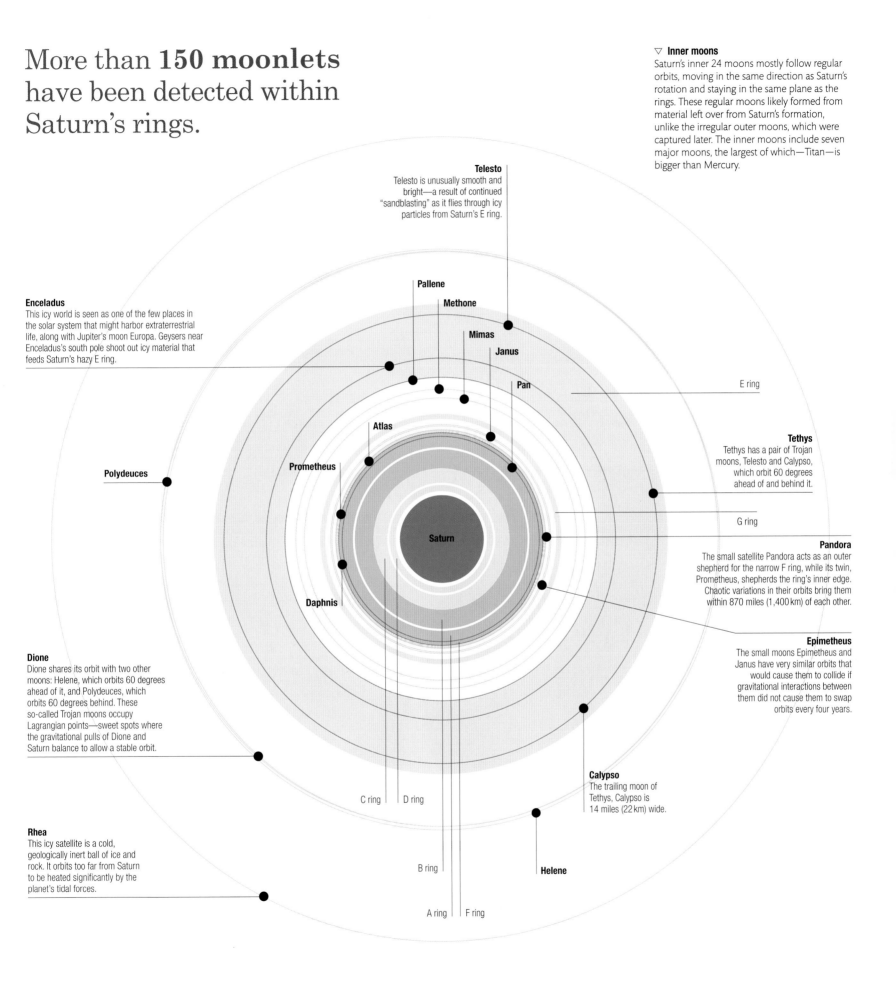

▽ **Inner moons**
Saturn's inner 24 moons mostly follow regular orbits, moving in the same direction as Saturn's rotation and staying in the same plane as the rings. These regular moons likely formed from material left over from Saturn's formation, unlike the irregular outer moons, which were captured later. The inner moons include seven major moons, the largest of which—Titan—is bigger than Mercury.

Telesto
Telesto is unusually smooth and bright—a result of continued "sandblasting" as it flies through icy particles from Saturn's E ring.

Pallene

Methone

Mimas

Janus

Pan

Enceladus
This icy world is seen as one of the few places in the solar system that might harbor extraterrestrial life, along with Jupiter's moon Europa. Geysers near Enceladus's south pole shoot out icy material that feeds Saturn's hazy E ring.

E ring

Atlas

Tethys
Tethys has a pair of Trojan moons, Telesto and Calypso, which orbit 60 degrees ahead of and behind it.

Prometheus

Polydeuces

Saturn

G ring

Pandora
The small satellite Pandora acts as an outer shepherd for the narrow F ring, while its twin, Prometheus, shepherds the ring's inner edge. Chaotic variations in their orbits bring them within 870 miles (1,400 km) of each other.

Daphnis

Epimetheus
The small moons Epimetheus and Janus have very similar orbits that would cause them to collide if gravitational interactions between them did not cause them to swap orbits every four years.

Dione
Dione shares its orbit with two other moons: Helene, which orbits 60 degrees ahead of it, and Polydeuces, which orbits 60 degrees behind. These so-called Trojan moons occupy Lagrangian points—sweet spots where the gravitational pulls of Dione and Saturn balance to allow a stable orbit.

C ring   D ring

Calypso
The trailing moon of Tethys, Calypso is 14 miles (22 km) wide.

Rhea
This icy satellite is a cold, geologically inert ball of ice and rock. It orbits too far from Saturn to be heated significantly by the planet's tidal forces.

B ring

Helene

A ring   F ring

SATURN'S
MAJOR MOONS

SEVEN OF SATURN'S MOONS ARE LARGE ENOUGH FOR GRAVITY TO HAVE PULLED THEM INTO ROUGHLY SPHERICAL SHAPES. SOME OF THESE HAVE BEEN DEAD WORLDS FOR BILLIONS OF YEARS, BUT OTHERS ARE GEOLOGICALLY ACTIVE.

Saturn's major moons are named after giants in Greek mythology. In order of distance from the planet, they are Mimas, Enceladus, Tethys, Dione, Rhea, Titan, and Iapetus. They range in size from Mimas, which is a mere 246 miles (396 km) wide, to mighty Titan, which at 3,200 miles (5,150 km) in diameter is 50 percent wider than Earth's moon. Titan was the first Saturnian moon to be discovered, in 1655. By 1789, all seven major moons had been located and named.

Dense atmosphere

The dark regions on Titan's surface may be dry seabeds.

▷ **Titan**
Larger than Mercury and Pluto, Titan is the only moon in the solar system with a significant atmosphere, and the only body besides Earth with nitrogen-rich "air." Composed of rock and ice, it has an average surface temperature of around −292°F (−180°C). Despite the cold, its dense atmosphere traps enough heat energy to drive a complex weather cycle, with liquid methane on the surface evaporating and raining back down like water on Earth. Titan also shows evidence of "cryovolcanic" eruptions of slushy ice onto the surface.

Light areas are regions of higher ground.

Impact basin

Engelier Crater

△ **Rhea**
Rhea is the second-largest Saturnian moon, but it is a great deal smaller than our Moon. It is a ball of ice and rock that has compressed under its own gravity to create an unusually dense form of ice. Rhea's heavily cratered surface, which includes two large impact basins, suggests it has been geologically inactive for billions of years.

△ **Iapetus**
The outermost of Saturn's major moons has a curious appearance, with a dark leading hemisphere and a much brighter trailing one. The dark pattern may be caused by deposits of carbon dust from Saturn's Phoebe ring. The sooty dust absorbs extra heat from the Sun, causing surface ice to evaporate and making the affected area even darker.

Cliffs, formed by the fracturing of crust.

Ithaca Chasma is a 1,250-mile- (2,000-km-) long canyon.

Long fractures known as sulci run across the surface of Enceladus.

Herschel Crater

△ **Dione**
This midsize icy moon has a heavily cratered surface and contains substantial amounts of denser rock within. Large-scale variations in the frequency of craters on Dione's surface suggest that some areas were smoothed out in the past by the eruption of cryovolcanoes. A network of faults across Dione's trailing hemisphere appears from a distance as bright streaks.

△ **Tethys**
Superficially similar to Dione, Tethys is lower in density, suggesting it consists of almost pure water ice. Although heavily cratered, it seems to have been active more recently than its neighbors and has large, smooth plains formed by cryovolcanic activity. Ithaca Chasma, a canyon-like surface crack, probably formed as Tethys's interior froze and expanded.

Enceladus △
This small moon is pulled in a gravitational tug of war between Saturn and Dione, which causes internal friction and heating. Melted ice erupts through the surface as vapor and water, forming spectacular geysers around the south pole. The fountains of ice are the source of the material in Saturn's faint E ring.

△ **Mimas**
Mimas is one of the smallest bodies in the solar system to have become spherical through its own gravity. Its pitted surface is dominated by the massive Herschel Crater, which measures 86 miles (130 km) wide. The impact that created this crater almost destroyed Mimas.

DESTINATION **LIGEIA MARE**

IN THE FAR NORTH OF SATURN'S LARGEST MOON, TITAN, IS A GLASSY LAKE SO HUGE THAT A PERSON STANDING ON THE SHORE WOULD SEE NO END TO IT. THIS IS LIGEIA MARE, WHICH IS NOT WATER BUT METHANE—A GAS THAT LIQUEFIES IN COLD AS INTENSE AS TITAN'S.

Several seas or large lakes of liquid hydrocarbon chemicals such as ethane and methane have been discovered near Titan's poles. Ligeia Mare is one of the biggest, larger than any of Earth's great freshwater lakes. Radar signals from NASA's Cassini orbiter have penetrated the lake and bounced back from its floor, revealing its depth and suggesting that it is composed of more or less pure methane. The surface is smooth and flat, though seasonal weather changes might stir up disturbances. Ligeia Mare's ragged shoreline is broken by bays and coves. Some areas of the shore flatten out into smooth beaches, or possibly methane mudflats; elsewhere the terrain is rougher and rises into hummocks.

The **amount of methane** in Ligeia Mare is estimated to be **40 times greater** than Earth's global reserves of liquid fuels.

Artist's impression based on Cassini radar and altimeter data

LOCATION

Latitude 80°N; longitude 248°W

560 FT (170 M)
**DEPTH OF LIGEIA MARE RECORDED BY
RADAR SIGNALS FROM CASSINI ORBITER**

LAKE PROFILE

Ligeia Mare is located close to Titan's north pole, along with the majority of the moon's lakes, and covers an area of approximately 48,650 sq miles (126,000 km²). The lake has a shoreline of some 1,240 miles (2,000 km) and, unlike the few lakes in the south, shows no signs of shrinkage due to evaporation of chemicals. Such differences may be linked to seasonal cycles in the opposing hemispheres.

Cassini radar image of Ligeia Mare. Smooth liquid areas are shown in blue.

Satellite view of Lake Superior (top) in North America to scale with Ligeia Mare.

NASA profile of the lake floor showing depth to an estimated maximum of 690 ft (210 m) in the center.

TITANIC RIVER

Dramatic evidence of methane rain and liquid runoff on Titan is provided by Cassini images of a 250-mile (400-km) river that drains into Ligeia Mare. It is named Vid Flumina after a poisonous river in Norse mythology.

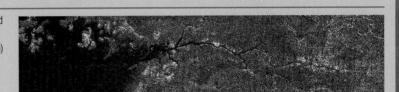

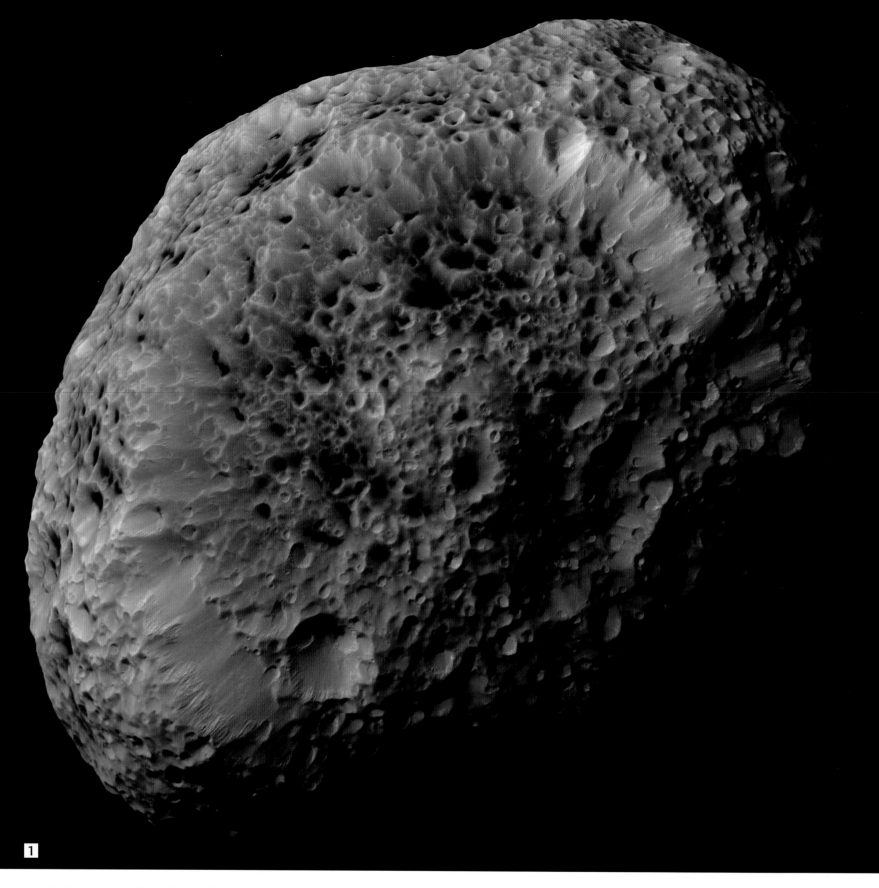

1

CASSINI'S **VIEW**

1 **Hyperion**
NASA's Cassini spacecraft has captured many spectacular images during its tour of the Saturn system, including this close-up of one of Saturn's oddest moons. Hyperion is not quite large enough for gravity to pull it into a sphere. Its odd shape, chaotic rotation, and spongelike surface suggest it is a fragment of a larger moon destroyed in a collision.

2 **Dione**
During a close encounter with Saturn's small, icy moon Dione, Cassini captured this breathtaking view across a sunlit crescent. Deep shadows starkly delineate the rims of impact craters on Dione's meteorite-scarred face. Much of the moon's surface is heavily pitted with such craters, some of the largest exceeding 60 miles (100 km) in diameter.

3 **Mimas**
Dwarfed by its parent, Saturn's innermost major moon Mimas hangs against the backdrop of the planet's northern hemisphere. The dark bands across Saturn's cloudscape are shadows cast by the rings onto the winter hemisphere. The scattering of sunlight through the relatively cloud-free northern sky tints the atmosphere blue.

4 **Titan**
Magnification diminishes the huge size difference between Saturn and its haze-covered moon Titan in this view. Titan's orbit is in the same plane as the rings of Saturn. The dense A and B rings cast a broad, dark band of shadow onto Saturn's southern hemisphere, with the Cassini division creating a bright split within it.

3

4

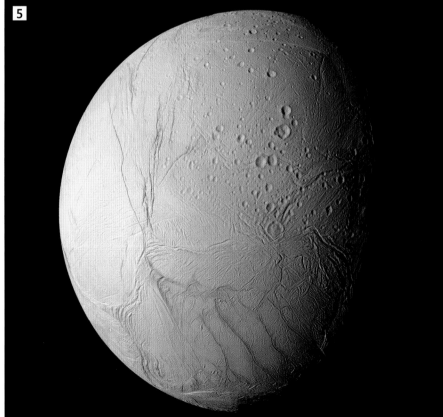

5 Enceladus

An enhanced-color image of Saturn's brightest satellite reveals striking differences in its landscape. The oldest terrain lies in the densely cratered north, while crater-free areas to the south indicate later resurfacing. Bluish ice marks the most recent features, including the distinctive "tiger stripes," where eruptions of subsurface water are still continuing.

Ice particles lofted into space by Enceladus's geysers have been detected moving at speeds of **39,000 mph (63,000 km/h).**

DESTINATION
ENCELADUS

A BRIGHT MOON WITHIN SATURN'S E RING, ENCELADUS IS SO TINY THAT A VISITOR COULD HIKE AROUND ITS EQUATOR IN TWO WEEKS. LOOKING SOUTH, THE TRAVELER WOULD SEE THE MOON'S VAST ICE GEYSERS SHOOTING ABOVE THE HORIZON.

The thick layer of bright snow that blankets Enceladus ensures that the surface of this tiny satellite remains cold as it reflects light and heat from the distant Sun. However, tidal forces, generated as Enceladus is pulled in different directions by Saturn and outer moons such as Dione, warm the interior enough to generate pockets of subterranean meltwater. Near the south pole, where tidal flexing creates deep fissures known as "tiger stripes," the water erupts to the surface, where it violently boils into the vacuum of space as a mix of vapor and ice crystals.

LOCATION

Latitude 4°N; **longitude** 209°W

COLD GEYSERS

Heat generated by tidal forces creates a reservoir of meltwater in Enceladus's interior. Under pressure, this erupts from the surface in jets of water vapor and ice particles.

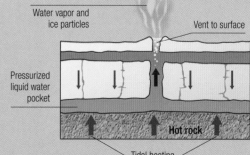

Water vapor and ice particles

Vent to surface

Pressurized liquid water pocket

Hot rock

Tidal heating

SPECTACULAR PLUMES

This image from NASA's Cassini spacecraft shows icy plumes from Enceladus's geysers streaming high into space. The view is color-enhanced to reveal the density of the plumes more clearly.

Artist's impression based on Cassini images

LORD OF
THE RINGS

BEAUTIFUL SATURN HAS ENCHANTED ASTRONOMERS EVER SINCE THE PLANET'S RINGS WERE FIRST SEEN THROUGH A TELESCOPE. MORE RECENTLY, SPACECRAFT HAVE REVEALED THE PLANET'S MOONS TO BE JUST AS FASCINATING.

Before spacecraft discovered rings around other planets, those around Saturn were thought to be unique. Although first seen in the 17th century, the rings remained a mystery for nearly 250 years before physicist James Clerk Maxwell explained their true nature. While few features on Saturn itself were seen until the mid-19th century, improving telescopes revealed structures and divisions within the rings, as well as a host of orbiting moons. However, it was not until the first interplanetary missions that astronomers began to grasp the complexity of the Saturnian system.

Ptolemy observing Saturn

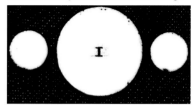

Galileo's interpretation of Saturn's rings

127
The outermost world
For early astronomers, Saturn has special significance as the slowest of the five known planets. Greek scholar Ptolemy places Saturn on the outermost of the crystal spheres he believes surround Earth, with only a shell of the fixed stars beyond it.

1610
Saturn's strange shape
The crude telescope of Galileo Galilei reveals that Saturn has a strange shape, leading the Italian scientist to suspect it has juglike "handles" or is orbited by two big moons. Unbeknownst to Galileo, he has seen a distorted view of the planet's rings.

Saturn's large outer moon, Phoebe

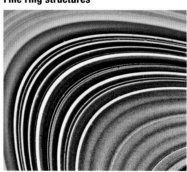

Fine ring structures

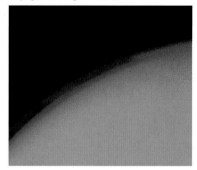

Voyager 1 image of Titan

2004
Phoebe flyby
After a long journey, NASA's Cassini spacecraft arrives and orbits Saturn. During its final approach, Cassini passes close to Phoebe, the mysterious outer moon. It sends back images of a cratered surface that suggests the moon is a captured comet or minor planet.

1981
Structure in the rings
Voyager 2 arrives at Saturn eight months after its sibling. Together, the Voyagers image all the major moons and reveal fine details within the rings, including individual ringlets and radial "spokes" of darker material rippling out across the ring system.

1980
First look at Titan
NASA's Voyager 1 spacecraft reaches Saturn. Its trajectory is revised to take it close to the giant moon Titan, allowing it to send back the first close-up images. Titan's thick atmosphere makes the surface beneath impossible to see.

Cassini image of Titan

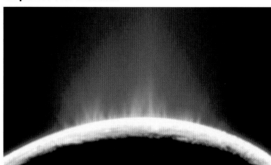

Ice plumes over Enceladus

2005
Piercing the veil
Cassini's infrared instruments peer through Titan's hazy atmosphere and photograph the surface. The images reveal a world whose features have apparently been smoothed by erosion processes, such as the flow of liquid methane across the landscape.

2005
Active Enceladus
During a close encounter with the small, bright moon Enceladus, Cassini sees plumes of icy material erupting hundreds of miles into space. Further studies reveal that the active geysers, powered by tidal heating of the moon's interior, emerge from surface faults near the south pole.

A sketch by Huygens showing Saturn's changing appearance

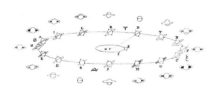

The Paris Observatory

The contrasting hemispheres of Iapetus

1655

Discovery of the rings
Dutch astronomer and instrument-maker Christiaan Huygens studies Saturn using a powerful telescope of his own design. He concludes that the planet is surrounded by a thin, flat ring. In the same year, Huygens discovers Saturn's largest moon, Titan.

1675

Splitting the rings
At the Paris Observatory, Italian-French astronomer Giovanni Cassini sees a dark circle within the rings—the boundary between the A and B rings, today called the Cassini division. This is the first hint that the rings have a complex internal structure.

1705

Two-tone moon
Having observed Iapetus on one side of Saturn since 1671, Cassini now detects the moon on the opposite side of the planet, finding that it is much fainter. He correctly concludes that Iapetus has a dark leading hemisphere and a brighter trailing one.

Pioneer 11 view of Saturn

Will Hay

1979

Pioneer 11
The first spacecraft to visit Saturn passes the planet at a distance of 13,000 miles (21,000 km) and beams back the most detailed images yet of Saturn's rings and atmospheric weather systems. Pioneer 11 also investigates flight paths for the later Voyager missions.

1933

The Great White Spot
British comic actor and amateur astronomer Will Hay discovers a huge white outburst on Saturn, later confirmed to be a storm similar to spots seen in 1876 and 1903. The Great White Spot is now recognized as Saturn's most prominent recurring weather feature.

1859

True nature of the rings
James Clerk Maxwell explains the true nature of Saturn's rings for the first time, showing through mathematics that they cannot be solid planes or ringlets, but must instead be made of countless tiny particles following independent, circular orbits.

Ontarius Lacus on Titan

The 2011 storm eruption

2005–2007

Lakes of Titan
Although Cassini's daughter probe, Huygens, lands in a dry equatorial region of Titan in 2005, Cassini's radar later finds lakes around Titan's poles. In 2007, Cassini uses infrared cameras to detect sunlight reflecting from a south polar lake called Ontarius Lacus.

2010

Fine ring structures
Images from Cassini reveal ripples and peaks at the outer edge of the B ring that cast their shadows across the mostly flat plane. These short-lived vertical structures are thought to be caused by the gravitational influence of small moonlets within Saturn's rings.

2011

A storm up close
Cassini charts the development of a huge white-spot storm that grows to cover an area more than eight times the size of Earth in Saturn's northern hemisphere. The storm seems to have been driven by warming from the onset of the northern spring.

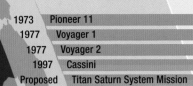

1973	Pioneer 11
1977	Voyager 1
1977	Voyager 2
1997	Cassini
Proposed	Titan Saturn System Mission

MISSIONS TO **SATURN**

SATURN AND ITS MOONS HAVE BEEN VISITED BY SEVERAL SPACECRAFT SINCE THE 1970S. THE FIRST MISSIONS WERE FLYBYS, BUT MORE RECENTLY A DECADE-LONG INVESTIGATION WAS UNDERTAKEN BY NASA'S CASSINI ORBITER.

Saturn was a key destination for the Pioneer missions that paved the way for the exploration of the outer solar system. While Pioneer 10 merely flew past Jupiter, Pioneer 11 used a "gravitational slingshot" from the giant planet to propel itself to Saturn in September 1979. The twin Voyager probes arrived in November 1980 and August 1981, and gave the first detailed views of Saturn's intriguing family of moons. Saturn was not revisited until 2004, when Cassini (and its companion, the Huygens Titan probe) became the first craft to orbit the ringed planet.

KEY

⬤ 🇺🇸	NASA (USA)
⬤ (🌐esa)	ESA (Europe)
	Joint NASA/ESA mission
••••••⬤	Destination
◻◯	Success

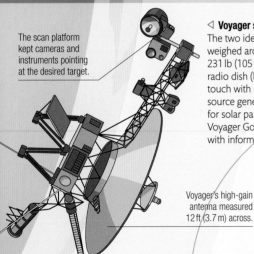

The scan platform kept cameras and instruments pointing at the desired target.

Voyager

Voyager's high-gain antenna measured 12 ft (3.7 m) across.

◁ **Voyager spacecraft**
The two identical Voyager spacecraft each weighed around 1,700 lb (773 kg) and carried 231 lb (105 kg) of scientific instruments. A large radio dish (high-gain antenna) kept the craft in touch with Earth, while a radioactive power source generated electricity without the need for solar panels. Each mission carried with it a Voyager Golden Record—a gold disc inscribed with information about Earth.

▷ **Discoveries**
The Voyager flybys confirmed the existence of countless individual ringlets within Saturn's main rings, as well as short-lived structures such as radial spokes. Although Titan's thick atmosphere proved impenetrable, the Voyagers discovered surface features on several of the other moons for the first time, as well as details of Saturn's own weather systems.

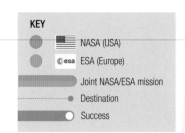

Color-enhanced Voyager 2 image of the rings

▷ **Mission profile**
Voyager 1 traveled considerably faster than Voyager 2, and overtook its sibling on the way to an encounter with Jupiter in 1979. Upon arrival at Saturn, Voyager 1 flew close to Titan before leaving the plane of the solar system. Voyager 2 moved on to visit Uranus and Neptune.

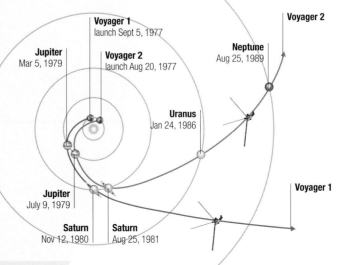

Voyager 1
launch Sept 5, 1977

Voyager 2
launch Aug 20, 1977

Jupiter
Mar 5, 1979

Jupiter
July 9, 1979

Saturn
Nov 12, 1980

Saturn
Aug 25, 1981

Uranus
Jan 24, 1986

Neptune
Aug 25, 1989

Voyager 2

Voyager 1

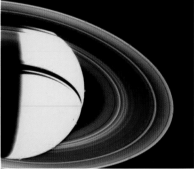

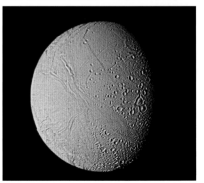

False-color Voyager 2 view of Enceladus

Saturn's moons **Titan** and **Enceladus** are key targets for **future missions**.

◁ **Cassini**
The enormous Cassini spacecraft is the size of a bus and has a mass of 4,740 lb (2,150 kg), making it the largest and most complex interplanetary craft sent into space so far. Onboard instruments include advanced radar, visible and infrared mapping cameras, magnetometers, and particle analysis tools. Cassini also transported the Huygens Titan probe, adding a further 770 lb (350 kg) to the overall payload.

Cassini is 22 ft (6.8 m) tall and contains over 8.7 miles (14 km) of cabling.

The Huygens probe was released over Titan in December 2004.

▷ **Launch**
In October 1997, Cassini blasted off from Cape Canaveral aboard a Titan-IVB/Centaur rocket. Its complex trajectory— including two flybys of Venus, one of Earth, and one of Jupiter, gaining speed with each encounter— meant that Cassini took nearly seven years to reach Saturn.

▽ **Discoveries**
Still in orbit after more than a decade at Saturn, Cassini has revolutionized our ideas about the planet and its satellites. Key breakthroughs include the confirmation of lakes on Titan and the discovery of ice plumes on Enceladus. Cassini has also revealed fine structure within Saturn's rings, and improved our understanding of the planet's complex weather systems.

Dark patches left by disappearing ice on the moon Iapetus

Huygens' view of its landing site on the surface of Titan

URANUS

ENIGMATIC URANUS KEEPS ITS SECRETS HIDDEN UNDER AN ALMOST CLOUDLESS FACE. UNIQUELY, URANUS SPINS ON ITS SIDE AS IT ORBITS THE SUN. ALTHOUGH NOT THE FARTHEST PLANET FROM THE SUN, IT IS THE COLDEST OF ALL.

"A curious nebulous star or perhaps a comet," noted William Herschel, a German-born British musician and amateur astronomer, on March 13, 1781. In fact, Herschel had discovered a new planet far beyond Saturn and, at a stroke, doubled the size of the known solar system.

Uranus is a giant world, but one so distant that it is barely visible to the naked eye. Even telescopes reveal little more: a handful of moons, whose orbits indicate that the planet is tipped sideways, and faint evidence of some dark rings. The Voyager 2 spacecraft flew past Uranus in 1986, but the images it returned proved disappointingly featureless, even under close scrutiny.

Over the following decades, as Uranus's orbit brought different parts of its face into the Sun, the planet has emerged from hibernation. Powerful telescopes are now revealing clouds swirling around this aquamarine world.

Data

Equatorial diameter	31,763 miles (51,118 km)
Mass (Earth = 1)	14.5
Gravity at equator (Earth = 1)	0.89
Mean distance from Sun (Earth = 1)	19.2
Axial tilt	82.2°
Rotation period (day)	17.2 hours (east to west)
Orbital period (year)	84.3 Earth days
Cloud-top temperature	−323°F (−197°C)
Moons	27

The amount of sunlight **Uranus** receives is only **0.25 percent** of that reaching **Earth.**

Unlike Saturn's rings of water ice, the rings around Uranus are made of dust and dark, rocky material.

▽ **Northern hemisphere**
Night and day at the poles each last for 42 years. The northern polar region is now coming into sight, brightening as Uranus moves around the Sun; the planet's changing seasons expose the region to more intense sunlight.

▽ **Tilt**
Uranus's axis is tilted at almost a right angle to its orbit, and the planet rotates the opposite way to all the other planets except Venus. This is probably because Uranus was "knocked over" by a giant impact soon after the planet's formation.

▽ **Southern hemisphere**
Voyager 2 sped directly toward the south pole of the tipped-up planet, which at the time was midway through its 42-year day. When the images of Uranus were enhanced, this was the brightest region; it is now fading.

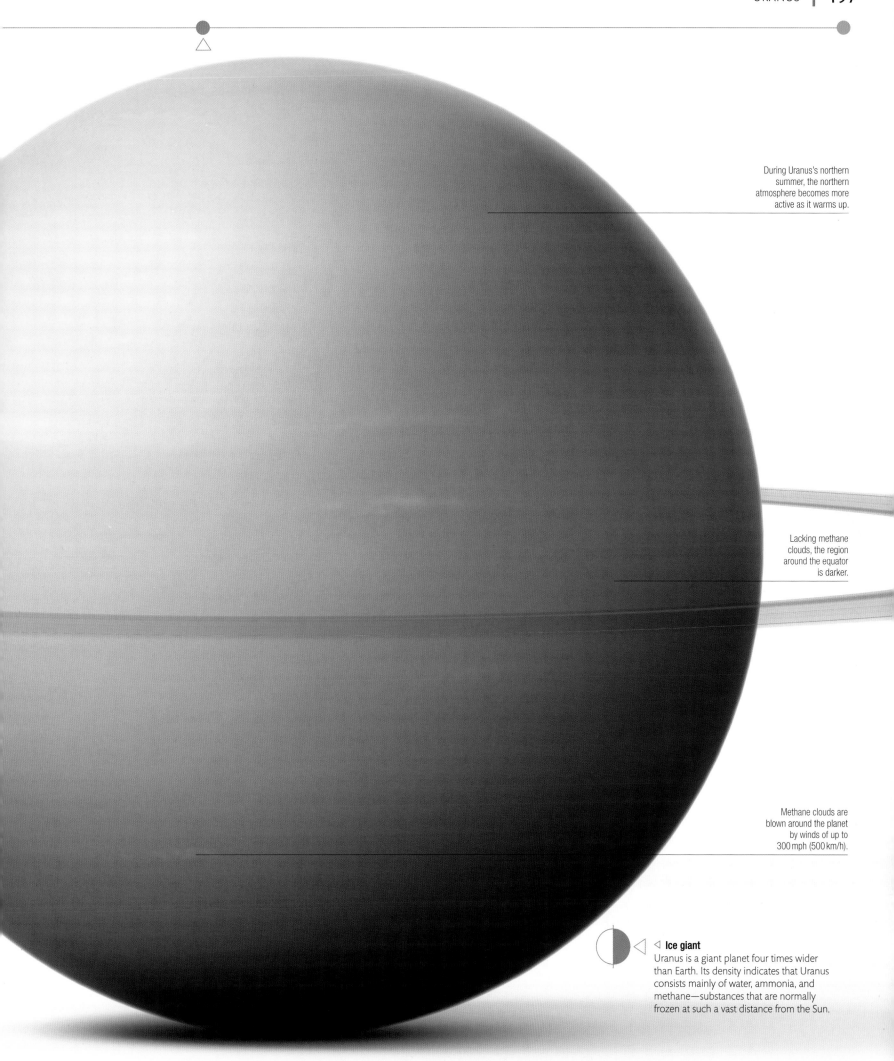

During Uranus's northern summer, the northern atmosphere becomes more active as it warms up.

Lacking methane clouds, the region around the equator is darker.

Methane clouds are blown around the planet by winds of up to 300 mph (500 km/h).

◁ **Ice giant**
Uranus is a giant planet four times wider than Earth. Its density indicates that Uranus consists mainly of water, ammonia, and methane—substances that are normally frozen at such a vast distance from the Sun.

Methane clouds

Like Neptune, Uranus may have a diamond sea around its core, with diamond hailstones raining into it.

URANUS **STRUCTURE**

BENEATH AN ATMOSPHERE TINGED BLUE-GREEN BY METHANE LIES A HUGE, SLUSHY OCEAN SURROUNDING A CORE OF HOT ROCK. THE PLANET'S LOPSIDED MAGNETIC FIELD MAY BE GENERATED BY A LAYER OF ELECTRICALLY CHARGED WATER.

If you were to descend into Uranus's aquamarine atmosphere, you would pass though successive cloud decks, the air becoming steadily thicker until you found yourself in a warm ocean with no distinct surface. This liquid ocean makes up most of the planet.

In the depths of Uranus's hidden ocean, water molecules break down to form a soup of hydrogen and oxygen ions. Currents in this sea of electrically charged particles are thought to generate Uranus's magnetic field, which is lopsided and off-center. If Earth had such a magnetic field, its poles would be as close to the equator as Cairo, Egypt or Brisbane, Australia.

Unlike the other giant planets, Uranus radiates less heat into space than it receives from the Sun. This may be because it was suddenly cooled by the immense impact that knocked the infant planet on its side.

Other planets **spin** like tops—Uranus rolls like a marble.

Core
Uranus's core is slightly less massive than planet Earth. A molten mixture of iron and magma, it has a temperature of more than 9,000°F (5,000°C) and is squeezed by pressure 10 million times greater than atmospheric pressure on Earth's surface.

Mantle
Astronomers call Uranus an ice giant because water, ammonia, and methane—the planet's main constituents—are normally frozen this far from the Sun. However, the high temperatures within the planet melt these substances to form a slushy ocean with a depth of 9,300 miles (15,000 km).

Atmosphere
The "air" on Uranus is mainly hydrogen and helium. There are layers of clouds at different depths in the atmosphere. Unique among the giant planets, Uranus has a tenuous outer atmosphere that is several times larger than the planet itself.

At the base of the mantle is a layer of superionic water—electrically charged hydrogen and oxygen—that glows yellow.

Clouds of ammonia form in the atmosphere. The deepest clouds consist of frozen water droplets.

Churning fluid within the mantle generates the planet's magnetic field.

The brightest and densest ring, Epsilon, is shepherded by two tiny moons, Cordelia and Ophelia, whose gravity helps maintain its shape.

◁ **Rings**
Uranus has a set of 13 rings. The first rings were discovered in 1977 when they unexpectedly blocked out the light of a distant star. Other rings were detected by Voyager 2 in 1986 and by the Hubble Space Telescope in 2003–05. All the rings of Uranus are narrow and, unlike Saturn's brilliant rings, as dark as coal.

Uranus has the coldest atmosphere of any planet, with temperatures plunging to –371°F (–224°C) in the troposphere—the densest part of the atmosphere.

THE **URANUS** SYSTEM

ASTRONOMERS DIVIDE URANUS'S 27 MOONS INTO THREE GROUPS: FIVE MAJOR MOONS, 13 SMALL INNER MOONS, AND NINE SMALL OUTER MOONS.

William Herschel discovered Uranus's two largest moons, Titania and Oberon, in 1787—six years after he had found the planet itself. Another amateur astronomer, the English brewer William Lassell, tracked down Umbriel and Ariel in 1851, and Dutch-American Gerard Kuiper located Miranda in 1948. Astronomers on a flying observatory, based aboard a Lockheed C-141A Starlifter transport plane, detected the narrow rings in 1977.

When Voyager 2 flew past Uranus in 1986, it captured detailed images of the moons and rings that were known at the time. Another 11 moons and two more rings turned up in the Voyager images. Since then, the Hubble Space Telescope and powerful instruments on Earth have identified the remainder of Uranus's moons and rings of which we are currently aware.

Moons to scale

Titania and Oberon rank among the solar system's top ten largest moons, but even all 27 moons combined would be no match for a single major moon of Jupiter, Saturn, or Neptune. Most of Uranus's moons are named after characters in plays by William Shakespeare, and a few from the poetry of Alexander Pope.

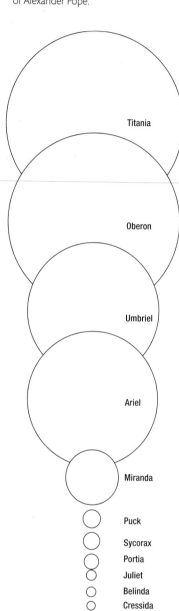

Titania

Oberon

Umbriel

Ariel

Miranda

Puck
Sycorax
Portia
Juliet
Belinda
Cressida
Rosalind
Caliban
Desdemona
Bianca
Prospero
Setebos
Ophelia
Cordelia
Stephano
Perdita
Mab
Francisco
Margaret
Ferdinand
Cupid
Trinculo

▷ Inner moons

The five largest moons, which orbit directly above the planet's tipped-up equator, were formed from the same spinning disk of gas and ice as Uranus itself. The 13 moons that lie closer to Uranus are in unstable orbits: past collisions have filled this region with rubble that still orbits Uranus, now corralled into narrow rings by the gravity of nearby moons.

Umbriel

The darkest of Uranus's moons, Umbriel is composed mainly of ice, coated in a layer of dark material perhaps made of organic (carbon-rich) compounds.

Oberon

This is the outermost of Uranus's five major moons. It is composed of a mixture of ice and rock, and its dark surface has a reddish tinge. Debris from space has smashed into Oberon, making it the most cratered of all Uranus's moons; one crater's central peak is 36,000 ft (11,000 m) high—taller than Mount Everest.

Titania

Titania is Uranus's largest moon and the eighth-biggest moon in the solar system. Titania's face is blemished by massive canyons and scarps, which formed when this moon expanded soon after its formation. Titania may have a very tenuous atmosphere of carbon dioxide.

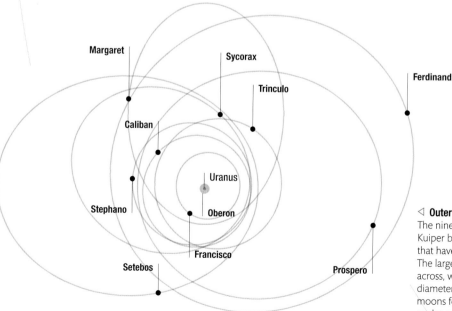

Margaret

Sycorax

Trinculo

Ferdinand

Caliban

Stephano

Uranus

Oberon

Setebos

Francisco

Prospero

◁ Outer moons

The nine outer moons are small icy worlds—Kuiper belt objects or the nuclei of comets—that have been captured by Uranus's gravity. The largest, Sycorax, is only 93 miles (150 km) across, while diminutive Trinculo has a diameter of less than 12 miles (20 km). These moons follow strange orbits, tilted at odd angles and looping in and out; Margaret has the highest eccentricity (least circular orbit) of any moon in the solar system.

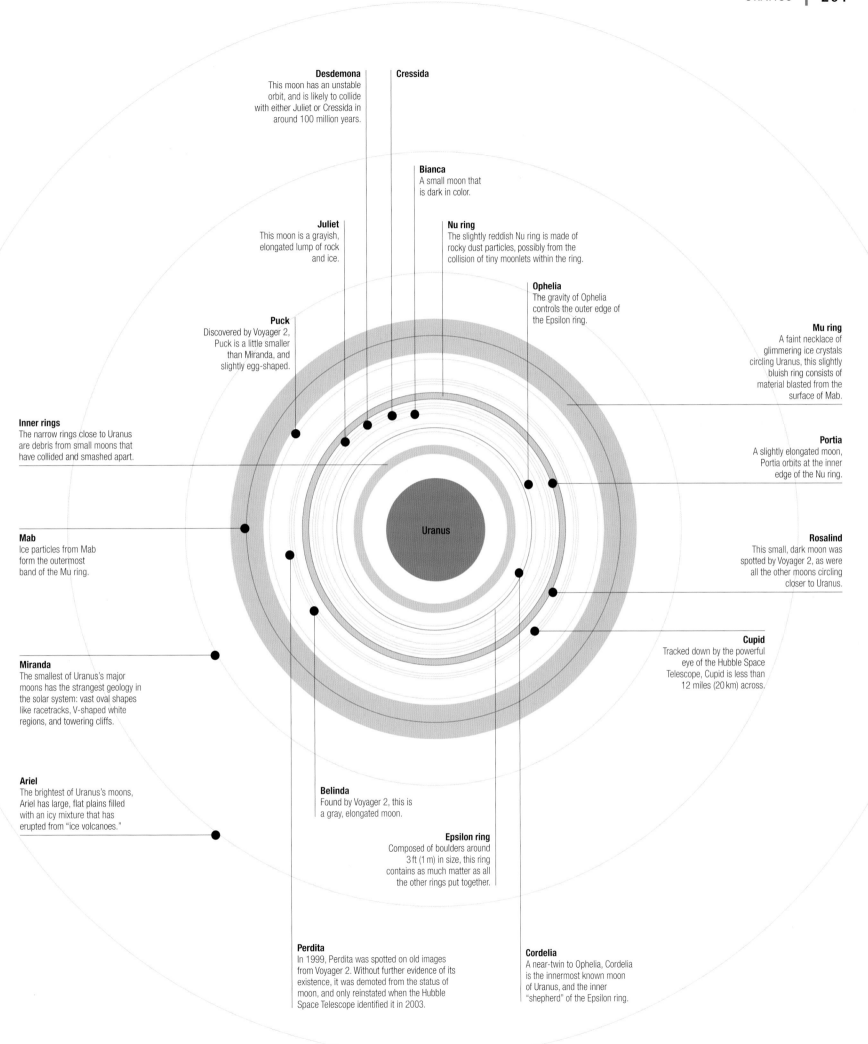

Desdemona
This moon has an unstable orbit, and is likely to collide with either Juliet or Cressida in around 100 million years.

Cressida

Bianca
A small moon that is dark in color.

Juliet
This moon is a grayish, elongated lump of rock and ice.

Nu ring
The slightly reddish Nu ring is made of rocky dust particles, possibly from the collision of tiny moonlets within the ring.

Ophelia
The gravity of Ophelia controls the outer edge of the Epsilon ring.

Puck
Discovered by Voyager 2, Puck is a little smaller than Miranda, and slightly egg-shaped.

Mu ring
A faint necklace of glimmering ice crystals circling Uranus, this slightly bluish ring consists of material blasted from the surface of Mab.

Inner rings
The narrow rings close to Uranus are debris from small moons that have collided and smashed apart.

Portia
A slightly elongated moon, Portia orbits at the inner edge of the Nu ring.

Uranus

Mab
Ice particles from Mab form the outermost band of the Mu ring.

Rosalind
This small, dark moon was spotted by Voyager 2, as were all the other moons circling closer to Uranus.

Miranda
The smallest of Uranus's major moons has the strangest geology in the solar system: vast oval shapes like racetracks, V-shaped white regions, and towering cliffs.

Cupid
Tracked down by the powerful eye of the Hubble Space Telescope, Cupid is less than 12 miles (20 km) across.

Ariel
The brightest of Uranus's moons, Ariel has large, flat plains filled with an icy mixture that has erupted from "ice volcanoes."

Belinda
Found by Voyager 2, this is a gray, elongated moon.

Epsilon ring
Composed of boulders around 3 ft (1 m) in size, this ring contains as much matter as all the other rings put together.

Perdita
In 1999, Perdita was spotted on old images from Voyager 2. Without further evidence of its existence, it was demoted from the status of moon, and only reinstated when the Hubble Space Telescope identified it in 2003.

Cordelia
A near-twin to Ophelia, Cordelia is the innermost known moon of Uranus, and the inner "shepherd" of the Epsilon ring.

DESTINATION
VERONA RUPES

THE TALLEST CLIFF IN THE SOLAR SYSTEM IS FOUND ON ONE OF THE SMALLER MOONS, URANUS'S MIRANDA. NAMED VERONA RUPES, THIS CLIFF IS SO HIGH—AND MIRANDA'S GRAVITY SO LOW—THAT A ROCK WOULD TAKE TEN MINUTES TO FALL FROM TOP TO BASE.

The near-vertical face of Verona Rupes, glistening with water ice like the rest of Miranda's surface, is almost 6 miles (10 km) high. It is not clear how such a huge structure was thrown up on so small a moon. The most likely explanation is tectonic activity early in Miranda's evolution. A more sensational theory suggests that Miranda was smashed to pieces in a colossal collision with another body and randomly reassembled itself, creating a scarred and fragmented surface pitted with craters, gouged with canyons, and crisscrossed by huge ridges.

**Artist's impression based
on NASA images**

LOCATION

Latitude −18°S; longitude 348°E

LAND PROFILE

Even the impressive walls of the Grand Canyon, which rise 1 mile (1.8 km) from the canyon floor, are dwarfed by Verona Rupes, which is about six times higher.

Elevation (km)

Miranda

Grand Canyon

Distance (km)

72 MILES (116 KM)
APPROXIMATE LENGTH OF VERONA RUPES RIDGE

FORMATION

Verona Rupes probably formed when a fault cracked Miranda's surface, and blocks of crust rose on one side of the fracture line and dropped on the other. Friction and erosion as the blocks rubbed against one another left grooves called slickensides on the cliff face.

Fault forms

Crust displaced vertically

NEPTUNE

YOU MIGHT EXPECT NEPTUNE TO BE A PLACID WORLD, SINCE IT IS THE MOST DISTANT PLANET FROM THE SUN. IN FACT, NEPTUNE HAS VIOLENT WEATHER SYSTEMS, HEAT WELLING UP FROM ITS INTERIOR, AND A MASSIVE ERUPTING MOON.

Neptune was discovered by deduction. In the 19th century, astronomers realized that Uranus was being pulled by the gravity of an unknown planet. French astronomer Urbain Leverrier calculated its position in 1846 (following a lead from John Couch Adams in England), and less than a year later, astronomers in Berlin found Neptune just where Leverrier had predicted.

A near twin to Uranus in size, Neptune is so far from the Sun that you need a telescope to see it at all. The eighth planet probably has the same internal structure as Uranus, along with a set of dark rings. When Voyager 2 passed Neptune in 1989, it showed an atmosphere in turmoil, with the fastest winds in the solar system. Even Neptune's most prominent feature, the Great Dark Spot, was short-lived.

Neptune Data

Equatorial diameter	49,528 km (30,775 miles)
Mass (Earth = 1)	17.1
Gravity at equator (Earth = 1)	1.1
Mean distance from Sun (Earth = 1)	30.1
Axial tilt	28.3°
Rotation period (day)	16.1 hours
Orbital period (year)	168.4 Earth years
Moons	14
Cloud-top temperature	−330°F (−201°C)

The near-supersonic winds of **Neptune's dark spots** can exceed **700 mph (1,200 km/h).**

Atmospheric methane absorbs red wavelengths in sunlight, giving the planet its characteristic blue color.

Neptune is surrounded by a system of thin and sparsely populated rings.

▽ **Tilt**

Neptune's axis is tilted at a similar angle to Earth's, so like Earth, the planet experiences seasons as it moves around the Sun. However, Neptune is so far from the Sun that each of its seasons lasts for more than 40 years.

▽ **Northern hemisphere**

It is currently winter in Neptune's northern hemisphere, so there is little activity in the region. Voyager 2 flew less than 3,000 miles (5,000 km) above the northern hemisphere's cloud tops—the closest of all its planetary encounters.

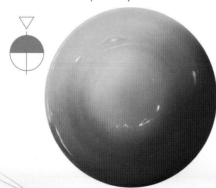

▽ **Southern hemisphere**

The southern hemisphere has been bathed in summer sunlight for the past 40 years. As a result, the south pole is the hottest spot on the planet, with temperatures rising to −310°F (−190°C).

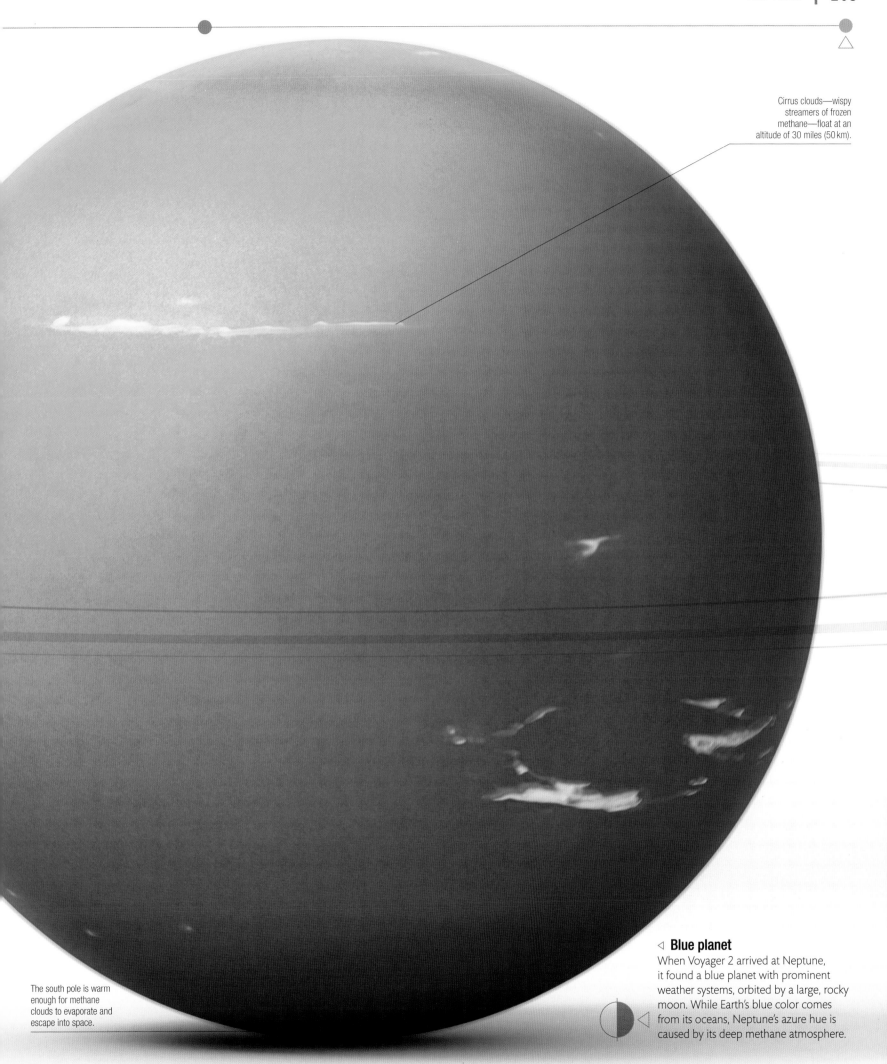

Cirrus clouds—wispy streamers of frozen methane—float at an altitude of 30 miles (50 km).

The south pole is warm enough for methane clouds to evaporate and escape into space.

◁ **Blue planet**

When Voyager 2 arrived at Neptune, it found a blue planet with prominent weather systems, orbited by a large, rocky moon. While Earth's blue color comes from its oceans, Neptune's azure hue is caused by its deep methane atmosphere.

A sea of liquid diamond may surround Neptune's core.

NEPTUNE STRUCTURE

NAMED AFTER THE GOD OF THE SEA IN ANCIENT ROMAN MYTHOLOGY, NEPTUNE IS LARGELY MADE OF WATER—JUST LIKE ITS TWIN, URANUS. DEEP INSIDE THE PLANET, THERE MAY BE A ROCKY CORE AND A SEA OF LIQUID DIAMOND.

Neptune is the third most massive planet, after Jupiter and Saturn. It is slightly smaller than neighboring Uranus because it has a thinner atmosphere, but its deeper liquid mantle makes it more massive overall.

Like Uranus, Neptune is sometimes called an ice giant because it formed from volatile compounds that existed as ices in the early solar system—mainly water, ammonia, and methane. Inside the planet's hot, dense interior, however, these compounds exist in a liquid form today.

Neptune's interior generates vast amounts of heat; around 60 percent more warmth wells up from deep inside the planet than arrives at its surface from the Sun. The heat and pressure in the lower mantle are so intense that methane may split into its constituent elements carbon and hydrogen, creating an ocean of liquid diamond around the core.

Diamond hailstones may rain down through Neptune's mantle.

Core
Neptune's core weighs 20 percent more than Earth and, like our planet, consists of rock and iron. Relative to Neptune's size, it's the most massive core of the giant planets. The core's central temperature probably exceeds 9,000°F (5,000°C).

Mantle
Most of Neptune's mass is in its mantle—a deep ocean of water, ammonia, and methane. Toward the bottom of the mantle, water molecules break up into oxygen and hydrogen ions. These electrically charged particles may be responsible for generating Neptune's magnetic field, which is tilted relative to the planet's axis of rotation.

Atmosphere
The turbulent cloud patterns in Neptune's atmosphere are only skin-deep, and the planet's dark-spot weather systems are short-lived. The deeper atmosphere extends one-fifth of the way to the core. It consists mainly of hydrogen and helium, with traces of methane providing the blue color.

◁ **Ring system**
Neptune has five very faint rings. Three are narrow, like the rings of Uranus, but two are broader bands of dust. The ring system was first detected from Earth during the 1980s, when it was noticed that something was blocking the light of the stars behind Neptune.

Galle is the innermost of Neptune's five rings. The existence of Neptune's rings was confirmed by the visit of Voyager 2 in 1989.

Le Verrier ring is shepherded by the tiny moon Despina; the moon's gravity helps to keep material within the ring.

Adams ring, the outermost of Neptune's rings, is unique in the solar system: its brightest regions are five distinct arcs following the same orbital path but separated from each other.

THE **NEPTUNE** SYSTEM

Moons to scale

Triton dominates Neptune's family of moons, accounting for 99.7 percent of the total mass of Neptune's entourage. At 1,700 miles (2,700 km) wide, it is the solar system's seventh-largest moon. Unlike Triton, which is spherical, Neptune's other moons are all probably irregular in shape.

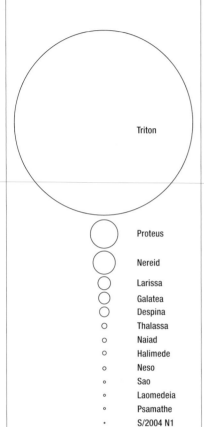

Triton

Proteus

Nereid

Larissa

Galatea

Despina

Thalassa

Naiad

Halimede

Neso

Sao

Laomedeia

Psamathe

S/2004 N1

LIKE ALL THE GAS GIANTS IN THE OUTER SOLAR SYSTEM, NEPTUNE IS SURROUNDED BY A FASCINATING, DYNAMIC ENVIRONMENT. HOST TO AT LEAST 14 MOONS, THE PLANET IS ALSO CIRCLED BY A SET OF FIVE VERY THIN RINGS.

The first moon to be identified was mighty Triton—only 17 days after Neptune itself was discovered in 1846. The astronomer who tracked it down was Englishman William Lassell. A fortune amassed as a brewer in the northern town of Bolton enabled Lassell to build giant telescopes and indulge his passion for astronomy.

Over a century passed before Nereid was discovered in 1949; a third moon, Larissa, followed in 1981. The rest of the moons were found more recently, either by the Voyager 2 spacecraft, which flew past Neptune in 1989, or by powerful, ground-based telescopes. The latest addition to the family—as yet unnamed—was spotted by the Hubble Space Telescope in 2013. All currently named moons of Neptune are named after water gods and spirits in Greek mythology.

Triton

This moon is an oddball, orbiting its planet backward—a characteristic not shared by any other large moon in the solar system. Like the outer moons, Triton was captured by the planet's gravity. The taming of such a large body wreaked havoc on the Neptune system, sending other moons into strange orbits. Triton's own orbit isn't stable: its destiny is to crash into Neptune.

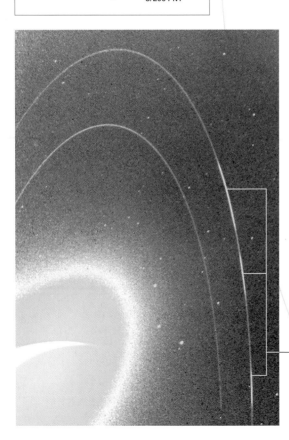

◁ Rings and arcs

Neptune is surrounded by five faint rings. Like Jupiter's rings, they consist largely of cosmic dust. The rings are named after astronomers who studied Neptune: Galle, Le Verrier, Lassell, Arago, and Adams. The outermost Adams ring has distinct clumps in it known as arcs, revealed in this image from Voyager 2. Ring particles normally spread out into a uniform circle, but astronomers believe the particles in the Adams ring are being confined by the gravity of Neptune's small moon Galatea, causing clumping.

Ring arcs

Neso

Halimede

Triton

Psamathe

Neptune

Nereid

Sao

Laomedeia

△ Outer moons

None of Neptune's outer moons has a circular orbit. Instead, they loop around the planet in great ellipses. Some of the orbits are highly inclined, and these orbits vary between prograde (forward) and retrograde (backward). All the outer moons, bar Nereid, are comparatively tiny. The majority of these moonlets were probably captured from the icy Kuiper belt by Neptune's gravity.

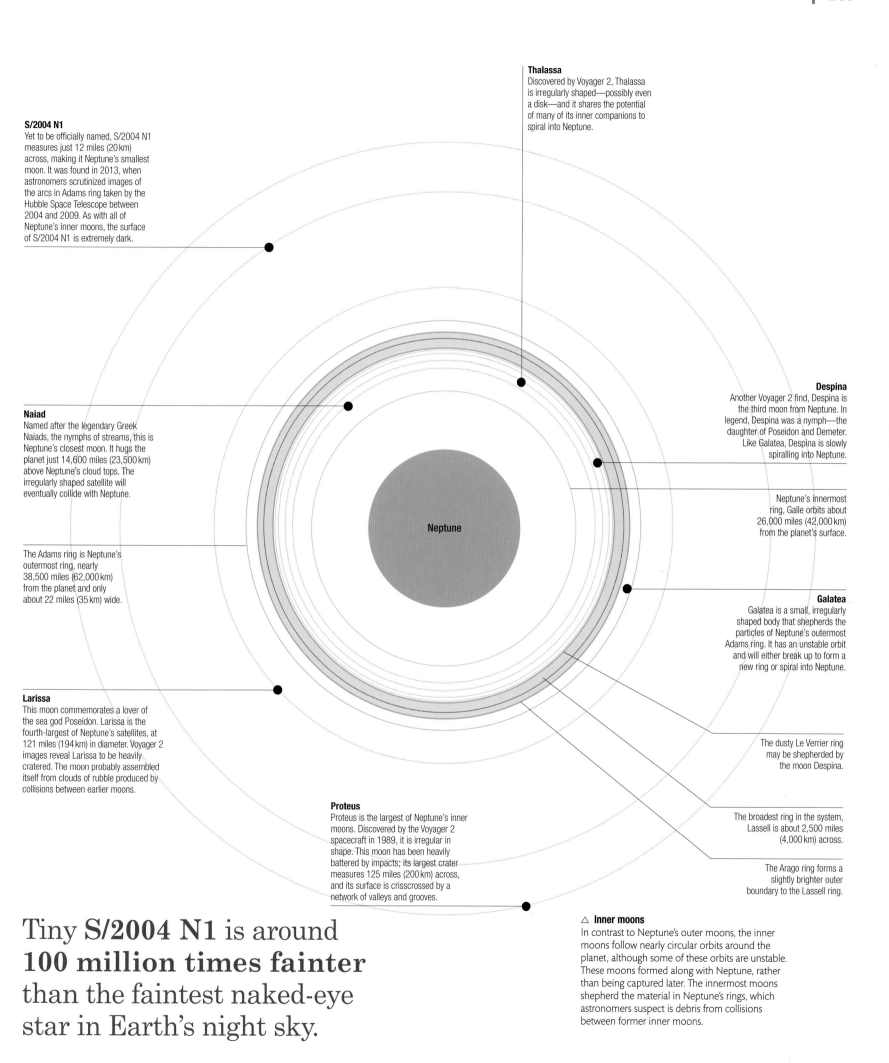

Thalassa
Discovered by Voyager 2, Thalassa is irregularly shaped—possibly even a disk—and it shares the potential of many of its inner companions to spiral into Neptune.

S/2004 N1
Yet to be officially named, S/2004 N1 measures just 12 miles (20 km) across, making it Neptune's smallest moon. It was found in 2013, when astronomers scrutinized images of the arcs in Adams ring taken by the Hubble Space Telescope between 2004 and 2009. As with all of Neptune's inner moons, the surface of S/2004 N1 is extremely dark.

Naiad
Named after the legendary Greek Naiads, the nymphs of streams, this is Neptune's closest moon. It hugs the planet just 14,600 miles (23,500 km) above Neptune's cloud tops. The irregularly shaped satellite will eventually collide with Neptune.

The Adams ring is Neptune's outermost ring, nearly 38,500 miles (62,000 km) from the planet and only about 22 miles (35 km) wide.

Larissa
This moon commemorates a lover of the sea god Poseidon. Larissa is the fourth-largest of Neptune's satellites, at 121 miles (194 km) in diameter. Voyager 2 images reveal Larissa to be heavily cratered. The moon probably assembled itself from clouds of rubble produced by collisions between earlier moons.

Proteus
Proteus is the largest of Neptune's inner moons. Discovered by the Voyager 2 spacecraft in 1989, it is irregular in shape. This moon has been heavily battered by impacts; its largest crater measures 125 miles (200 km) across, and its surface is crisscrossed by a network of valleys and grooves.

Despina
Another Voyager 2 find, Despina is the third moon from Neptune. In legend, Despina was a nymph—the daughter of Poseidon and Demeter. Like Galatea, Despina is slowly spiralling into Neptune.

Neptune's innermost ring, Galle orbits about 26,000 miles (42,000 km) from the planet's surface.

Galatea
Galatea is a small, irregularly shaped body that shepherds the particles of Neptune's outermost Adams ring. It has an unstable orbit and will either break up to form a new ring or spiral into Neptune.

The dusty Le Verrier ring may be shepherded by the moon Despina.

The broadest ring in the system, Lassell is about 2,500 miles (4,000 km) across.

The Arago ring forms a slightly brighter outer boundary to the Lassell ring.

Neptune

△ **Inner moons**
In contrast to Neptune's outer moons, the inner moons follow nearly circular orbits around the planet, although some of these orbits are unstable. These moons formed along with Neptune, rather than being captured later. The innermost moons shepherd the material in Neptune's rings, which astronomers suspect is debris from collisions between former inner moons.

Tiny **S/2004 N1** is around **100 million times fainter** than the faintest naked-eye star in Earth's night sky.

DESTINATION
TRITON

WITH A SURFACE TEMPERATURE OF –391°F (–235°C), NEPTUNE'S MOON TRITON IS ONE OF THE COLDEST PLACES IN THE SOLAR SYSTEM. YET THIS FRIGID WORLD IS VOLCANICALLY ACTIVE.

Triton's "retrograde" orbit—which runs in the opposite direction of the planet's rotation—suggests this moon is probably a captured Kuiper belt object from the icy outer limits of the solar system. Images from Voyager 2 reveal the surface to be a jumble of rocky outcrops, ridges, furrows, and craters. All of these tell us that Triton's surface is very young—just a few million years old.

Triton has a tenuous atmosphere of nitrogen and a reddish surface coated in methane and nitrogen ice. But its most famous features are the geysers discovered by Voyager in 1989. They spew out plumes of nitrogen gas mixed with dark dust and can reach heights of 5 miles (8 km) before falling back to stain the surface. An eruption can last a whole year.

Triton's shiny surface of **methane ice** and **nitrogen frost** reflects **70 percent** of the sunlight it receives.

Artist's impression based on images from Voyager 2 spacecraft

LOCATION

Latitude 31°S; longitude 37°E

SOUTH POLAR CAP

Titan is so cold that its air freezes on the ground. The highly reflective south polar cap is made of frozen nitrogen and methane. Cosmic rays striking the methane have created other organic compounds, giving the frost a pinkish hue. The polar cap is also peppered with dark spots and streaks from geysers.

Not imaged

Polar cap

WIND DIRECTION

Voyager images reveal the speed and direction of winds in the south polar region. The winds carry dark material from geysers northeast before it falls back, leaving black streaks. Scientists estimate the southwest winds reach speeds of 25 mph (40 km/h).

Deposit

Deposit

Wind

Geyser

Geyser

THE BLUE PLANETS

FOR MILLENNIA, PEOPLE KNEW ONLY THE INNERMOST FIVE PLANETS, SO IT WAS A GREAT SURPRISE WHEN WILLIAM HERSCHEL STUMBLED UPON URANUS IN 1781—A DISCOVERY THAT TRIGGERED THE HUNT FOR MORE HIDDEN WORLDS.

The discovery of Uranus and, later, Neptune was all the more surprising because these planets were giants, four times wider than Earth. Ever since, astronomers have continued to scour the skies for new planets. Many smaller worlds have been found, including Pluto, but these are now classed as dwarf planets or Kuiper belt objects. Careful study of the orbits of these distant icy bodies may yet reveal another giant lurking in the dark depths of the outer solar system.

John Flamsteed

1612
Galileo spots Neptune
Galileo observes Jupiter's moons and draws Neptune—which lies behind Jupiter in 1612—but thinks it is a star. Had he checked its motion, Galileo would have found Neptune before Uranus was known—and preempted its discovery by over 230 years.

1690
Observation of Uranus
The first Astronomer Royal, John Flamsteed, enters Uranus into his star catalog, naming it 34 Tauri. It is the planet's first recorded observation. Uranus is seen a further 22 times before its discovery, but astronomers dismiss it as a star.

Natural-color view of Uranus

False-color view of Uranus

Clyde Tombaugh

1986
Voyager 2 visits Uranus
The first close-up images from Voyager 2's trip to Uranus reveal a bland planet with 11 dark rings and ten previously unknown moons. The highlight is the contorted surface of the moon Miranda, with high cliffs and strange, racecourse-shaped markings.

1977
Uranus's rings discovered
Astronomers aboard a flying observatory over the Pacific Ocean are amazed as they watch a distant star disappearing behind Uranus. The star dims briefly, five times in all. They deduce that the planet must have a set of dark, very narrow rings that block the star's light.

1930
Discovery of Pluto
Amateur astronomer Clyde Tombaugh continues the search at Lowell Observatory. In February 1930, he photographs a faint moving object. Tombaugh calculates that it lies beyond Neptune. British schoolgirl Venetia Burney suggests the name Pluto.

Neptune's Great Dark Spot

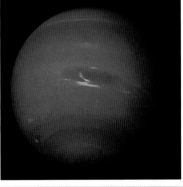

Pair of rings around Uranus

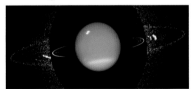

1989
The Great Dark Spot
Voyager 2 reveals Neptune's violent weather, with speeding clouds and a huge weather system, the Great Dark Spot. It confirms Neptune has a set of patchy rings. It also finds geysers erupting from the frozen surface of the planet's giant moon Triton.

1994
The Great Dark Spot disappears
The Hubble Space Telescope views Neptune and discovers that the Great Dark Spot has vanished; it was a transitory weather system, unlike Jupiter's 300-year-old Great Red Spot. The next year, Hubble views a large dark spot on the opposite side of Neptune.

2005
Extra rings for Uranus
Long-exposure images from Hubble reveal two faint rings around Uranus, farther out than the known ring system. The outer ring consists of dust ejected from the moon Mab, while the other ring may be the remains of a moon shattered in a collision.

William Herschel

Herschel's telescope

John Couch Adams

1781

Discovery of Uranus
British astronomer and musician William Herschel spots Uranus in his telescope, at first suspecting it is a star or comet. When astronomers calculate its orbit, it becomes clear that Herschel has discovered a new planet. He is the first person ever to do so.

1787

Two moons of Uranus seen
Using a large telescope, Herschel discovers Titania, Oberon, and four spurious moons and rings. Herschel notes their orbits are at "a considerable angle"—a clue to the planet's tilt. For 50 years, no one else has a telescope powerful enough to see them.

1843

Uranus's orbit
Astronomers find Uranus is straying from its orbit, probably pulled by the gravity of an unknown planet. Mathematician John Couch Adams calculates the location of the object responsible, but his work is ignored by Astronomer Royal George Airy.

Lowell's observatory

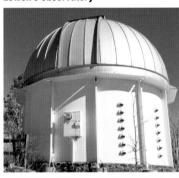

George III

Urbain Leverrier

1906

Search for Planet X
Astronomers note that Uranus and Neptune seem to feel the tug of another world. Boston businessman Percival Lowell had established an observatory in Arizona to study the supposed canals of Mars, and here he begins to search for the mystery "Planet X."

1850

Naming of Uranus
Herschel had called his new planet Georgium Sidus, "George's Star" (after King George III). This clashes with the other planets' mythological names and is unpopular. Johann Bode suggests Uranus, father of Saturn. In 1850, Britain's Nautical Almanac Office agrees.

1846

Discovery of Neptune
French astronomer Urbain Leverrier comes up with the same position for the planet as Adams. He sends the prediction to the Berlin Observatory, which has a new star chart for that region of sky. On the first night he looks, Johann Galle sees Neptune.

Pluto and its moons

Eris and its moon Dysnomia

2006

Pluto is demoted
The International Astronomical Union reclassifies Pluto as a dwarf planet. Astronomers have now discovered more than 1,000 similar icy bodies beyond Neptune. These include Eris, which is about the same size as Pluto.

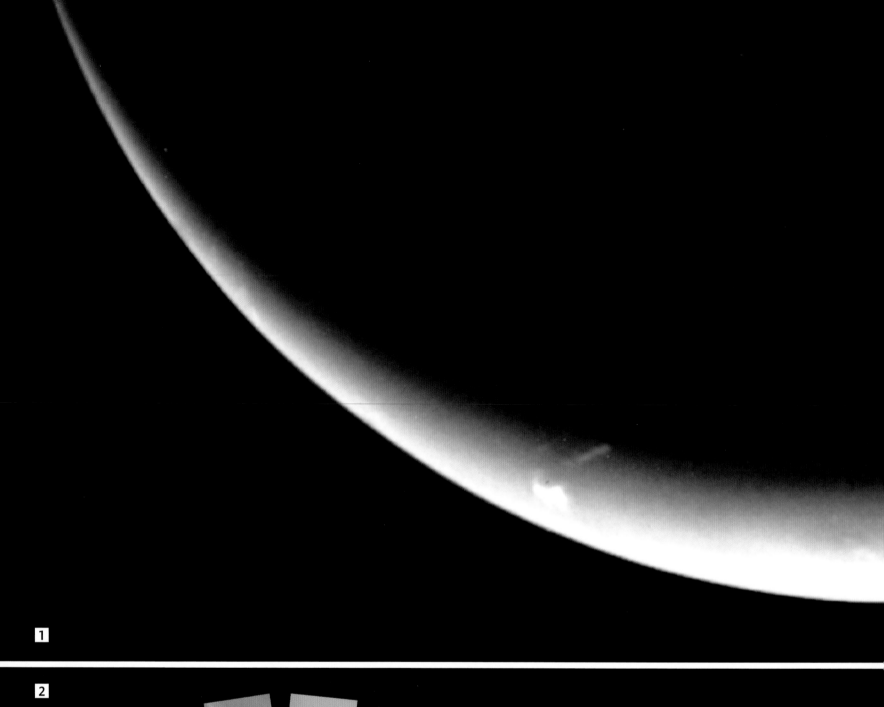

1

2

Jupiter

Venus

Earth

The Sun

Saturn

Uranus

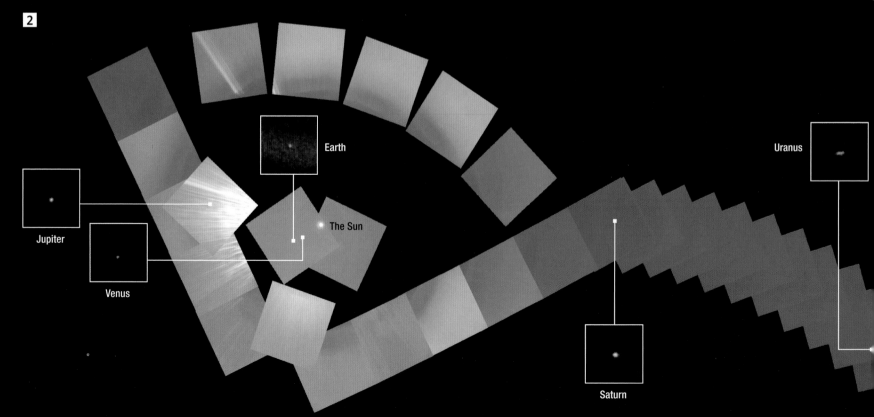

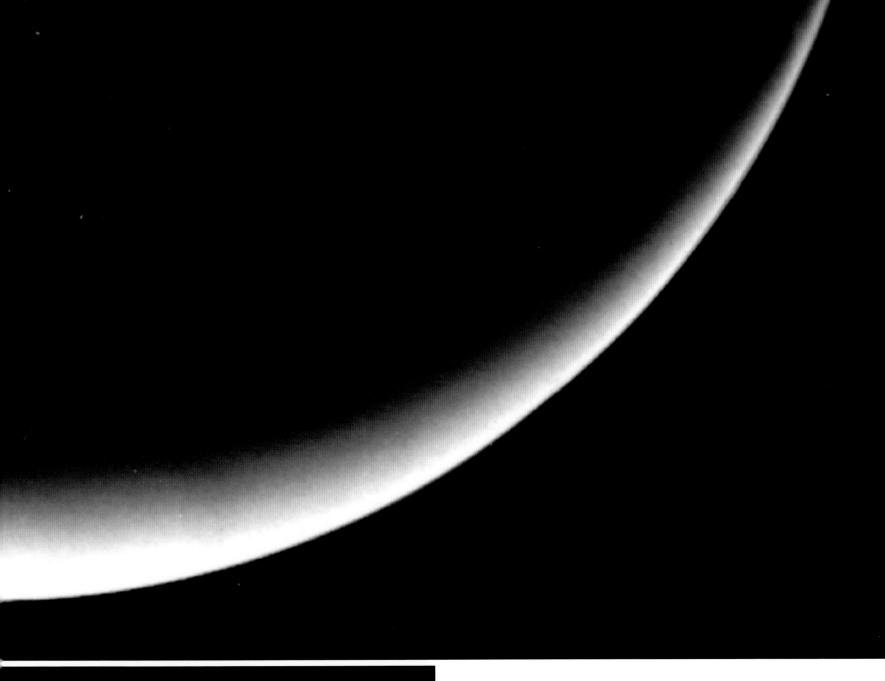

Neptune

VOYAGERS' **GRAND TOUR**

1 Goodbye to the planets

This stunning arc of the crescent Neptune was captured by the outward-bound Voyager 2 in 1989 as it departed from its final encounter with a planet. The twin Voyager spacecraft were launched in 1977 to explore the giant planets. Voyager 1 flew past Jupiter and Saturn, but Voyager 2 visited all four gas giants.

2 Looking back

Voyager 1's portrait of the solar system, captured in 1990 when the spacecraft was 3.7 billion miles (6 billion km) from Earth, was the first-ever image of our planetary system taken from outside. It was also the last image taken by either Voyager. The mosaic comprises 60 wide-angle frames; insets show the planets magnified many times. From Voyager's great distance, Earth was a point of light measuring only 0.12 pixels across.

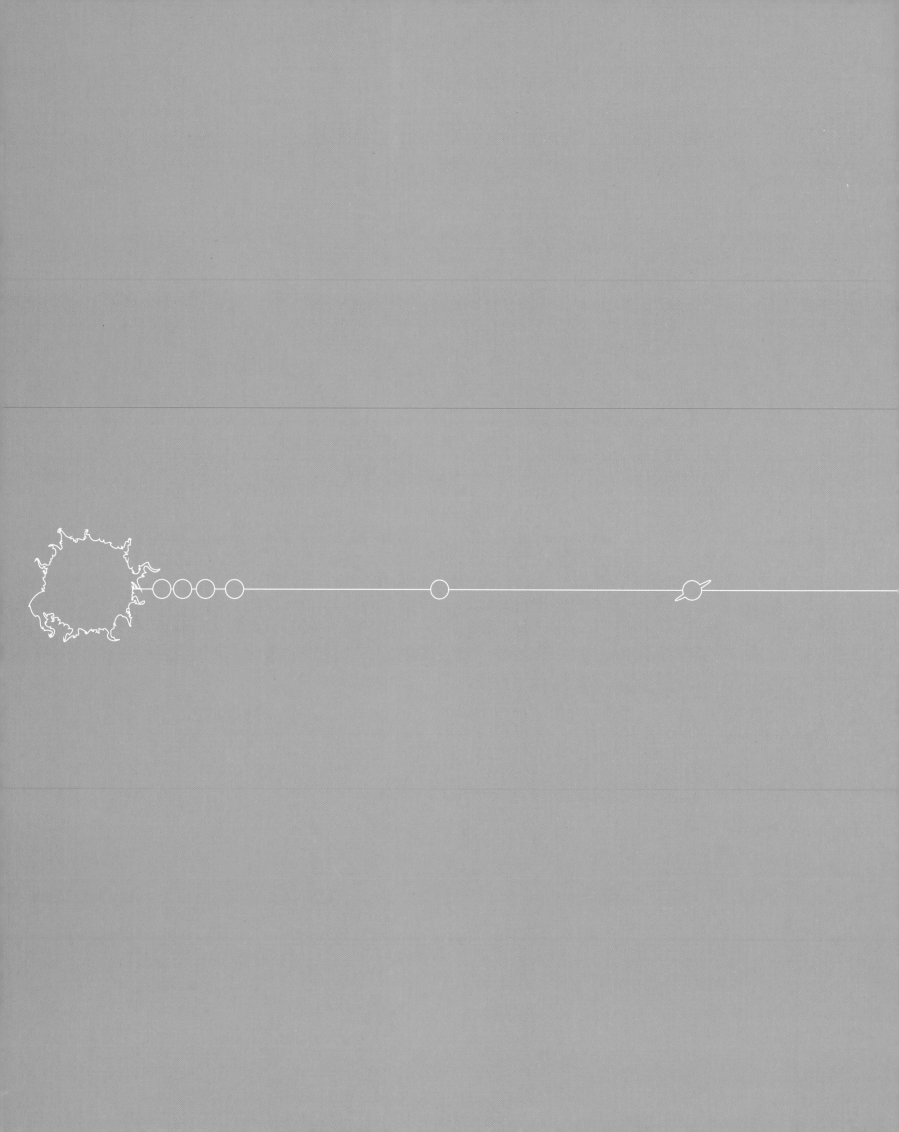

OUTER LIMITS

THE **KUIPER BELT**

A BIG QUESTION AT THE END OF THE 20TH CENTURY WAS WHETHER ANYTHING LAY BEYOND PLUTO. THE EXISTENCE OF A BELT OF ICY OBJECTS WAS PREDICTED BUT NOT CONFIRMED UNTIL THE 1990S, WHEN THE FIRST OBJECTS WERE FOUND.

The Kuiper belt begins about 30 times farther from the Sun than Earth (30 AU) and stretches to 50 AU. More than 100,000 Kuiper belt objects (KBOs) larger than around 60 miles (100 km) wide are believed to exist in the belt. They formed at the dawn of the solar system and were thrown into their present eccentric orbits by the gravitational fields of the giant planets. Typical KBOs are classed as cubewanos (pronounced "cue-be-one-oh"), which are found throughout the belt. The name comes from 1992 QB1, the first cubewano discovered. KBOs on the far edge of the belt follow eccentric orbits. This region, termed the scattered disk, is the source of short-period comets.

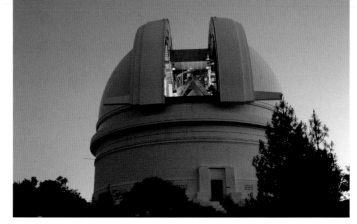

△ **Search**
A typical telescope used to search for KBOs and scattered disk bodies is the Samuel Oschin 48-in (1.2-m) telescope on Mount Palomar, California. It was used to find the KBO Orcus and the dwarf planet Eris. Two images are made of a region of sky, one week apart. Anything that moves from one image to the next is a solar system body. The stationary objects are stars.

Pluto has an eccentric orbit inclined at 17.1° and ranging from 48.9 to 29.7 AU.

The main belt is a flattened disk measuring about 2 billion miles (3 billion km) from edge to edge.

Neptune

The inner part of the belt contains plutinos—bodies in a 3:2 orbital resonance with Neptune (for every two orbits of the plutino, Neptune makes three).

▷ **Ice ring**
We know of more than a thousand KBOs. Made of rock and ice, they are similar in composition to the nuclei of comets, but the larger ones are more dense. Their surfaces, which measure less than −360°F (−220°C) in temperature, are covered with ices including water, carbon dioxide, methane, and ammonia, and are colored by interactions with cosmic rays. Originally termed the Edgeworth–Kuiper Belt after Kenneth Edgeworth, who predicted its existence in 1943, and Gerard Kuiper, who in 1951 hypothesized that it had formed and dissipated, the name has been shortened to the Kuiper Belt.

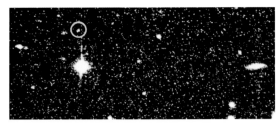

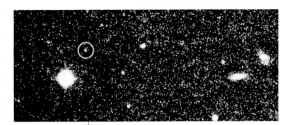

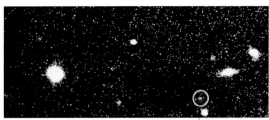

△ Discovery

After five years of searching, David Jewitt and Jane Luu discovered the first KBO in August 1992 using the 87-in (2.2-m) University of Hawaii telescope on Mauna Kea. Named 1992 QB1, it was located about 4 billion miles (6 billion km) from the Sun and was about 100,000 million times fainter than Jupiter. These European Southern Observatory images were taken one month after discovery.

In the center of the circle is 1992 QB1, imaged on September 27, 1992, four hours after the image at left.

A day later, 1992 QB1 had moved position against the background stars, traveling at a few seconds of arc per hour.

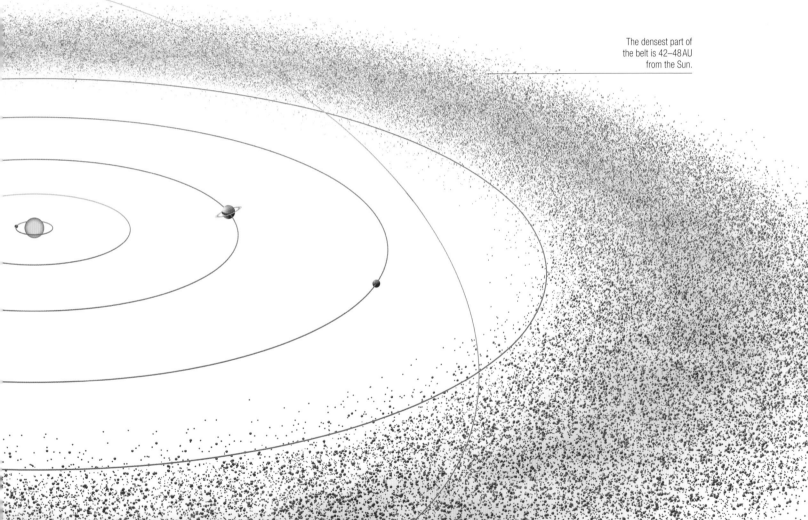

The densest part of the belt is 42–48 AU from the Sun.

The outer edge contains scattered disk objects. Their eccentric orbits stretch to about 9 billion billion miles (15 billion billion km) from the Sun.

DWARF PLANETS

LIKE TRUE PLANETS, DWARF PLANETS HAVE ENOUGH MASS TO BECOME SPHERICAL UNDER THEIR OWN GRAVITY. HOWEVER, THEY LACK THE GRAVITATIONAL FORCE TO SWEEP THEIR ORBITS CLEAR OF OTHER BODIES.

As they form, planets clear their orbits of minor objects, such as asteroids, either by pulling them in and amalgamating with them or by flinging them elsewhere. Dwarf planets cannot do this, though they may have sufficient gravity to capture their own moons.

The definition of a dwarf planet was agreed upon by the International Astronomical Union in 2006. The most famous example is Pluto, which orbits far from the Sun in the freezing Kuiper belt at the edge of the solar system. Once referred to as the ninth planet, Pluto was assigned to the new category along with several similar bodies found in the outer solar system. Among these are Eris (the largest known dwarf planet), Haumea, and Makemake. The asteroid Ceres, located in the asteroid belt between Mars and Jupiter, was also given dwarf planet status in 2006.

Eris
Diameter 1,445 miles (2,326 km)

Pluto
Diameter 1,433 miles (2,306 km)

Haumea
Diameter 1,218 miles (1,960 km)

Makemake
Diameter 895 miles (1,440 km)

Quaoar (possible dwarf planet)
Diameter 665 miles (1,070 km)

Sedna (possible dwarf planet)
Diameter 618 miles (995 km)

Ceres
Diameter 592 miles (952 km)

Orcus (possible dwarf planet)
Diameter 570 miles (917 km)

Ixion (possible dwarf planet)
Diameter 404 miles (650 km)

Earth
Diameter
7,917 miles
(12,742 km)

Discovering Pluto

US astronomer Clyde Tombaugh discovered Pluto in 1930 while searching for "Planet X"—a hypothetical ninth planet thought to be responsible for irregularities in the orbits of Neptune and Uranus. Pluto was named the ninth planet, though it turned out to have too little mass to exert gravitational pull on the gas giants. Its eccentric and tilted orbit is typical of Kuiper belt bodies.

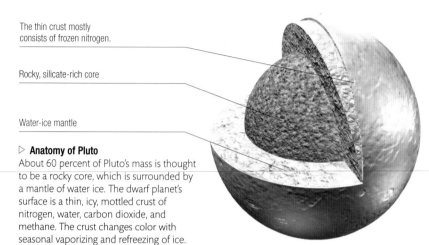

The thin crust mostly consists of frozen nitrogen.

Rocky, silicate-rich core

Water-ice mantle

▷ **Anatomy of Pluto**
About 60 percent of Pluto's mass is thought to be a rocky core, which is surrounded by a mantle of water ice. The dwarf planet's surface is a thin, icy, mottled crust of nitrogen, water, carbon dioxide, and methane. The crust changes color with seasonal vaporizing and refreezing of ice.

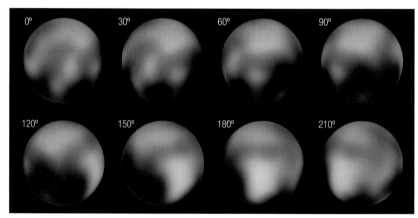

△ **Hubble view of Pluto**
Pluto is so small and distant that sharp images are impossible to obtain. Even the Hubble Space Telescope is unable to resolve details smaller than a few hundred miles wide. Here we see Pluto in rotation. The dark areas are carbon-rich residues that have formed where ultraviolet radiation and solar wind particles caused methane to react with carbon dioxide ice.

▽ **New Horizons mission**
In January 2006, NASA launched the New Horizons mission to Pluto. After a nine-year interplanetary journey, the spacecraft is due to fly past Pluto and its various small moons at 6 miles (11 km) per second on July 14, 2015. It will obtain detailed colored images of the dwarf planet's sunlit surface and will use scientific instruments to measure its surface temperature and analyze its atmosphere.

Moons of Pluto

Pluto has five known moons, all with names linked to the underworld in classical mythology. The New Horizons spacecraft is expected to find more. Charon, the largest moon, was discovered in 1978 by American astronomer James Christy and is named after the ferryman of Hades in Greek mythology. The four smaller moons were discovered in the 21st century using Hubble Space Telescope data. Nix and Hydra are about 60 miles (100 km) in diameter, and Styx and Kerberos are a mere 12 miles (20 km).

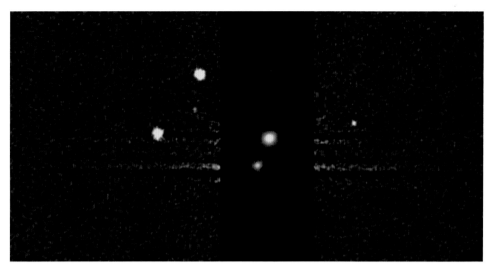

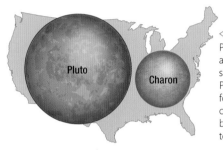

◁ **Size comparison**
Pluto is barely twice as wide as its moon Charon. The large size of Charon relative to Pluto suggests the two bodies formed at the same time out of the same material, Charon breaking away from Pluto due to spin-induced instability.

△ **Pluto and moons**
This Hubble image shows Pluto with its five known moons. The brightness of Pluto and Charon (in the dark band) has been reduced to make the other moons visible. From left, the objects are: Hydra, Styx, Nix (top), Charon, Pluto, and Kerberos. All the moons have circular orbits close to Pluto and in the same plane, indicating they are not captured objects. They may have formed after a collision between Pluto and another body.

▽ **Artist's impression**
A visitor standing on the icy surface of Pluto would see a faint, distant Sun, the moon Charon, and traces of the hazy nitrogen-methane atmosphere. The roughness of the surface is caused by cratering by smaller Kuiper belt bodies, cryovolcanic activity, and seasonal variations in temperature during which upper layers of ice and snow turn to vapor and then refreeze.

COMETS

COMETS ARE MOUNTAIN-SIZED, DIRTY SNOWBALLS THAT FORMED AT THE DAWN OF THE SOLAR SYSTEM. OCCASIONALLY ONE APPROACHES THE SUN, CHANGES RADICALLY IN SIZE AND APPEARANCE, AND BECOMES BRIGHT ENOUGH TO BE SEEN.

An estimated 1 trillion comets exist in the freezing outer reaches of the solar system. Unchanged since the planets formed, each is a lump of snow, ice, and rocky dust: a cometary nucleus. These icy bodies are too small to be seen from Earth, but if one ventures into the planetary part of the solar system, it can develop a spectacular glowing halo and tails, making it bright enough to be detected. Many comets are found using telescopes, often accidentally by asteroid hunters, but many also pass unnoticed. Some revisit us regularly, with return periods ranging from a few years to hundreds. Others are unexpected and may not pass our way again for thousands or millions of years—or ever. Newly discovered comets take the name of the discoverer. The greatest number of discoveries, over 2,500, has been made by the SOHO spacecraft.

Close to the Sun

When a comet gets closer to the Sun than the asteroid belt, solar heating causes its nucleus to lose mass. This material forms a coma— a huge cloud of gas and dust around the nucleus—and two tails that continuously disperse into space. Each time the nucleus passes the Sun, a layer of surface material about 3 ft (1 m) deep is used up in a fresh coma and tails. A comet such as Halley, which orbits the Sun every 76 or so years, will eventually run out of material and vanish.

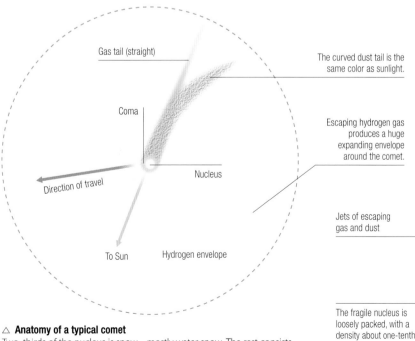

Anatomy of a typical comet

Gas tail (straight)

The curved dust tail is the same color as sunlight.

Coma

Escaping hydrogen gas produces a huge expanding envelope around the comet.

Direction of travel

Nucleus

To Sun — Hydrogen envelope

△ **Anatomy of a typical comet**
Two-thirds of the nucleus is snow—mostly water snow. The rest consists of small rock particles and dust. Released gas and dust form the coma, which may grow to 60,000 miles (100,000 km) wide, and two tails, both of which are pushed back by the solar wind. The gas tail is straight but the dust tail curves back toward the comet's orbital path.

△ **Hale–Bopp**
Bright, naked-eye comets occur at a rate of one per decade. Comet Hale–Bopp, one of the 20th century's brightest comets, was visible to the naked eye in 1996 and 1997. Here, the material in its two tails is being pushed out of the solar system. The white dust tail shines as sunlight reflects off its dust particles. The blue ionized gas tail actually emits its own light and is more structured—the paths of the particles within it are determined by magnetic fields in the solar wind.

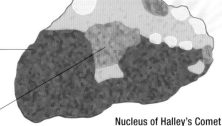

Jets of escaping gas and dust

The fragile nucleus is loosely packed, with a density about one-tenth that of ice; much of the interior is void.

Surface depression

Nucleus of Halley's Comet

◁ **Cometary nucleus**
The nucleus of a comet is irregularly shaped and typically about half a mile (1 km) across, with a black, dusty surface. Where the dust is thinnest, the transmission of solar heat causes the snow beneath to change into gas. The gas escapes, taking some of the overlying dust with it and leaving depressions on the surface of the comet.

◁ Sungrazer
Some comets fly close enough to the Sun to pass through its outer atmosphere, the solar corona. Others, such as Comet SOHO 6 (left), get so close that they dive into the Sun and are destroyed. Called sungrazers, many such comets are seen by the SOHO spacecraft as it studies the Sun. In this SOHO image, the Sun is blocked out by a disk to reveal the Sun's corona and Comet SOHO 6's final moments (top left).

▷ Meteor shower
Larger dust particles do not get pushed into the comet tail but slowly gain on, or fall behind, the cometary nucleus. They eventually form an annulus—a ring of dust around the comet's orbital path. If Earth travels through a comet's annulus, individual dust particles form meteors—shooting stars—as they speed through Earth's atmosphere. The meteors in a shower radiate from a specific spot in the sky.

COMET ORBITS

MOST COMETS EXIST WITHIN THE OORT CLOUD, FAR BEYOND THE PLANETS AND OUR VISION. WE KNOW THEY EXIST BECAUSE OCCASIONALLY ONE OF THESE BODIES IS DIVERTED INTO THE INNER SOLAR SYSTEM AND FORMS A COMA AND TAILS.

Cometary orbits, unlike planetary ones, are highly elliptical (oval), so a comet's distance from the Sun varies greatly over time. Comets that leave the Oort cloud and travel toward the Sun are classed according to how long one orbit takes. Short-period comets hug the plane of the planets and have periods of less than 20 years. Intermediate-period comets pass close to the Sun every 20–200 years and their orbits have a wide range of inclinations. Long-period comets, which are also randomly inclined, have periods ranging from 200 years to tens of million of years. Some travel so far from the Sun that they may fly halfway to nearby stars.

Cometary orbits are affected by the gravitational fields of the planets. Short-period comets have become trapped in the inner solar system by Jupiter's gravity, and Jupiter can easily flip a comet from a short orbit back to a longer one. Some long-period comets are ejected from the solar system altogether and sail off into the galaxy. Others are pulled closer to the Sun, giving astronomers a chance of detecting them.

Straight, blue-white tail of ionized gas

Curved, white dust tail

The Sun is located at one focal point of a comet's elliptical orbit.

The tails are at their longest when the comet is closest to the Sun.

The tails always face away from the Sun.

The tails shorten and disappear as the comet moves away from the Sun.

△ **Around the Sun**
Comets follow elliptical orbits—oval-shaped loops around two focal points. They do not travel along these orbits at a constant speed, but accelerate as they move closer to the Sun, then slow down again as they move away from it. Comets are visible from Earth only when they fly close to the Sun, as what we see of them is their tails, which are created when the Sun's heat vaporizes material on the comet's surface, creating a streak of debris.

Comet Halley's orbital period varies from 76 to 79.3 years. It was last seen in 1986 and will next appear in 2061.

▷ **Comets in the inner solar system**
By the end of 2013, astronomers had detected about 5,000 comets passing through the planetary part of the solar system. Around 500 are short-period comets, such as Comet Tempel 1. First recorded in 1867, Tempel 1 returned in 1873 and 1879 but then did not reappear until 1967, due to a change in its orbit. Comet Halley is an intermediate-period comet first recorded in 240 BCE and seen 30 times since. The long-period Comet Hyakutake appeared brightly in Earth's sky in 1996. It previously visited the Sun 17,000 years before, but on its 1996 orbit, gravitational interaction with the giant planets disturbed its orbit so much that it will not return for 70,000 years.

Comet Tempel 1 currently orbits every 5.5 years within the orbits of Mars and Jupiter; its proximity to Jupiter means that its orbital path and period will change.

Sun

Mercury

Venus

Earth

Mars

Jupiter

Saturn

Uranus

Neptune

Short-period comets have less elliptical orbits and are regular visitors to the inner solar system.

Long-period comets have extremely elliptical orbits and only rarely make an appearance in the inner solar system.

Discovered in 1996, Hyakutake made one of the closest approaches to Earth in the 20th century and was one of the brightest objects in the night sky.

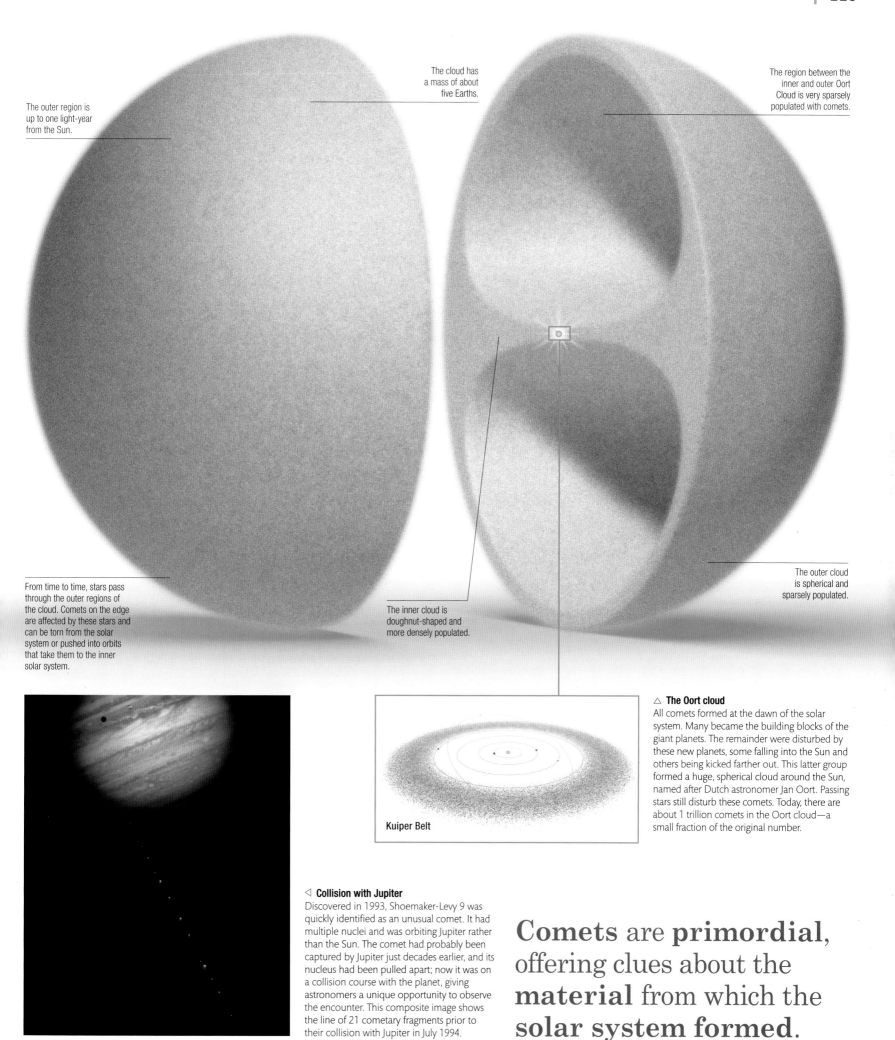

The outer region is up to one light-year from the Sun.

The cloud has a mass of about five Earths.

The region between the inner and outer Oort Cloud is very sparsely populated with comets.

From time to time, stars pass through the outer regions of the cloud. Comets on the edge are affected by these stars and can be torn from the solar system or pushed into orbits that take them to the inner solar system.

The inner cloud is doughnut-shaped and more densely populated.

The outer cloud is spherical and sparsely populated.

Kuiper Belt

△ **The Oort cloud**

All comets formed at the dawn of the solar system. Many became the building blocks of the giant planets. The remainder were disturbed by these new planets, some falling into the Sun and others being kicked farther out. This latter group formed a huge, spherical cloud around the Sun, named after Dutch astronomer Jan Oort. Passing stars still disturb these comets. Today, there are about 1 trillion comets in the Oort cloud—a small fraction of the original number.

◁ **Collision with Jupiter**

Discovered in 1993, Shoemaker-Levy 9 was quickly identified as an unusual comet. It had multiple nuclei and was orbiting Jupiter rather than the Sun. The comet had probably been captured by Jupiter just decades earlier, and its nucleus had been pulled apart; now it was on a collision course with the planet, giving astronomers a unique opportunity to observe the encounter. This composite image shows the line of 21 cometary fragments prior to their collision with Jupiter in July 1994.

Comets are **primordial**, offering clues about the **material** from which the **solar system formed**.

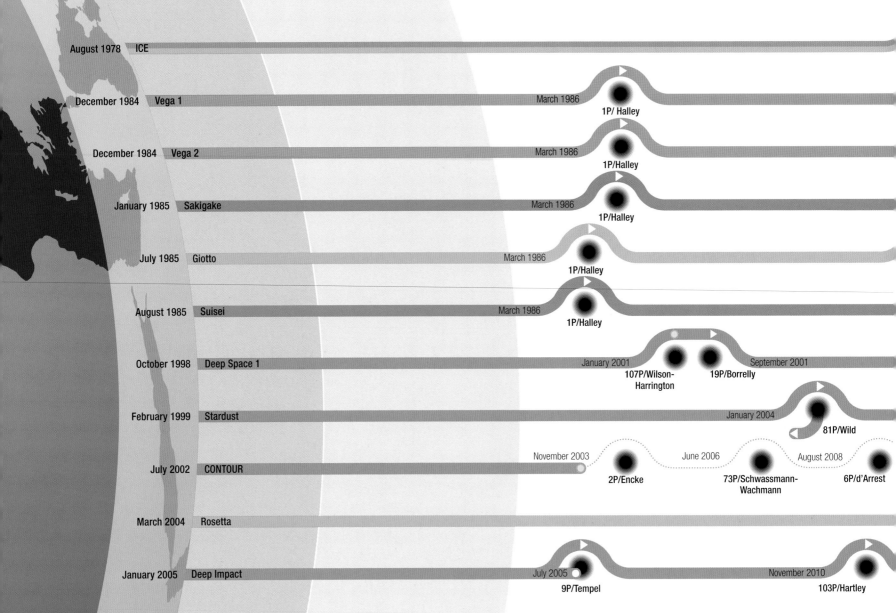

EARTH ORBIT

August 1978	ICE
December 1984	Vega 1	March 1986	1P/ Halley
December 1984	Vega 2	March 1986	1P/Halley
January 1985	Sakigake	March 1986	1P/Halley
July 1985	Giotto	March 1986	1P/Halley
August 1985	Suisei	March 1986	1P/Halley
October 1998	Deep Space 1	January 2001	107P/Wilson-Harrington	19P/Borrelly	September 2001
February 1999	Stardust	January 2004	81P/Wild
July 2002	CONTOUR	November 2003	2P/Encke	June 2006	73P/Schwassmann-Wachmann	August 2008	6P/d'Arrest
March 2004	Rosetta
January 2005	Deep Impact	July 2005	9P/Tempel	November 2010	103P/Hartley

KEY

- Joint mission— NASA/ESA
- NASA (USA)
- JAXA (Japan)
- RFSA (Russia)
- esa — ESA (Europe)
- Destination
- ▷ Flyby
- ◉ Orbit
- Sample return
- Lander/impactor
- Failure

▷ Giotto

Prior to 1986, astronomers had no idea what a comet nucleus looked like. Their first view came on March 13 of that year when ESA's Giotto imaged the nucleus of Halley's Comet. It revealed a 9.5-mile- (15.3-km-) long, potato-shaped mass with bright jets of gas and dust erupting from its surface. Hills and valleys could be seen on the nucleus's generally smooth surface. Giotto was inside the comet's coma and only just survived the battering from its dust. Its mission extended, Giotto flew by Comet Grigg-Skjellerup in 1992.

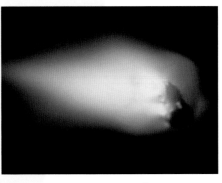

Giotto image of nucleus of Halley's Comet

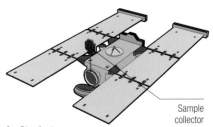

Sample collector

▷ Stardust

The first sample of comet material was obtained by NASA's Stardust spacecraft. To capture dust particles from the comet without vaporizing them, Stardust used an incredibly lightweight, porous material known as aerogel, which was mounted on a collecting device shaped like a tennis racket. The collector and its precious cargo separated from Stardust and returned to Earth in January 2006.

September 1985 **21P/Giacobini-Zinner**

July 1992 **26P/Grigg-Skjellerup**

November 1998
21P/Giaobini-Zinner

February 2011
9P/Tempel

August 2014
**67P/Churyumov
-Gerasimenko**

MISSIONS TO **COMETS**

IN THE PAST 30 YEARS OUR KNOWLEDGE OF COMETS HAS IMPROVED ENORMOUSLY, THANKS TO A SMALL NUMBER OF SPACECRAFT THAT HAVE SAILED THROUGH THE GLOWING COMAS AROUND COMETS TO VISIT THEIR ICY NUCLEI.

Hidden in the glare of their brilliant comas, and too small to be viewed with telescopes, comet nuclei can be seen clearly only by spacecraft. The first craft to return detailed images of a comet nucleus was Giotto, which launched in 1985 and passed within 375 miles (600 km) of Halley's Comet less than a year later. Its images confirmed the theory that comets are made of dirt and snow. More ambitious missions followed, including NASA's Stardust, which scooped a sample of dust from Comet Wild 2 and brought it back to Earth, and ESA's Rosetta, the first craft designed to land on a comet nucleus.

In the **ten-year** journey to its comet target, the Rosetta spacecraft orbited the Sun **five times**.

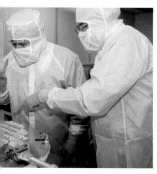

Scientists analyzing the Stardust sample

▷ **Deep Impact**
With the aim of studying a comet's interior, Deep Impact fired a self-guided impactor into the nucleus of Comet Tempel 1 in 2005. A cloud of debris obscured the craft's view, but the Stardust craft was redirected to Tempel 1 and later imaged the impact crater. Deep Impact moved off to meet Comet Hartley 2, getting within 450 miles (700 km) of its peanut-shaped, 1.2-mile- (2-km-) long nucleus.

Deep Impact image of nucleus of Comet Hartley 2

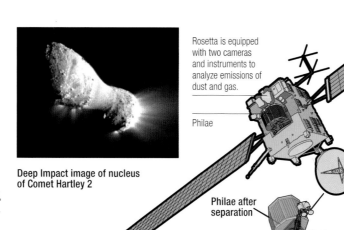

Rosetta is equipped with two cameras and instruments to analyze emissions of dust and gas.

Philae

Philae after separation

◁ **Rosetta**
ESA's Rosetta is the most ambitious comet mission to date. Launched in 2004, Rosetta is designed to orbit the 2.5-mile- (4-km-) wide nucleus of Comet Churyumov-Gerasimenko for more than a year, monitoring the comet as its coma and tails form. Rosetta carries a small lander, Philae, designed to land on the nucleus.

COSMIC SNOWBALLS

1 McNaught
In early 2007, McNaught became the brightest comet since 1965, easily visible with the naked eye—even, for a while, in daylight. Here, McNaught and the Sun are seen setting over the Pacific Ocean. It is not a sight that will be repeated; McNaught is a single-apparition comet that will never return to the inner solar system.

2 Hyakutake
In March 1996, this comet, named after the Japanese amateur astronomer who discovered it, came within 9 million miles (15 million km) of Earth. In May, ESA's Ulysses spacecraft unexpectedly detected Hyakutake's gas tail 355 million miles (570 million km) from the nucleus—the longest comet tail ever detected. Hyakutake was also the first comet observed to emit X-rays.

3 C/2001 Q4 (NEAT)
Discovered in 2001 by NASA's Near-Earth Asteroid Tracking (NEAT) system in Pasadena, CA, C/2001 Q4 was first visible in the southern hemisphere. The comet reached full brightness in May 2004, about 30 million miles (48 million km) from Earth. It will not return; its eccentric orbit will eject it from the solar system.

4 Hale-Bopp
The most widely observed comet of the 20th century, Hale-Bopp was present in the sky for 18 months, its brightness peaking in April 1997. Hale-Bopp's nucleus is unusually large, at 19–25 miles (30–40 km) across. Jupiter's gravity altered the comet's orbital path, reducing its orbital period from around 4,200 years to about 2,500.

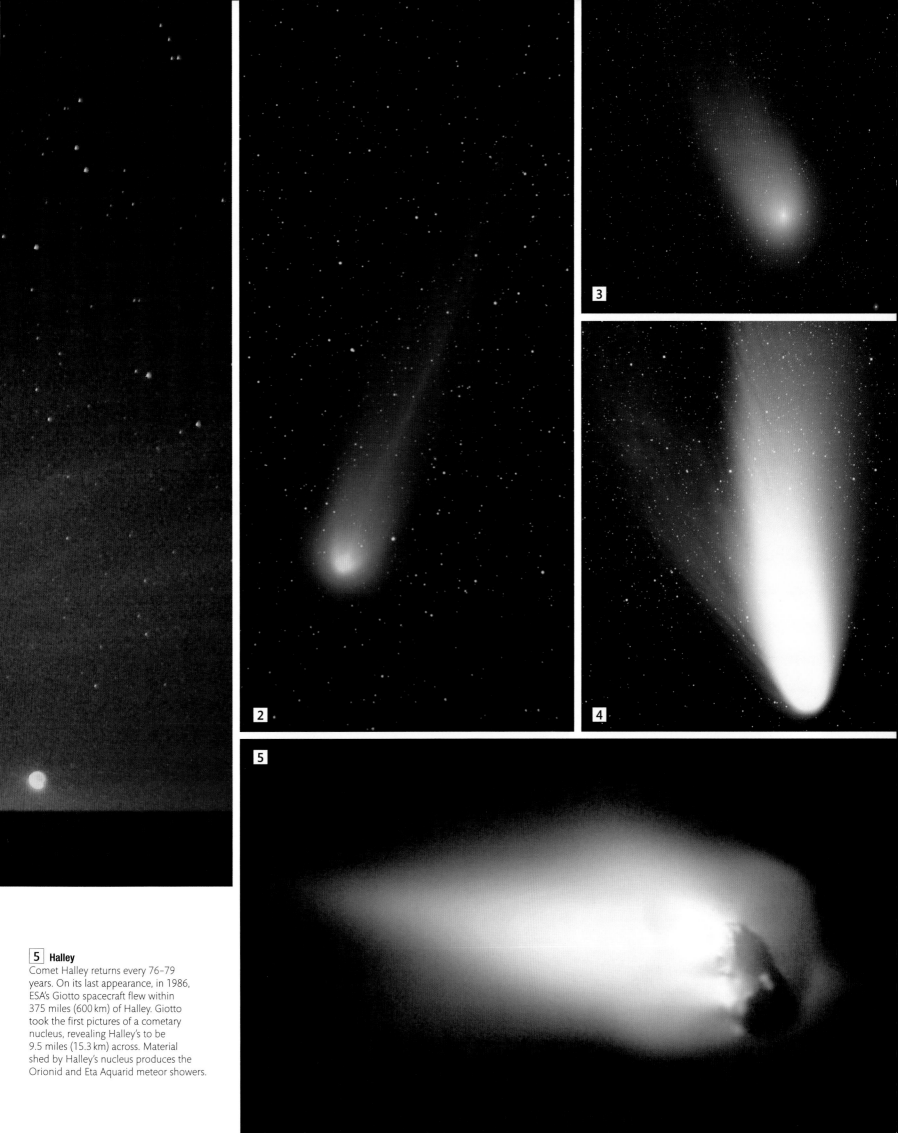

5 | Halley

Comet Halley returns every 76–79 years. On its last appearance, in 1986, ESA's Giotto spacecraft flew within 375 miles (600 km) of Halley. Giotto took the first pictures of a cometary nucleus, revealing Halley's to be 9.5 miles (15.3 km) across. Material shed by Halley's nucleus produces the Orionid and Eta Aquarid meteor showers.

PROPHETS
OF DOOM

ONCE SEEN AS MYSTERIOUS CELESTIAL APPARITIONS OF ILL OMEN, COMETS ARE NOW KNOWN TO BE PRIMORDIAL PLANETARY BUILDING BLOCKS LEFT OVER FROM THE FORMATION OF THE SOLAR SYSTEM.

It was only after English astronomer Edmond Halley realized in the 1690s that certain comets are permanent members of our solar system that astronomers began hunting for comets in the night sky. Cometary masses were found to be insignificant, so the source of the gas and dust in their comas and tails was a mystery. In 1950, American astronomer Fred Whipple proposed that comets have a "dirty snowball" nucleus that loses mass with each orbit of the Sun. A nucleus was seen for the first time in 1986, and in July 2005 the Deep Impact spacecraft became the first craft to make physical contact with a comet nucleus.

Silk Atlas of Comets

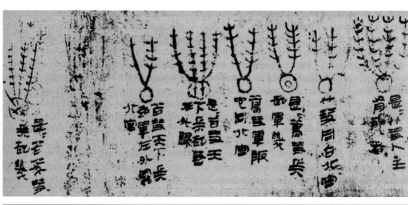

2500 BCE

Earliest observations
Chinese astronomers are convinced that comets are astrologically significant. They monitor the sky for these "broom stars," said to bring bad luck. The 185 BCE *Silk Atlas of Comets* (above), from a tomb in Mawangdui, shows the oldest representations of comets.

5 BCE

The Star of Bethlehem
The biblical star said to have led the Magi to the infant Jesus could have been a planet or a comet. For his nativity fresco in the Arena Chapel, Padua, Italian artist Giotto de Bondone bases his Star of Bethlehem on the 1301 apparition of Comet Halley.

Great Comet photograph by Gill

1900

Formation of tails
Swedish physicist Svante Arrhenius proposes that solar radiation pressure pushes cometary dust into the tail. Fifty years later, astronomers realize the gas tail takes shape as magnetic field lines in the solar wind become draped around this tail, sometimes disconnecting part of it.

1882

Great Comet photographed
Scottish astronomer David Gill takes the first photograph of the Great Comet of 1882, showing background stars through the spectacular tail. American astronomer Edward E. Barnard makes the first cometary discovery by photography—Comet 1892 V.

1868

Chemical make-up
English astronomer William Huggins uses spectroscopy to prove that comets contain hydrocarbon compounds. Spectroscopy also shows that curved comet tails contain dust particles, while straight, bluish tails are ionized, excited molecules from cometary snow.

Jan Oort

Nucleus of Comet Halley

1932

Oort cloud
Estonian astrophysicist Ernst Öpik suggests that long-period comets come from a huge comet cloud surrounding the solar system—now known as the Oort cloud, after Dutch astronomer Jan Oort.

1950

Comet nucleus
American astronomer Fred Whipple suggests that the heart of a comet is a "dirty snowball" nucleus, just a few miles across, made of water ice, snow, and dust. An image of a comet nucleus is later taken in 1986, when the Giotto spacecraft visits Comet Halley.

1979

Comets and life
British astronomers Chandra Wickramasinghe and Fred Hoyle suggest that life arrived on Earth via comets, but others disagree. However, comets often collide with planets: in 1994, astronomers watch Comet Shoemaker-Levy 9 impacting with Jupiter's atmosphere.

Bayeux Tapestry

Edmond Halley

1066

Battle of Hastings

Comets are believed to foretell doom, disease, death, and disaster. Comet Halley is in the sky six months before the death of England's King Harold at the Battle of Hastings. In this scene from the Bayeux Tapestry, soldiers point at the bad omen.

1531

Comet tails

In *Astronomicum Caesareum*, the German astronomer Petrus Apianus shows that comet tails always point away from the Sun. They grow longer as a comet approaches the Sun, then die away as the comet travels out to the colder reaches of the solar system.

1680

Comet orbits

English mathematical genius Isaac Newton is the first to calculate a comet's path. Edmond Halley later calculates more cometary orbits, realizing that a comet he saw in 1682 had been seen before. Halley's Comet returns about every 76 years.

1833 Leonid meteor shower

Caroline Herschel

1866

Comets and meteors

The Italian astronomer Giovanni Schiaparelli realizes that comets and meteoroid streams are related. As a comet decays, dust slowly spreads around its orbit, forming a meteoroid stream. If Earth intersects this stream, we get a meteor storm, such as the Leonids.

1786

Caroline Herschel

British astronomer Caroline Herschel becomes the first woman to discover a comet, using a special telescope made by her astronomer brother William Herschel. Comet Herschel-Rigollet, discovered in 1788, is named after her.

1755

Origin and mass

Prussian philosopher Immanuel Kant suggests that comets are remnants of the planetary formation process. Lexell's Comet comes within 1.4 million miles (2.3 million km) of Earth in 1770. Its mass is calculated as less than 0.02 of Earth's.

Impactor strikes Comet Tempel 1

ESA's Rosetta team celebrate the spacecraft's reawakening

2005

Deep Impact mission

A 815-lb (370-kg) impactor from NASA's Deep Impact spacecraft is launched at Comet Tempel 1 and strikes the nucleus. In February 2011, the comet is visited by the Stardust spacecraft, which takes images of the 500-ft-(150-m-) wide crater formed by the impactor.

2014

Rosetta reawakening

After 31 months in hibernation mode, the ESA's Rosetta spacecraft, launched in 2004 on a trip to Comet Churyumov-Gerasimenko, is successfully reawakened. Rosetta is set to orbit Churyumov-Gerasimenko for 17 months as the comet journeys around the Sun.

WORLDS BEYOND

The Milky Way galaxy, seen here arching over Cape Palliser in New Zealand, may be home to hundreds of billions of planets, but most are impossible to see. The first extrasolar planets (beyond the solar system) were detected in 1992; since then, over 2,000 have been found. Usually invisible to even the most powerful telescopes, they reveal their presence by pulling on their parent star, making it wobble, or by causing a tiny diminution in the star's light as they pass in front. Since large exoplanets close to stars are easiest to detect, most discovered to date are "hot Jupiters"—gas giants that orbit their star in just a few days, typically on a wild, elliptical path. Nevertheless, astronomers have begun to capture faint but tantalizing images of exoplanets, and just a few appear to have water in their atmospheres, making them possible habitats for life. The search is now on for Earth's twin—a small, rocky world similar to our own.

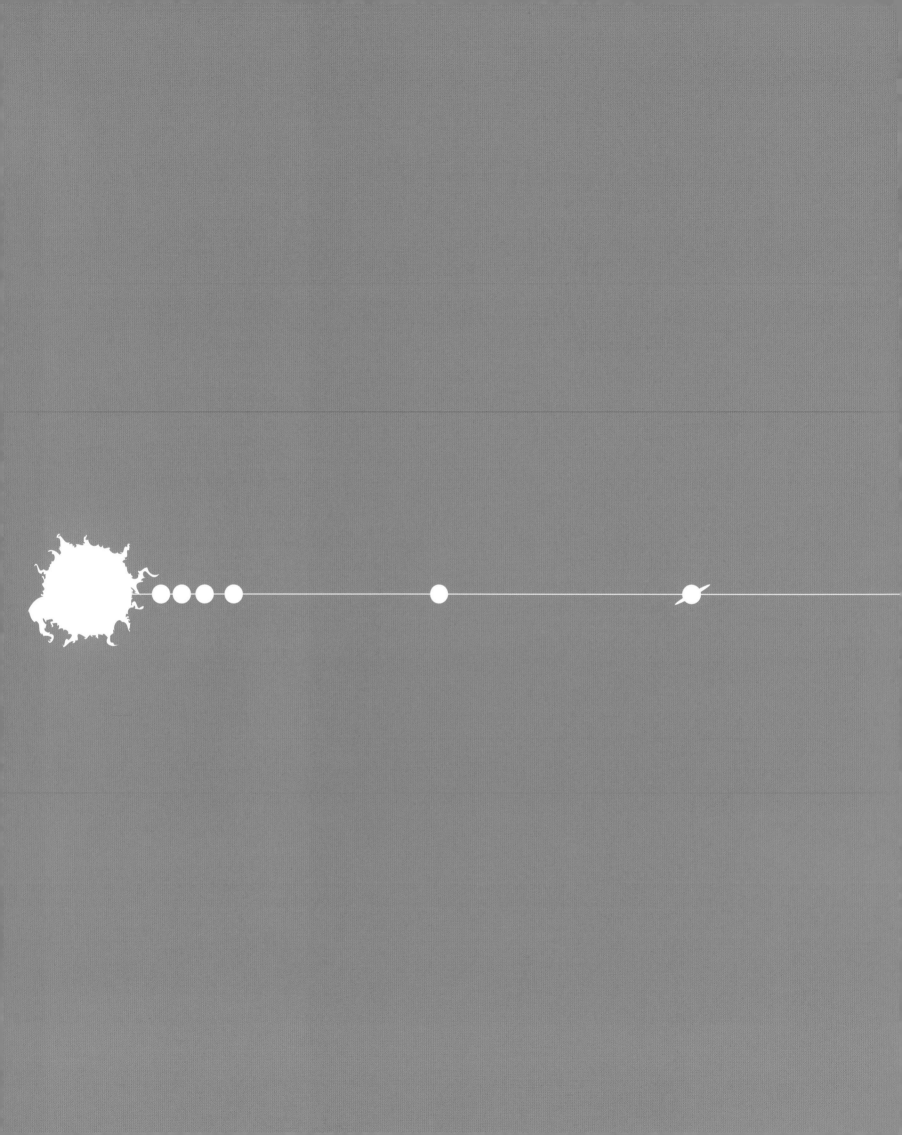

REFERENCE

THE **SOLAR SYSTEM** AND **ITS PLANETS**

Elements in the solar system

In the Sun, the elements exist as atoms. Moving away from the Sun, temperatures fall, and the atoms combine to form larger molecules. Hydrogen and oxygen combine to form water; carbon and oxygen form carbon dioxide; carbon and hydrogen form methane; and iron, silicon magnesium, and oxygen form various rock minerals. Hydrogen and helium dominate the solar system. Other elements make up only 1.9 percent of the solar system's total mass.

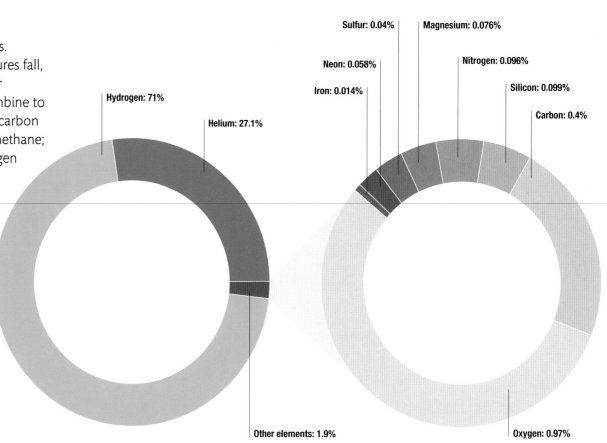

Sulfur: 0.04% Magnesium: 0.076%

Neon: 0.058% Nitrogen: 0.096%

Iron: 0.014% Silicon: 0.099%

Hydrogen: 71% Carbon: 0.4%

Helium: 27.1%

Other elements: 1.9% Oxygen: 0.97%

Planet data

There are eight planets in the solar system. They range in size from Mercury, with a diameter a third of Earth's, to Jupiter, which is 11 times wider than Earth. The four planets closest to the Sun are Mercury, Venus, Earth, and Mars. They are small, dense, rocky bodies with solid surfaces and very few satellites. Earth is the only one with a wet surface. The outer four planets—Jupiter, Saturn, Uranus, and Neptune—are large bodies with substantial cores made of rock and metal, surrounded by very thick atmospheres and many satellites. The cold surfaces of the gas giants are essentially their cloud tops. In the table below, the radius given is the mean for the rocky planets and equatorial for the gas giants. Gravity is surface gravity for the rocky planets and gravity at the equator for the gas giants.

	Mercury	Venus	Earth	Mars	Jupiter	Saturn	Uranus	Neptune
Radius	1,516 miles (2,440 km)	3,761 miles (6,052 km)	2,963 miles (6,378 km)	2,110 miles (3,396 km)	44,423 miles (71,492 km)	37,448 miles (60,268 km)	15,881 miles (25,559 km)	15,387 miles (24,764 km)
Topographic range	6 miles (10 km)	9 miles (15 km)	12 miles (20 km)	19 miles (30 km)	–	–	–	–
Mass (Earth equals 1)	0.06	0.82	1	0.11	317.83	95.16	14.54	17.15
Density	5,427 kg/m³	5,243 kg/m³	5,514 kg/m³	3,933 kg/m³	1,326 kg/m³	687 kg/m³	1,271 kg/m³	1,638 kg/m³
Flattening	0	0	0.00335	0.00589	0.06487	0.09796	0.0229	0.0171
Rotation period	1,407.6 hours	5,832.5 hours	23.9 hours	24.6 hours	9.9 hours	10.7 hours	17.2 hours	16.1 hours
Solar day (sunrise to sunrise)	4,222.6 hours	2,802.0 hours	24.0 hours	24.7 hours	9.9 hours	10.7 hours	17.2 hours	16.1 hours
Gravity (Earth = 1)	0.38	0.91	1	0.38	2.36	1.02	0.89	1.12
Axial tilt	0.01°	2.6°	23.4°	25.2°	3.1°	26.7°	82.2°	28.3°
Escape velocity	9,619 mph (15,480 km/h)	23,174 mph (37,296 km/h)	25,022 mph (40,270 km/h)	11,251 mph (18,108 km/h)	133,098 mph (214,200 km/h)	79,411 mph (127,800 km/h)	47,646 mph (76,680 km/h)	52,568 mph (84,600 km/h)
Apparent magnitude	−2.6 to 5.7	−4.9 to −3.8	–	+1.6 to −3	−1.6 to −2.94	+1.47 to −0.241	5.9 to 5.32	8.02 to 7.78
Average temperature	333°F (167°C)	880°F (470°C)	59°F (15°C)	−81°F (−63°C)	−162°F (−108°C)	−218°F (−139°C)	−323°F (−197°C)	−328°F (−201°C)
Number of moons	0	0	1	2	67+	62+	27+	14+

Planetary orbits

The planet's orbits are governed by the Sun's gravitational field. Initially it was thought that the planets followed circular orbits around the Sun, but in the early 17th century, German mathematician and astronomer Johannes Kepler discovered that they follow non-circular, elliptical orbits, with two focal points.

The time it takes each planet to go around the Sun once is known as the orbital period. This period increases greatly with distance—Mercury, the innermost planet, takes only 88 days to orbit the Sun, while Neptune, the most distant from the Sun, takes 165 years. The closest point of each planet's orbit to the Sun is known as perihelion, and the farthest as aphelion.

The solar system is almost flat, with all the planets orbiting the Sun in approximately the same plane. Each planet's orbital plane is slightly tilted relative to Earth's, and the angle between the two is known as orbital inclination.

While all planets follow elliptical orbits, they are not all exactly the same elliptical shape. The extent to which an orbit deviates from a circle is known as eccentricity. An eccentricity of zero indicates a perfectly circular orbit.

	Mercury	Venus	Earth	Mars	Jupiter	Saturn	Uranus	Neptune
Perihelion	28.6 million miles (46.0 million km)	66.8 million miles (107.5 million km)	91.4 million miles (147.1 million km)	128.4 million miles (206.6 million km)	460.2 million miles (740.5 million km)	840.5 million miles (1,352.6 million km)	1,703.4 million miles (2,741.3 million km)	2,761.7 million miles (4,444.5 million km)
Aphelion	43.4 million miles (69.8 million km)	67.7 million miles (108.9 million km)	94.5 million miles (152.1 million km)	154.8 million miles (249.2 million km)	507.4 million miles (818.6 million km)	941.1 million miles (1,514.5 million km)	1,866.4 million miles (3,003.6 million km)	2,824.6 million miles (4,545.7 million km)
Orbital period	87.969 days	224.701 days	365.256 days	686.980 days	4,332.589 days	10,759.22 days	30,685.4 days	60,189 days
Orbital velocity	29.75 miles/sec (47.87 km/sec)	21.76 miles/sec (35.02 km/sec)	18.50 miles/sec (29.78 km/sec)	14.99 miles/sec (24.13 km/sec)	8.12 miles/sec (13.07 km/sec)	6.02 miles/sec (9.69 km/sec)	4.23 miles/sec (6.81 km/sec)	3.37 miles/sec (5.43 km/sec)
Orbital inclination	7.0°	3.39°	0°	1.850°	1.304°	2.485°	0.772°	1.769°
Eccentricity	0.206	0.007	0.017	0.094	0.049	0.057	0.046	0.011

MERCURY
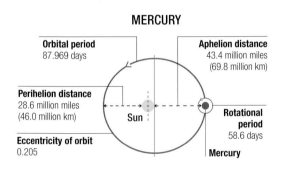

Orbital period
87.969 days

Aphelion distance
43.4 million miles
(69.8 million km)

Perihelion distance
28.6 million miles
(46.0 million km)

Sun

Rotational period
58.6 days

Eccentricity of orbit
0.205

Mercury

VENUS
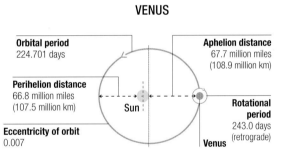

Orbital period
224.701 days

Aphelion distance
67.7 million miles
(108.9 million km)

Perihelion distance
66.8 million miles
(107.5 million km)

Sun

Rotational period
243.0 days
(retrograde)

Eccentricity of orbit
0.007

Venus

EARTH
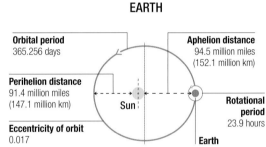

Orbital period
365.256 days

Aphelion distance
94.5 million miles
(152.1 million km)

Perihelion distance
91.4 million miles
(147.1 million km)

Sun

Rotational period
23.9 hours

Eccentricity of orbit
0.017

Earth

MARS
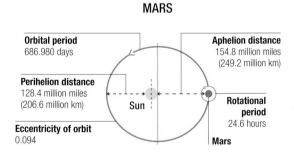

Orbital period
686.980 days

Aphelion distance
154.8 million miles
(249.2 million km)

Perihelion distance
128.4 million miles
(206.6 million km)

Sun

Rotational period
24.6 hours

Eccentricity of orbit
0.094

Mars

JUPITER
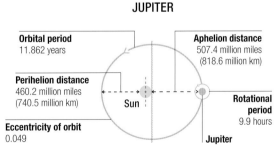

Orbital period
11.862 years

Aphelion distance
507.4 million miles
(818.6 million km)

Perihelion distance
460.2 million miles
(740.5 million km)

Sun

Rotational period
9.9 hours

Eccentricity of orbit
0.049

Jupiter

SATURN
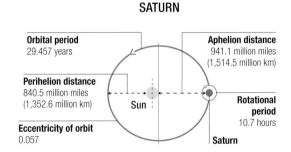

Orbital period
29.457 years

Aphelion distance
941.1 million miles
(1,514.5 million km)

Perihelion distance
840.5 million miles
(1,352.6 million km)

Sun

Rotational period
10.7 hours

Eccentricity of orbit
0.057

Saturn

URANUS
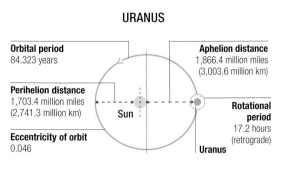

Orbital period
84.323 years

Aphelion distance
1,866.4 million miles
(3,003.6 million km)

Perihelion distance
1,703.4 million miles
(2,741.3 million km)

Sun

Rotational period
17.2 hours
(retrograde)

Eccentricity of orbit
0.046

Uranus

NEPTUNE
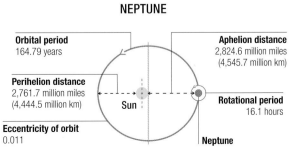

Orbital period
164.79 years

Aphelion distance
2,824.6 million miles
(4,545.7 million km)

Perihelion distance
2,761.7 million miles
(4,444.5 million km)

Sun

Rotational period
16.1 hours

Eccentricity of orbit
0.011

Neptune

Scientists are hunting for **new planets** in our solar system, **beyond Neptune.**

ANATOMY OF
THE PLANETS

Interiors

Temperature, density, and pressure increase toward the center of a planet. While the crust of a rocky planet is solid, deeper materials are more viscous or even molten. Circulation of these fluid materials allows heavier compounds and elements, such as metals, to sink toward the center, forming a core, while more buoyant materials, such as rocky minerals, rise. In some planets, such as Earth, the great pressure in the center produces a solid core, inside a liquid-metal outer core. Currents in this outer core generate the magnetic field that surrounds the planet. The gas giants consist mostly of gases such as hydrogen, helium, methane, and ammonia, but they may also have cores of rock and metal, compressed into solid form by the colossal weight of the material surrounding them.

	Mercury	Venus	Earth	Mars	Jupiter	Saturn	Uranus	Neptune
Radius	1,516 miles (2,440 km)	3,761 miles (6,052 km)	3,963 miles (6,378 km)	2,110 miles (3,396 km)	44,423 miles (71,492 km)	37,449 miles (60,268 km)	15,882 miles (25,559 km)	15,389 miles (24,766 km)
Mean density	5,427 kg/m³	5,204 kg/m³	5,515 kg/m³	3,396 kg/m³	1,326 kg/m³	687 kg/m³	1,318 kg/m³	1,638 kg/m³
Crust thickness	93 miles (150 km)	31 miles (50 km)	19 miles (30 km)	28 miles (45 km)	– –	– –	– –	– –
Central pressure	0.4Mbar	3Mbar	3.6Mbar	0.4Mbar	80Mbar	50Mbar	20Mbar	20Mbar
Central temperature	2,000K	5,000K	6,000K	2,000K	20,000K	11,000K	7,000K	7,000K

Magnetic fields

For a planet to have a significant magnetic field, it must have a molten, metallic inner region that is spinning rapidly. Earth has a larger magnetic field than Venus because the planet rotates more than 200 times faster than Venus. Mars and Mercury have very small magnetic fields because their metallic cores are solid. Jupiter and Saturn have very large magnetic fields; they spin twice as fast as Earth and have a large zone of liquid metallic hydrogen around the core. Magnetic moment is a measure of the magnetic field's strength.

	Mercury	Venus	Earth	Mars	Jupiter	Saturn	Uranus	Neptune
Magnetic moment (Earth = 1)	0.0007	<0.0004	1	<0.000025	20,000	600	50	25
Angle between magnetic and rotation axis	14°	–	10.8°	–	–9.6°	–1°	–59°	–47°
Magnetic axis offset from planet center (planetary radii)	–	–	0.08	–	0.12	0.04	0.3	0.55
Distance to nearest edge of magnetic field (planetary radii)	1.5	–	11	–	80	20	20	25

Ring systems

All the gas giants have ring systems, but only Saturn's rings are bright enough to see through a small telescope. The rings of Jupiter, Uranus, and Neptune are extremely faint and can be observed only by using large infrared telescopes or spacecraft. Ring systems are found in a planet's equatorial plane and are usually close to the planet, where the powerful gravitational field disrupts the accretion of debris into a significant satellite.

	Jupiter	Saturn	Uranus	Neptune
Radius (planet radius = 1)	1.4–3.8	1.09–8	1.55–3.82	1.7–2.54
Radius in miles/km	62,137–167,770 miles (100,000–270,000 km)	41,570–298,755 miles (66,900–480,000 km)	24,606–60,708 miles (39,600–97,700 km)	26,098–39,084 miles (42,000–62,900 km)
Thickness	18.6–186 miles (30–300 km)	<0.6 miles (<1 km)	0.09 miles (0.15 km)	Unknown
Particle size	< 0.00004 in (< 0.001 mm)	0.0004 in – 30 ft (0.01–10 m)	0.00004 in – 30 ft (0.001–10 m)	Unknown
Diameter of moon of equivalent mass	6 miles (10 km)	280 miles (450 km)	6 miles (10 km)	6 miles (10 km)

Saturn's rings from Cassini spacecraft

Atmospheres of rocky planets

All planets have atmospheres, though Mercury's is very thin and is continually blown away by the solar wind. The atmospheres of the terrestrial planets formed from gases released by the crust. High surface temperatures can cause atmospheric gases to escape into space, but the rate of escape is lower on more massive planets, which have a more powerful gravitational field. The low mass of Mercury, coupled with its high temperature, explains why it has such a small atmosphere. The water that once existed on Venus has escaped because the planet is so hot. Earth's atmosphere is influenced by plant life, which removes carbon dioxide and releases oxygen in a cycle not known to occur elsewhere.

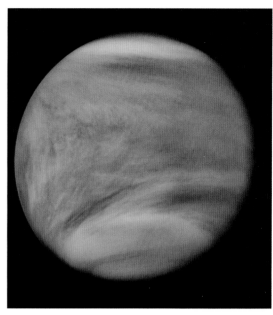

Venus's cloudy atmosphere, seen from Pioneer orbiter

	Mercury	Venus	Earth	Mars
Surface pressure	<0.00001mbar	92,000mbar	1,014mbar	6.4mbar
Pressure variability	0mbar	0mbar	870–1,085mbar	4.0–8.7mbar
Surface density	–	65 kg/m³	1.217 kg/m³	0.02 kg/m³
Average temperature	332.6°F (167°C)	867.2°F (464°C)	59°F (15°C)	–81.4°F (–63°C)
Temperature range	–290 to 800°F (–180 to 430°C)	32°F (0°C)	–130 to 122°F (–90 to 50°C)	–225 to 95°F (–143 to 35°C)
Wind speeds	– –	0.98 to 3.28 ft/s (0.3 to 1.0 m/s)	0 to 328 ft/s (0 to 10 m/s)	6.56 to 98.4 ft/s (2 to 30 m/s)

Atmospheric gases

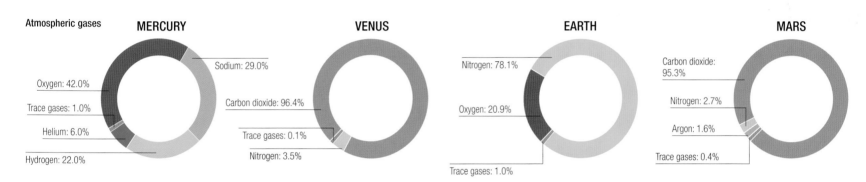

MERCURY
Sodium: 29.0%
Oxygen: 42.0%
Trace gases: 1.0%
Helium: 6.0%
Hydrogen: 22.0%

VENUS
Carbon dioxide: 96.4%
Trace gases: 0.1%
Nitrogen: 3.5%

EARTH
Nitrogen: 78.1%
Oxygen: 20.9%
Trace gases: 1.0%

MARS
Carbon dioxide: 95.3%
Nitrogen: 2.7%
Argon: 1.6%
Trace gases: 0.4%

Atmospheres of gas giants

The gas giant planets captured their huge hydrogen and helium atmospheres during their formation, and they were cold enough and massive enough to retain these atmospheres. The apparent surfaces of Jupiter and Saturn are upper-atmosphere clouds colored by ammonia and other compounds. Uranus and Neptune have a few percent of methane in their atmospheres, which gives them their bluish green color.

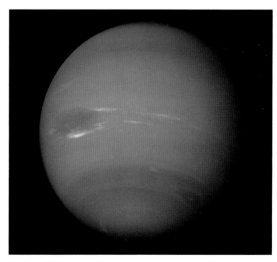

Neptune from Voyager 2

	Jupiter	Saturn	Uranus	Neptune
Temperature at 1 bar pressure	–162.4°F (–108°C)	–218.2°F (–139°C)	–322.6°F (–197°C)	–329.8°F (–201°C)
Temperature at 0.1 bar	–257.8°F (–161°C)	–308.2°F (–189°C)	–364°F (–220°C)	–360.4°F (–218°C)
Density at 1 bar	0.16 kg/m³	0.19 kg/m³	0.42 kg/m³	0.45 kg/m³
Wind speed	0–492 ft/sec (0–150 m/sec)	0–1,312 ft/sec (0–400 m/sec)	0–820 ft/sec (0–250 m/sec)	0–1,902 ft/sec (0–580 m/sec)

Atmospheric gases

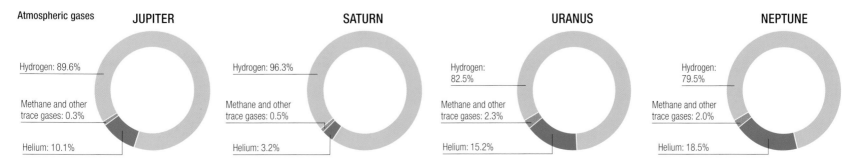

JUPITER
Hydrogen: 89.6%
Methane and other trace gases: 0.3%
Helium: 10.1%

SATURN
Hydrogen: 96.3%
Methane and other trace gases: 0.5%
Helium: 3.2%

URANUS
Hydrogen: 82.5%
Methane and other trace gases: 2.3%
Helium: 15.2%

NEPTUNE
Hydrogen: 79.5%
Methane and other trace gases: 2.0%
Helium: 18.5%

MOONS, ASTEROIDS, AND COMETS

Major moons

Mercury and Venus are the only planets without satellites. The other rocky planets have very few moons—Earth has the Moon, which is one-eightieth of Earth's mass, and Mars has two very small satellites, thought to be asteroids captured from the nearby asteroid belt. In contrast, the gas giants have large numbers of satellites and more are likely to be discovered. Most are small, but the four main moons of Jupiter are all large and bright enough to be seen from Earth through binoculars. Orbital inclination is the difference between the plane of a moon's orbit and that of the planet's equator. Eccentricity is the extent to which an elliptical orbit deviates from a circle, where zero is a perfect circle.

Planet	Moon	Diameter	Density	Escape velocity	Orbital period	Surface temperature	Orbital inclination	Orbital eccentricity	Date of discovery	Discoverer
Earth	The Moon	2,157 miles (3,472 km)	3,346 kg/m³	2.38 km/sec	27.322 days	−274 to 248°F (−170 to 120°C)	5.145°	0.0549	–	–
Mars	Deimos	7.8 miles (12.6 km)	1,471 kg/m³	0.0056 km/sec	1.2624 days	−40°F (−40°C)	0.93°	0.00033	1877	Hall
	Phobos	13.8 miles (22.2 km)	1,876 kg/m³	0.0114 km/sec	0.3189 days	−40°F (−40°C)	1.093°	0.0151	1877	Hall
Jupiter	Io	2,264 miles (3,644 km)	3,528 kg/m³	2.558 km/sec	1.769 days	−292 to −220°F (−180 to −140°C)	0.050°	0.0041	1610	Galileo
	Europa	1,940 miles (3,122 km)	3,010 kg/m³	2.025 km/sec	3.551 days	−364 to −238°F (−220 to −150°C)	0.471°	0.0094	1610	Galileo
	Ganymede	3,270 miles (5,262 km)	1,936 kg/m³	2.741 km/sec	7.154 days	−328 to −184°F (−200 to −120°C)	0.204°	0.0011	1610	Galileo
	Callisto	2,996 miles (4,821 km)	1,834 kg/m³	2.440 km/sec	16.689 days	−310 to −166°F (−190 to −110°C)	0.205°	0.0074	1610	Galileo
	Amalthea	104 miles (167 km)	857 kg/m³	0.058 km/sec	0.49818 days	−238°F (−150°C)	0.374°	0.0032	1892	Barnard
	Himalia	106 miles (170 km)	2,000 kg/m³	0.1 km/sec	250.2 days	−238°F (−150°C)	30.486°	0.1513	1904	Perrine
Saturn	Mimas	246 miles (396 km)	1,148 kg/m³	0.159 km/sec	0.942 days	−346°F (−210°C)	1.566°	0.0202	1789	Herschel
	Enceladus	313 miles (504 km)	1,609 kg/m³	0.239 km/sec	1.370 days	−400 to −202°F (−240 to −130°C)	0.010°	0.0047	1789	Herschel
	Tethys	660 miles (1,062 km)	984 kg/m³	0.394 km/sec	1.887 days	−310°F (−190°C)	0.168°	0.0001	1684	Cassini
	Dione	698 miles (1,123 km)	1,478 kg/m³	0.51 km/sec	2.737 days	−301°F (−185°C)	0.002°	0.0022	1684	Cassini
	Rhea	949 miles (1,527 km)	1,236 kg/m³	0.635 km/sec	4.518 days	−364 to −283°F (−220 to −175°C)	0.327°	0.00126	1672	Cassini
	Titan	3,201 miles (5,151 km)	1,880 kg/m³	2.639 km/sec	15.945 days	−292°F (−180°C)	0.3485°	0.0288	1655	Huygens
	Iapetus	912 miles (1,468 km)	1,088 kg/m³	0.573 km/sec	79.32 days	−292 to −220°F (−180 to −140°C)	15.47°	0.286	1671	Cassini
	Phoebe	132 miles (213 km)	1,638 kg/m³	0.1 km/sec	−545.09 days	Unknown	173.04°	0.156	1899	Pickering
Uranus	Miranda	293 miles (471 km)	1,200 kg/m³	0.079 km/sec	1.4135 days	−346°F (−210°C)	1.232°	0.0013	1948	Kuiper
	Ariel	720 miles (1,158 km)	1,660 kg/m³	0.558 km/sec	2.5204 days	−346°F (−210°C)	0.260°	0.0012	1851	Lassell
	Umbriel	726 miles (1,169 km)	1,390 kg/m³	0.52 km/sec	4.1442 days	−328°F (−200°C)	0.205°	0.0039	1851	Lassell
	Titania	980 miles (1,577 km)	1,711 kg/m³	0.773 km/sec	8.7059 days	−328°F (−200°C)	0.340°	0.0011	1787	Herschel
	Oberon	946 miles (1,523 km)	1,630 kg/m³	0.726 km/sec	13.463 days	−328°F (−200°C)	0.058°	0.0014	1787	Herschel
	Portia	84 miles (135 km)	1,300 kg/m³	0.058 km/sec	0.5132 days	−346°F (−210°C)	0.059°	0.00005	1986	Synnott
Neptune	Proteus	261 miles (420 km)	1,300 kg/m³	0.17 km/sec	1.122 days	−364°F (−220°C)	0.075°	0.0005	1989	Voyager team
	Triton	1,682 miles (2,707 km)	2,061 kg/m³	1.455 km/sec	−5.877 days	−391°F (−235°C)	156.885°	0.00006	1846	Lassell
	Nereid	211 miles (340 km)	1,500 kg/m³	0.156 km/sec	360.14 days	−364°F (−220°C)	7.090°	0.7507	1949	Kuiper
	Larissa	121 miles (194 km)	0.076 kg/m³	0.076 km/sec	0.555 days	−364°F (−220°C)	0.205°	0.0014	1981	Reitsema
	Galatea	109 miles (176 km)	0.0556 kg/m³	0.0556 km/sec	0.429 days	−364°F (−220°C)	0.34°	0.0001	1989	Voyager team

Asteroid Belt

Between the orbits of Mars and Jupiter lie a huge number of rocky and metallic bodies called asteroids. Asteroids orbit the Sun in the same way as the planets, but they are smaller and mostly irregular in shape.

This table lists the largest asteroids in the Asteroid Belt, starting with Ceres, which has sufficient mass to form a spherical shape and is therefore classed as a dwarf planet as well as an asteroid. Brightness is apparent magnitude from Earth.

Name	Diameter	Dimensions	Rotation period	Brightness
1 Ceres	592 miles (952 km)	606 x 606 x 565 miles (975 x 975 x 909 km)	0.3781 days	6.64 – 9.34
2 Pallas	338 miles (544 km)	362 x 345 x 311 miles (582 x 556 x 500 km)	0.3256 days	6.49 – 10.65
4 Vesta	326 miles (525 km)	356 x 346 x 277 miles (573 x 557 x 446 km)	0.2226 days	5.1 – 8.48
10 Hygiea	268 miles (431 km)	329 x 253 x 230 miles (530 x 407 x 370 km)	1.15 days	9.0 – 11.97
704 Interamnia	203 miles (326 km)	217 x 189 miles (350 x 304 km)	0.364 days	9.9 – 13.0
52 Europa	196 miles (315 km)	236 x 205 x 155 miles (380 x 330 x 250 km)	0.2347 days	–
511 Davida	180 miles (289 km)	222 x 183 x 144 miles (357 x 294 x 231 km)	0.2137 days	9.5 – 12.98
87 Sylvia	178 miles (286 km)	239 x 165 x 143 miles (385 x 265 x 230 km)	0.2160 days	–
65 Cybele	170 miles (273 km)	188 x 180 x 144 miles (302 x 290 x 232 km)	0.1683 days	10.67 –13.64
15 Eunomia	167 miles (268 km)	222 x 158 x 132 miles (357 x 255 x 212 km)	0.2535 days	7.9 – 11.24
3 Juno	160 miles (258 km)	199 x 166 x 124 miles (320 x 267 x 200 km)	0.3004 days	7.4 – 11.55
31 Euphrosyne	159 miles (256 km)	Unknown	0.2305 days	10.16 – 13.61
624 Hector	150 miles (241 km)	230 x 166 x 124 miles (370 x 267 x 200 km)	0.2884 days	13.79 – 15.26
88 Thisbe	144 miles (232 km)	137 x 125 x 104 miles (221 x 201 x 168 km)	0.2517 days	–
324 Bamberga	142 miles (229 km)	Unknown	1.226 days	–
451 Patientia	140 miles (225 km)	Unknown	0.4053 days	–
532 Herculina	138 miles (222 km)	Unknown	0.3919 days	8.82 – 11.99
48 Doris	138 miles (222 km)	173 x 88 miles (278 x 142 km)	0.4954 days	–

Periodic comets

Comets that have been observed returning to the Sun more than once are known as periodic comets. Short-period comets are those that orbit the Sun in fewer than 200 years. Those with orbital periods less than 20 years are known as Jupiter-family comets; they are kept in the inner solar system by Jupiter's gravity and reach their aphelion (maximum distance from the Sun) close to Jupiter's orbit. They also have low inclinations, while longer-period comets have random inclinations.

Name	Orbital period	Sightings	Next due
1P/Halley	75.32 years	30	Jul 2061
2P/Encke	3.30 years	62	Mar 2017
3D/Beila	6.619 years	6	–
6P/d'Arrest	6.54 years	20	Mar 2015
9P/Tempel 1	5.52 years	12	Aug 2016
17P/Holmes	6.883 years	10	Mar 2014
21P/Giacobini-Zinner	6.621 years	15	Sep 2018
29P/Schwassmann-Wachmann 1	14.65 years	7	Mar 2019
39P/Oterma	19.43 years	4	Jul 2023
46P/Wirtanen	5.44 years	10	Dec 2018
50P/Arend	8.27 years	8	Feb 2016
55P/Tempel-Tuttle	33.22 years	5	May 2031
67P/Churyumov Gerasimenko	6.45 years	7	Aug 2015
81P/Wild 2	6.408 years	6	Jul 2016
109P/Swift-Tuttle	133.3 years	5	Jul 2126

Great comets

Every ten years or so, a comet appears in the night sky that is so bright it is easily visible to the naked eye for a period of several weeks. These comets are unpredictable because they are long-period comets with orbital periods of hundreds to many thousands of years. In this table, brightness represents apparent magnitude, and the comet's closest approach to Earth is given in astronomical units, where 1 AU is equal to the distance between Earth and the Sun.

Name	Year	Brightness	Closest to Earth
Great Comet	1811	0	1.22 AU
Great March Comet	1843	< −3	0.84 AU
Donati's Comet	1858	0.5	0.54 AU
Great Comet	1861	0	0.13 AU
Coggia	1874	0.5	0.29 AU
Great September Comet	1882	< −3	0.99 AU
Great Comet	1901	1	0.83 AU
Great January Comet	1910	1.5	0.86 AU
Skjellerup-Maristany	1927	1	0.75 AU
Arend-Roland	1957	−0.5	0.57 AU
Seki-Lines	1962	−2	0.62 AU
Ikeya-Seki	1965	2	0.91 AU
Bennett	1970	0.5	0.69 AU
West	1976	−1	0.79 AU
Hyakutake	1996	1.5	0.10 AU
Hale-Bopp	1997	−0.7	1.32 AU
McNaught	2007	-6	0.82 AU

Meteor showers

Showers of meteors are seen regularly each year as Earth travels through the trail of debris left behind by any of several decaying "parent" comets. Dust particles, or meteoroids, from the comets burn up in the upper 60–45 miles (100–75 km) of the atmosphere, producing long, thin, short-lived tubes of excited, ionized gas molecules. These brilliant streaks of light are informally known as shooting stars.

The meteors from a specific shower all appear to radiate from the same point in the sky. This spot is known as the radiant, and meteor showers are named after the constellation in which the radiant is located.

The "most meteors" column in the table below gives the number of meteors that would be seen per hour during the shower's peak, if the radiant was directly overhead.

Name	Peak	Speed	Most meteors	Parent comet
Quadrantids	Jan 4	25.5 miles/sec (41 km/sec)	120 per hour	C/1490 Y1
Lyrids	Apr 22	30 miles/sec (48 km/sec)	10 per hour	C/1861 G1
Eta Aquarids	May 5	41 miles/sec (66 km/sec)	30 per hour	Halley
Arietids	Jun 7	23 miles/sec (37 km/sec)	54 per hour	69PMachholz
Zeta Perseids	Jun 9	18 miles/sec (29 km/sec)	20 per hour	2P/Encke
Delta Aquarids	Jul 29	25.5 miles/sec (41 km/sec)	16 per hour	Marsden/Kracht
Perseids	Aug 13	36 miles/sec (58 km/sec)	80 per hour	Swift-Tuttle
Draconids	Oct 8	12.4 miles/sec (20 km/sec)	Variable	Unknown
Orionids	Oct 21	41.6 miles/sec (67 km/sec)	25 per hour	Halley
Leonids	Nov 17	41.1 miles/sec (71 km/sec)	Variable	55P/Tempel-Tuttle
Geminids	Dec 13	21.7 miles/sec (35 km/sec)	75 per hour	3200 Phaethon
Ursids	Dec 23	20.5 miles/sec (33 km/sec)	10 per hour	8P/Tuttle

The Leonid meteor shower over Joshua Tree National Park, California

EXPLORING **SPACE**

Landmark missions

The Space Age has revolutionized our understanding of the solar system. Distant planets that were once no more than fuzzy disks in a telescope have now been scrutinized from close quarters and mapped in detail. The earliest space missions were merely brief flybys. Then came orbiting spacecraft, atmospheric probes, landers, and finally rovers. This table lists key missions, some of which are ongoing.

Mission	Country of origin	Launch date	Target	Mission type	Achievement
Mariner 2	USA	Aug 2, 1962	Venus	Flyby	First data from Venus's atmosphere
Mariner 4	USA	Nov 28, 1964	Mars	Flyby	First close-up photographs of Martian surface
Venera 7	USSR	Aug 17, 1970	Venus	Lander	First soft landing
Mariner 9	USA	May 30, 1971	Mars	Orbiter	Extensive photography
Pioneer 10	USA	Mar 3, 1972	Jupiter	Flyby	Extensive data and photographs
Pioneer 11	USA	Apr 6, 1973	Jupiter/Saturn	Flyby	First flyby of Saturn; discovery of F ring
Mariner 10	USA	Nov 3, 1973	Mercury	Flyby	First close-up images of Mercury's surface
Venera 9	USSR	June 8, 1975	Venus	Lander/orbiter	First images returned from Venus's surface
Viking 1	USA	Aug 20, 1975	Mars	Lander	Surface experiments and search for life
Voyager 2	USA	Aug 20, 1977	Outer planets	Flyby	Passes Jupiter and Saturn; makes first flybys of Uranus (Jan 24, 1986) and Neptune (Aug 24, 1989)
Voyager 1	USA	Sept 5, 1977	Jupiter/Saturn	Flyby	Detailed investigation
Venera 11/12/13/14	USSR	1978–1981	Venus	Landers/flybys	Images from surface of Venus
Magellan	USA	May 5, 1989	Venus	Orbiter	Maps entire surface of Venus using radar
Galileo	USA	Oct 18, 1989	Jupiter	Orbiter	Long-term observations of Jovian system; investigation of atmosphere with descent probe
Mars Pathfinder	USA	Dec 2, 1996	Mars	Lander	Releases Sojourner, the first rover to operate on Mars
Cassini	USA/others	Oct 15, 1997	Saturn	Orbiter	Ongoing; extensive collection of data and images
Huygens	ESA	Oct 15, 1997	Titan	Lander	First probe to land on a gas giant moon
Mars Express	ESA	June 2, 2003	Mars	Orbiter	Finds evidence for previous existence of surface water
Mars Exploration	USA	June 10, 2003	Mars	Lander	Releases Spirit rover; first drilling of Martian rock
Mars Exploration	USA	July 7, 2003	Mars	Lander	Releases Opportunity rover; close-up soil investigations
MESSENGER	USA	Aug 3, 2004	Mercury	Orbiter	First spacecraft to orbit Mercury
Mars Reconnaissance	USA	Aug 12, 2005	Mars	Orbiter	Investigates history of water on Mars
Venus Express	ESA	Nov 9, 2005	Venus	Orbiter	Investigates Venus's atmosphere
New Horizons	USA	Jan 19, 2006	Pluto	Flyby	First attempted flyby of Pluto
Mars Phoenix	USA	Aug 4, 2007	Mars	Lander	Polar exploration, searches for life and water
Mars Science Lab	USA	Nov 26, 2011	Mars	Rover	Curiosity rover investigates climate and geology

Launch at Cape Canaveral of Atlas V rocket carrying Curiosity rover, November 2011

World's largest rockets

Saturn V, used in the Apollo project of the 1960s and 70s that sent astronauts to the Moon, was the largest rocket ever built. Launches of its Soviet rival, the N1, were attempted multiple times but each ended in disaster.

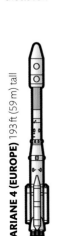
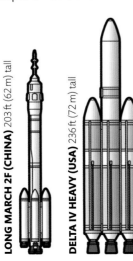
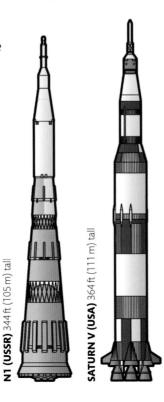

ARIANE 4 (EUROPE) 193 ft (59 m) tall

LONG MARCH 2F (CHINA) 203 ft (62 m) tall

DELTA IV HEAVY (USA) 236 ft (72 m) tall

N1 (USSR) 344 ft (105 m) tall

SATURN V (USA) 364 ft (111 m) tall

Launch sites

Around 20 spaceflight launch sites have been constructed worldwide. Among the most important are Kennedy Space Center on Cape Canaveral and Baikonur in Kazakhstan. Sites closer to the equator can launch heavier cargo, because rockets taking off from there are given a final boost by Earth's spin.

Vandenberg Air Force Base, USA

Cape Canaveral, USA

Guiana Space Centre, French Guiana

Baikonur, Kazakhstan

Tanegashima Space Center, Japan

Xichang, China

Satish Dhawan Space Centre, India

Major launch sites

Satellites

Since the 1950s, thousands of satellites have been launched into orbit around Earth for observation and research. Many of these are now no longer functional. Some of the most important are listed below.

Name	Country of origin	Launch date	Achievement
Sputnik 1	USSR	Oct 5, 1957	First artificial satellite to orbit Earth
Sputnik 2	USSR	Nov 3, 1957	Carries a dog, Laika, into orbit
Explorer 1	USA	Jan 31, 1958	Discovers Van Allen radiation belts around Earth
SMM	USA	Feb 14, 1980	Observes Sun at solar maximum
Ulysses	ESA/USA	Oct 6, 1990	Observes Sun's polar regions
SOHO	ESA	Dec 2, 1995	X-ray and extreme UV observations of Sun; detects many Sun-grazing and impacting comets
POLAR	USA	Feb 24, 1996	Observes Earth's aurorae from polar orbit
TRACE	USA	Apr 2, 1998	Observes coronal loops in Sun's atmosphere
CLUSTER II	ESA	July/Aug 2000	Four spacecraft investigate Earth's magnetosphere
STEREO	USA	Oct 25, 2006	Two spacecraft produce 3D images of Sun

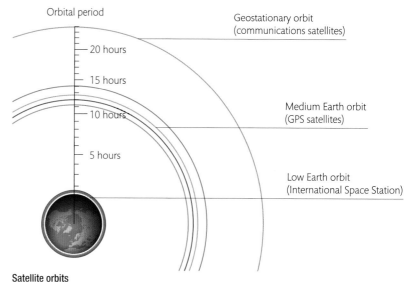

Satellite orbits

Space stations

Orbiting low above Earth, these crewed satellites serve as laboratories and workplaces. They stay within reach of reusable spacecraft that serve as ferries, restocking supplies and replacing crews. Space stations are used for low gravity experiments, Earth observation, and studies of the effects of long-term space exposure on the human body. It is very rare that anyone remains on board a station for more than a year.

Skylab (USA)   Salyut 1 (USSR)   Tiangong–1 (China)

Name	Launch date	Crew size	Days occupied	Crewed visits	Uncrewed visits	Mass (kg)
Salyut 1	Apr 19, 1971	3	24	2	0	18,400
Skylab	May 14, 1973	3	171	3	0	77,000
Salyut 6	Sept 29, 1977	2	683	16	14	9,000
Salyut 7	Apr 19, 1982	3	861	10	15	19,000
Mir	Feb 7, 1986	3	4,594	39	68	130,000
International Space Station	Nov 20, 1998	6	Ongoing	74	69	470,700
Tiangong-1	Sept 29, 2011	3	Ongoing	2	1	8,500

MIR (USSR)

International Space Station

Observing the skies

For centuries, astronomers have observed the heavens with the naked eye or through simple magnifying telescopes. But the visible light we see is just one part of a much bigger spectrum of electromagnetic rays that reaches Earth from space. Stars and other objects such as galaxies emit invisible radio waves, X-rays, and infrared and ultraviolet rays. Modern telescopes can detect all of these, and each type of radiation provides different information.

Telescopes, regardless of type and construction, all do essentially the same thing. They collect electromagnetic radiation and focus it to create an image or spectrum. Because of absorption and turbulence in Earth's atmosphere, telescopes are mainly located on high mountains or launched into space.

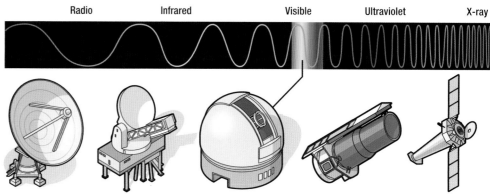

Radio   Infrared   Visible   Ultraviolet   X-ray

Radio telescopes
The huge dishes on these telescopes focus radio waves generated by such sources as galaxies, pulsars, and black holes.

Microwave telescopes
Microwaves are short-wavelength radio waves. Microwave telescopes allow astronomers to study radiation left by the Big Bang.

Optical telescopes
Using large lenses or bowl-shaped, segmented mirrors, optical telescopes gather faint visible light to see beyond naked-eye range.

Ultraviolet telescopes
As little ultraviolet light reaches Earth, ultraviolet telescopes are used from space to detect radiation from the Sun, stars, and galaxies.

X-ray telescopes
These telescopes are used in space to capture high-energy rays from very hot sources, such as the Sun and supernova explosions.

GLOSSARY

A

Accretion
(1) The colliding and sticking together of small, solid particles and bodies to make progressively larger ones. (2) The process whereby a celestial body grows in mass by accumulating matter from its surroundings.

Aphelion
The point in its elliptical orbit at which a body such as a planet, asteroid, or comet is at its greatest distance from the Sun. See also *perihelion*.

Apogee
The point in its elliptical orbit around Earth at which a body such as the Moon or a spacecraft is at its greatest distance from Earth. See also *perigee*.

Arachnoid
A volcanic structure on the surface of Venus that consists of a series of concentric ridges, resembling a spiderweb.

Asteroid
A small, irregular solar system object, with a diameter of less than 600 miles (1,000 km). Asteroids are made of rock and/or metal, and are thought to be detritus left over from the formation of the planets. Most asteroids occur in the asteroid belt, which lies between the orbits of Mars and Jupiter, but asteroids are found throughout the solar system. See also *asteroid belt, near-Earth asteroid*.

Asteroid belt
A doughnut-shaped region of the solar system, lying between the orbits of Mars and Jupiter, that contains a high concentration of asteroids.

Astronomical unit (AU)
A unit of distance, defined as the average distance between Earth and the Sun. 1 AU = 92,956,000 miles (149,598,000 km).

Atom
A building block of ordinary matter. It consists of a central nucleus surrounded by a cloud of electrons.

Aurora (plural: aurorae)
A glowing display of light in Earth's upper atmosphere (or the atmosphere of another planet), caused by particles from the Sun's solar wind becoming trapped in the planet's magnetic field and being drawn toward the magnetic poles. As the particles collide with gases in the atmosphere, they excite atoms and cause them to emit light. See also *solar wind*.

B

Background radiation
Remnant radiation from the Big Bang, which is still detectable as a faint distribution of microwave radiation across the whole sky. See also *Big Bang*.

Big Bang
The event in which the universe was born. According to the Big Bang theory, the universe originated a finite time ago in an extremely hot, dense initial state and ever since then has been expanding. The Big Bang was the origin of space, time, and matter.

C

Caldera
A bowl-shaped depression caused by the collapse of a volcanic structure into an empty magma chamber. See also *crater*.

Celestial poles
The celestial equivalent of Earth's poles. The night sky appears to rotate on an axis through the two celestial poles.

Celestial sphere
An imaginary sphere, surrounding Earth, on which all celestial objects appear to lie.

Centaur
A solar system body that occupies the same region as the gas giant planets. Centaurs are smaller than planets and have features in common with asteroids and comets.

Center of mass
The balance point within a system of bodies around which those bodies revolve. Where the system consists of two bodies, it is located on a line joining their centers.

Chondrite
A stony meteorite that contains many small spherical objects called chondrules. Carbonaceous chondrites are thought to be some of the least-altered remnants of the protoplanetary disk from which the solar system originally formed. See also *meteorite, protoplanetary disk*.

Chromosphere
A thin layer in the Sun's atmosphere that lies between the photosphere and the corona. See also *corona, photosphere*.

CNSA (China National Space Administration)
The national space agency of the People's Republic of China.

Coma
The cloud of gas and dust surrounding the nucleus of a comet, forming its glowing head. See also *comet*.

Comet
A small body, composed mainly of dust-laden ice, that orbits the Sun, typically following an elongated, elliptical path. When a comet enters the inner solar system, heating causes gas and dust to evaporate from its solid nucleus, forming an extensive cloud called a coma and one or more tails. See also *coma, tail*.

Conjunction
A close alignment in the sky of two or more celestial bodies, which occurs when they lie in the same direction as viewed from Earth. When a planet lies directly on the opposite side of the Sun from Earth, it is said to be at superior conjunction. When either Mercury or Venus passes between Earth and the Sun, the planet is said to be at inferior conjunction. See also *opposition*.

Convection
The transport of heat by rising bubbles or plumes of hot liquid or gas. In a convection cell, rising streams of hot material cool, spread out, and then sink down to be reheated, so maintaining a continuous circulation. Convection in Earth's mantle drives the movement of tectonic plates over Earth's surface.

Convective zone
An internal region of the Sun, below the photosphere and above the radiative zone, in which pockets of hot gas expand and rise toward the solar surface. See also *photosphere, radiative zone*.

Core
The central region of a star or planet.

Corona
The outermost part of the Sun's atmosphere. The solar corona has a very low density and a very high temperature of 2–9 million °F (1–5 million °C). From Earth's surface the corona can only be seen in detail during an eclipse.

Coronal mass ejection
A huge, rapidly expanding bubble of plasma that is ejected from the Sun's corona. Containing billions of tons of material in the form of ions and electrons, together with associated magnetic fields, a typical coronal mass ejection propagates outward through interplanetary space at a speed of a few hundred miles per second. See also *corona, ion, plasma*.

Cosmic rays
Highly energetic subatomic particles, such as electrons, protons, and atomic nuclei, that hurtle through space at velocities close to the speed of light.

Crater
A bowl- or saucer-shaped depression in the surface of a planet or satellite. An impact crater is one excavated by a meteorite, asteroid, or comet impact, whereas a volcanic crater develops around the vent of a volcano.

Crust
The thin, rocky or icy, cold, solid, outermost layer of a planet or moon.

D

Dwarf planet
A body that orbits the Sun and is massive enough to have formed a round shape but is not sufficiently massive to clear its orbital path of other objects.

E

Eccentricity
The extent to which a body's orbit deviates from a perfect circle. An orbit with a high eccentricity is a very elongated ellipse; an orbit of low eccentricity is almost circular. See also *ellipse*.

Eclipse
The passage of one celestial body into the shadow cast by another. A lunar eclipse occurs when the Moon passes into Earth's shadow. A total lunar eclipse takes place when the whole of the Moon lies within the dark cone of Earth's shadow, and a partial lunar eclipse when only part of the Moon is in the shadow. A solar eclipse is when part of Earth's surface enters the shadow cast by the Moon. In a total solar eclipse, the Sun is completely obscured by the dark disk of the Moon. A partial solar eclipse occurs when only part of the Sun's surface is hidden. If the Moon passes directly between the Sun and Earth when it is close to apogee, it will appear smaller than the Sun, and its dark disk will be surrounded by a ring, or annulus, of sunlight; such an event is called an annular eclipse.

Ecliptic
(1) The plane of Earth's orbit around the Sun. (2) The track along which the Sun travels around the celestial sphere, relative to the background stars, in the course of a year. See also *celestial sphere*.

Ejecta
Material thrown outward by the blast of an impact. Sometimes the material, which may be much brighter than the adjacent surface, forms extensive streaks, or rays, radiating out from the impact point.

Electromagnetic radiation
Oscillating electric and magnetic disturbances that propagate energy through space in the form of waves (electromagnetic waves). Examples include light and radio waves.

Electromagnetic (EM) spectrum
The entire range of energy emitted by different objects in the universe, from the shortest wavelengths (gamma rays) to the longest (radio waves). Our eyes can see a specific range within the spectrum called visible light.

Electron
A lightweight fundamental particle with a negative electrical charge. A cloud of electrons surrounds the nucleus of an atom. See also *atom*.

Ellipse
A shape like a flattened circle, or oval. See also *eccentricity*, *orbit*.

Equinox
An occasion when the Sun is vertically overhead at a planet's equator, and day and night have equal duration for the whole planet.

ESA (European Space agency)
An international space exploration organization with 20 European member countries.

Escape velocity
The minimum speed at which a projectile must be launched in order to recede forever from a massive body and not fall back. Earth's escape velocity is 7 miles (11.2 km) per second.

Extrasolar planet (exoplanet)
A planet that orbits a star other than the Sun. Since the first confirmed detection of one in 1992, more than 2,000 exoplanets have been detected.

F

Frequency
The number of crests of a wave that pass a given point in one second. See also *electromagnetic radiation*, *wavelength*.

Fusion (nuclear fusion)
A process whereby atomic nuclei join to form heavier atomic nuclei. Stars are powered by fusion reactions that take place in their cores and release large amounts of energy.

G

Galaxy
A large aggregation of stars and clouds of gas and dust, held together by gravity. Galaxies may be elliptical, spiral, or irregular in shape. They may contain from a few million to several trillion stars. See also *Milky Way*.

Galilean moon
Any of the four largest of Jupiter's moons (Io, Europa, Ganymede, and Callisto). They were discovered by Italian astronomer Galileo Galilei.

Gamma radiation
Electromagnetic radiation with extremely short wavelengths (shorter than X-rays) and very high frequencies. See also *electromagnetic radiation*, *electromagnetic spectrum*.

Gas giant
A large planet like Jupiter or Saturn that consists mainly of hydrogen and helium. See also *rocky planet*.

Geocentric
(1) Treated as being viewed from the center of the Earth. (2) Having the Earth at the center (of a system). A satellite that is traveling around the Earth is in a geocentric orbit. Geocentric cosmology was the theory that the Sun, Moon, planets, and stars revolved around a central Earth. See also *heliocentric*.

Gravity
An attractive force between all objects that have mass or energy, experienced on Earth as weight. The force of gravity keeps moons in orbit around planets and planets in orbit around the Sun.

Greenhouse effect
The process by which atmospheric gases make the surface of a planet hotter than it would otherwise be. Incoming sunlight is absorbed at the planet's surface and re-radiated as infrared radiation, which is then absorbed by greenhouse gases, such as carbon dioxide. Part of this trapped radiation is re-radiated back toward the ground, so raising its temperature.

H

Heliocentric
Having the Sun at the center. A body that travels around the Sun has a heliocentric orbit. The heliocentric model of the solar system proposed in 1543 by Polish astronomer Nicolaus Copernicus overturned the previously dominant geocentric model. See also *geocentric*.

Heliosphere
The region of space around the Sun within which the solar wind and interplanetary magnetic field are confined by the pressure of the interstellar medium. See also *interstellar medium*, *magnetic field*, *solar wind*.

Helium burning
The generation of energy in the cores of red giant stars by means of fusion reactions that convert helium into other elements. See also *fusion*.

Hydrogen burning
The generation of energy by means of fusion reactions that convert hydrogen into helium. Hydrogen burning takes place in the core of the Sun. See also *fusion*.

I

Infrared radiation
Electromagnetic radiation with wavelengths longer than visible light but shorter than microwaves or radio waves. It is the main form of radiation emitted from many cool astronomical objects. See also *electromagnetic radiation*.

Interstellar medium
The gas and dust that permeates the space between the stars within a galaxy.

Ion
A particle or group of particles with a net electrical charge. The process by which ions form from atoms is called ionization. See also *electron*, *plasma*.

Isotope
One of two or more forms of a chemical element, the atoms of which contain the same number of protons but different numbers of neutrons. For example, helium-3 and helium-4 are isotopes of helium; a nucleus of helium-4 (the heavier, more common isotope) has two protons and two neutrons, but a nucleus of helium-3 contains two protons and one neutron. See also *atom*, *nucleus*.

J

JAXA (Japan Aerospace Exploration Agency)
Japan's national aerospace agency.

K

Kepler's laws of planetary motion
Three laws that describe the orbits of planets around the Sun. The first states that each planet's orbit is an ellipse; the second shows how a planet's speed varies as it travels around its orbit; and the third links its orbital period to its average distance from the Sun.

Kuiper belt
A region of the solar system beyond Neptune containing icy-rocky bodies. See also *Oort cloud*.

Kuiper belt object
An icy body in the Kuiper belt region beyond the orbit of Neptune.

L

Leading hemisphere
The hemisphere of a moon in a synchronous orbit around a planet that faces forward, into the direction of motion. See also *trailing hemisphere*, *synchronous rotation*.

Light-year
The distance that light travels through a vacuum in one year: 1 light-year = 5,878 billion miles (9,460 billion km).

Limb
The outer edge of the observed disk of the Sun, a moon, or a planet.

Lithosphere
The physically solid, hard, rigid outer layer of a planet or satellite. See also *crust*, *mantle*, *tectonic plate*.

Lunar eclipse
See *eclipse*.

M

Magma
Subsurface molten or semi-molten rock, often containing dissolved gas or gas bubbles. When magma erupts onto the surface of a planet, it is called lava.

Magnetic field
The region around a magnetized body within which magnetic forces affect the motion of electrically charged particles.

Magnetosphere
The region of space around a planet within which the planet's magnetic field is sufficiently strong to deflect the solar wind, preventing most solar wind particles from reaching the planet. See also *magnetic field*, *solar wind*.

Main belt
See *asteroid belt*.

Mantle
The warm, slightly viscous, rocky layer that lies between the core and the crust of a planet or moon. See also *core, crust*.

Mare (plural: maria)
A dark, low-lying area of the Moon, filled with lava.

Mass-energy
A measure of the energy possessed by anything from a subatomic particle to the entire universe, taking into account that mass is convertible into energy and so has an energy equivalence.

Meteor
The short-lived streak of light, also called a shooting star, seen when a meteoroid hits Earth's atmosphere and is heated by friction. See also *meteorite, meteoroid*.

Meteorite
A meteoroid that reaches the ground and survives impact. Meteorites are usually classified according to their composition as stony, iron, or stony-iron. See also *meteor, meteoroid*.

Meteoroid
A lump or small particle of rock, metal, or ice orbiting the Sun in interplanetary space. See also *asteroid, comet, meteor, meteorite*.

Microwave
Electromagnetic radiation with wavelengths longer than infrared and visible light but shorter than radio waves.

Milky Way
The barred spiral galaxy that contains the solar system. The Milky Way is visible to the naked eye as a band of faint light across the night sky. See also *galaxy*.

Molecular cloud
A cool, dense cloud of dust and gas, within which the temperature is low enough for atoms to join together to form molecules, such as molecular hydrogen or carbon monoxide, and within which conditions are suitable for stars to form.

Moon
A natural satellite orbiting a planet. The Moon is Earth's natural satellite.

N

NASA (National Aeronautics and Space Administration)
The agency of the United States government with responsibility for the nation's space program.

Near-Earth asteroid
An asteroid whose orbit comes close to, or intersects, Earth's orbit. Formally, it is defined as an asteroid with a perihelion distance of less than 1.3 times Earth's mean distance from the Sun.

Nebula (plural: nebulae)
A cloud of gas and dust in interstellar space, visible because it is illuminated by embedded or nearby stars or because it obscures more distant stars. See also *planetary nebula, solar nebula*.

Neutrino
A fundamental particle, of exceedingly low mass and with zero electrical charge, that travels at close to the speed of light.

Neutron
A particle with zero electrical charge, found in all atomic nuclei except those of hydrogen. See also *atom, nucleus*.

Nucleus
(1) The compact central core of an atom. (2) The solid, ice-rich body of a comet.

O

Occultation
The passage of one body in front of another, which causes the more distant one to be wholly or partially hidden.

Oort cloud (Oort–Öpik cloud)
A spherical distribution of trillions of icy bodies such as cometary nuclei that surrounds the solar system and extends out to a radius of about 1.6 light-years from the Sun. It provides the reservoir from which long-period and "new" comets originate. Its existence was proposed in 1950 by Dutch astronomer Jan H. Oort (a similar idea had also been suggested by Estonian astronomer Ernst J. Öpik). See also *comet*.

Opposition
The time when Mars, or one of the giant planets, lies on the opposite side of Earth from the Sun and is highest in the sky at midnight. The planet is closest to Earth, and appears at its brightest, at this time. See also *conjunction*.

Orbit
The path a celestial body takes in space under the influence of the gravity of other, relatively nearby, objects. The orbits of the planets are elliptical in shape, although some are nearly circular.

Orbital period
The time an orbiting body takes to travel once around the object it is orbiting.

P

Penumbra
(1) The lighter, outer part of the shadow cast by an opaque body. An observer within the penumbra can see part of the illuminating source. (2) The less-dark and less-cool outer region of a sunspot. See also *eclipse, sunspot, umbra*.

Perigee
The point in its orbit where a body orbiting Earth is at its closest to Earth. See also *apogee*.

Perihelion
The point in its orbit where a planet or other solar system body is at its closest to the Sun. See also *aphelion*.

Phase
The proportion of the hemisphere of the Moon or a planet that is illuminated by the Sun and visible from Earth, at any particular instant.

Photon
A particle of electromagnetic radiation. See also *electromagnetic radiation*.

Photosphere
The thin, gaseous layer at the base of the Sun's atmosphere, from which visible light is emitted and which forms the apparent visible surface of the Sun. See also *chromosphere, corona*.

Planet
A celestial body that orbits a star, is sufficiently massive to have cleared its orbital path of debris, and is roughly spherical. See also *dwarf planet*.

Planetary nebula
A glowing shell of gas ejected by a star of similar mass to the Sun toward the end of its evolutionary development. In a small telescope, it resembles a planet's disk. See also *nebula*.

Planetesimal
One of the large number of small bodies, composed of rock or ice, that formed in the early solar system and from which the planets were eventually assembled through the process of accretion. See also *solar nebula*.

Plasma
A mixture of positively charged ions and negatively charged electrons that behaves like a gas, but conducts electricity and is affected by magnetic fields. Examples include the solar corona and solar wind. See also *corona, solar wind*.

Precession
A slow change in the orientation of a body's rotational axis, caused by the gravity of neighboring bodies.

Prominence
A vast, flamelike plume of plasma emerging from the Sun's photosphere.

Proton
A positively charged particle that is a constituent of every atomic nucleus. See also *atom, nucleus*.

Protoplanet
A precursor of a planet, which forms through the gradual aggregation of planetesimals. Protoplanets collide to form planets. See also *planetesimal, protoplanetary disk*.

Protoplanetary disk
A flattened disk of dust and gas around a newly formed star, within which matter may be aggregating to form the precursors of planets. See also *planetesimal, protoplanet*.

Protostar
A star in the early stages of formation, consisting of the center of a collapsed cloud that is heating up and growing through the addition of surrounding matter, but inside which hydrogen fusion has not yet begun.

R

Radiative zone
An internal region of the Sun, below the convective zone and above the core, in which light energy works its way slowly upward, colliding with atomic nuclei and being re-radiated billions of times. See also *convective zone*.

Radio telescope
An instrument designed to detect radio waves from astronomical sources. The most familiar type is a concave dish that collects radio waves and focuses them onto a detector.

Red giant star
A large, highly luminous star with a low surface temperature and a reddish color. A red giant "burns" helium in its core rather than hydrogen and is approaching the final stages of its life.

Regolith
A layer of dust and loose rock fragments that covers the surface of a planet, moon, or asteroid.

Relativity
Two theories developed in the early 20th century by Albert Einstein. The special theory of relativity describes how the relative motion of observers affects their measurements of mass, length, and time. One consequence is that mass and energy are equivalent. The general theory of relativity treats gravity as a distortion of space-time. See also *space-time*.

Resonance
A gravitational interaction between two

orbiting bodies that occurs when the orbital period of one is an exact, or nearly exact, multiple of the orbital period of the other. For example, Jupiter's moon Io is in a 1:2 resonance with the moon Europa (Io's period is half of Europa's period). When a small object is in resonance with a more massive one, it experiences a periodic gravitational tug each time one of the bodies overtakes the other, the cumulative effect of which gradually changes its orbit.

Retrograde motion
(1) An apparent temporary reversal in the direction of motion of a planet, such as Mars, when it is being overtaken in its orbital motion by Earth. (2) Orbital motion in the opposite direction of that of Earth and the other planets of the solar system. (3) The motion of a satellite along its orbit in the opposite direction of the rotation of its parent planet.

Retrograde rotation
The rotation of a planet or moon in the opposite direction of its orbit. All the planets orbit the Sun in the same direction that the Sun rotates. Most planets also rotate (spin) in the same direction, but Venus and Uranus have retrograde rotation.

Ring
A flat belt of small particles and lumps of material that orbits a planet, usually in the plane of the planet's equator. Jupiter, Saturn, Uranus, and Neptune each have many rings.

Rocky planet
A planet composed mainly of rock, with basic characteristics similar to Earth's. The four rocky planets in the solar system are Mercury, Venus, Earth, and Mars. See also *gas giant*.

Rupes
Scarps or cliffs on the surface of a planet or a satellite.

S

Satellite
A body that orbits a planet, otherwise known as a moon. An artificial satellite is an object deliberately put in orbit around Earth or another solar system body.

Shepherd moon
A small moon whose gravitational pull "herds" orbiting particles into a well-defined ring around a planet.

Solar cycle
A cyclical variation in solar activity (for example, the production of sunspots and flares), which reaches a maximum at intervals of about 11 years. The sunspot

cycle is the 11-year variation in the number and distribution of sunspots. See also *solar flare, sunspot*.

Solar eclipse
See *eclipse*.

Solar flare
A localized brightening of the Sun's surface, accompanied by a violent release of huge amounts of energy in the form of electromagnetic radiation, subatomic particles, and shock waves.

Solar mass
A unit of mass equal to the mass of the Sun.

Solar nebula
The cloud of gas and dust from which the solar system formed. As the cloud collapsed, most of its mass accumulated at the center to form the Sun. The rest flattened out into a disk, within which planets assembled by accretion. See also *accretion, protoplanetary disk*.

Solar system
The Sun together with the eight planets, smaller bodies (dwarf planets, moons, asteroids, comets, Kuiper belt objects, trans-Neptunian objects), dust, and gas that orbit the Sun.

Solar wind
A continuous stream of fast-moving charged particles, mainly electrons and protons, that escapes from the Sun and flows outward through the solar system.

Space-time
The combination of the three dimensions of space (length, breadth, height) and the single time dimension. See also *relativity*.

Spectral line
A bright or dark line that appears in an object's spectrum, due to emission or absorption of radiation by that object at a particular wavelength. Patterns of spectral lines serve as fingerprints of different chemical elements, allowing astronomers to identify the composition of distant objects by analyzing their light.

Spectroscopy
The science of obtaining and studying the spectra of objects. Because the appearance of a spectrum is influenced by factors such as chemical composition, temperature, velocity, and magnetic fields, spectroscopy can reveal a wealth of information about the properties of various celestial bodies. See also *spectrum*.

Spectrum
The full range of wavelengths of light emitted by a celestial object. The spectrum, and the presence of any

spectral lines, gives clues about the chemical and physical properties of the object. See also *spectral line*.

Spiral galaxy
A galaxy that consists of a spheroidal central concentration of stars (the nuclear bulge) surrounded by a flattened disk composed of stars, gas, and dust, within which the major visible features are clumped together into a pattern of spiral arms. See also *galaxy*.

Star
A huge sphere of glowing plasma that generates, or has generated, energy by means of nuclear fusion reactions in its core. Our Sun is a star of medium size. See also *fusion, plasma*.

Stellar wind
An outflow of charged particles from the atmosphere of a star. See also *solar wind*.

Sunspot
A region of intense magnetic activity in the Sun's photosphere that appears dark because it is cooler than its surroundings. See also *photosphere, solar cycle*.

Synchronous rotation
The rotation of a body around its axis in the same period of time that it takes to orbit another body. The orbiting body always keeps the same face turned toward the object around which it is orbiting. Earth's Moon displays synchronous rotation. See also *orbital period, satellite*.

T

Tail (of a comet)
A stream of ionized gas or dust that is swept out of the head (coma) of a comet when the comet is near the Sun. See also *comet*.

Tectonic plate
One of the large, rigid sections into which Earth's lithosphere is divided. Driven by convection currents in the mantle, tectonic plates drift slowly across the surface of the planet. Their collision and production give rise to phenomena such as earthquakes, volcanic activity, and mountain building. The term tectonic is sometimes also used to refer to large-scale geological structures, and features resulting from their movement, on planets other than Earth. See also *convection, crust, lithosphere, mantle*.

Tidal forces
Tidal forces occur when gravity does not pull equally strongly on both sides of a celestial body. Tidal forces between Earth and the Moon cause Earth's oceans to swell as tides and trigger moonquakes in the lunar crust. Tidal forces also cause

friction inside celestial bodies, generating internal heat that, for example, produces volcanoes on Jupiter's moon Io.

Trailing hemisphere
The hemisphere of a moon in a synchronous orbit around a planet that faces backward, away from the direction of motion. See also *leading hemisphere, synchronous rotation*.

Trans-Neptunian object
A body orbiting the Sun beyond the orbit of Neptune.

Transit
The passage of a smaller body in front of a larger on—for example, the passage of Venus across the face of the Sun.

Trojan
An object such as an asteroid or moon that shares the same orbit as a larger body but stays at one of two gravitationally stable points which are about 60° ahead of or behind the larger body.

U

Ultraviolet radiation
Electromagnetic radiation with wavelengths shorter than visible light but longer than X-rays.

Umbra
(1) The dark central part of the shadow cast by an opaque body. The illuminating source will be completely hidden from view at any point within the umbra. (2) The darker, cooler, central region of a sunspot. See also *eclipse, penumbra, sunspot*.

W

Wavelength
The distance between two successive crests in a wave motion. See also *electromagnetic radiation, frequency*.

X

X-ray
Electromagnetic radiation with wavelengths shorter than ultraviolet radiation but longer than gamma rays.

Z

Zenith
The point in the sky directly above an observer.

INDEX

Bold page numbers refer to main entries.

A

absorption lines 37
Acidalia Planitia (Mars) 111
Adams, John Couch 204, 213
Adams ring (Neptune) 207, 208, 209
Addams Crater (Venus) 58
Adrastea 157
aerogel 226
Africa 72, 73
African Plate 76
Agassiz, Louis 86
Airy, George 213
Alba Mons (Mars) 110
Aldrin, Buzz 108
ALH84001 meteorite 131
Alpha Regio (Venus) 58
Alta Regio (Venus) 59
Alvarez, Luis 87
Amalthea 157
Amalthea group (Jupiter) 156-7
Amazon Basin 73
Ammisaduqa, King of Babylon 68
ammonia 14, 16, 153, 167, 170, 172, 173, 197, 198, 199, 206
    clouds 168, 172, 173, 178
    ice 155, 172, 178
ammonium hydrosulfide 155
Amor asteroids 142
Ananke group (Jupiter) 156
Anaxagoras 105
Anaximander 86
Andes 73, 76
animals 82, 83
annular solar eclipses 35
annulus, comets 223
Antarctic Plate 77
Antarctica 72
Antoniadi Dorsum (Mercury) 48
antumbra 35
Aphrodite Terra (Venus) 64
Apianus, Petrus 231
apogee 92
Apollo (god) 36, 54
Apollo 7 107
Apollo 9 107
Apollo 11 20, 89, 108-9
Apollo 15 96-7, 109
Apollo 16 101
Apollo 17 39, 103, 104, 108, 109
Apollo 20 101
Apollo asteroids 142
Apollo project 90, 104, 107, 108-9
Arabia Terra (Mars) 42
arachnoid volcanoes 63
Arago ring (Neptune) 208, 209
arcs 208
Arecibo radio telescope (Puerto Rico) 54, 69
Arecibo Valles (Mercury) 48
argon 75, 113
Argyre Planitia (Mars) 111
Ariane 4 rocket 242
Ariel 200, 201
Aristarchus 21

Aristarchus Crater (Moon) 88
Aristotle 86
Armstrong, Neil 20, 108
Arrhenius, Svante 230
Arsia Mons (Mars) 110, 121
Artemis Corona (Venus) 63
Ascraeus Mons 121
Asgard impact basin (Callisto) 163
ash flows 62
ashen light (Venus) 69
Association for the Advancement of Science 36
asteroid belt 12, 138, 140-41, 142
asteroid hunters 222
asteroids 11, 12, 16, 128, 138-45, 240
    capture of 145
    discovery of 20
    as dwarf planets 220
    evolution 139
    family of 141
    impact 49, 82, 87, 92, 116, 138, 142, 143, 163, 165
    melting 139
    missions to 144-5
    near-Earth 140, 142-3
    origins and collisions 141
    samples from 145
    size and scale 18-19, 138
    types of 138
astronauts, Apollo project 108-9
astronomical calendar 36
Astronomicum Caesareum (Apianus) 231
astronomy 243
Atalanta Planitia (Venus) 58, 59
Aten asteroids 142
Atiras asteroids 142
Atlantic Ocean 72, 73
atmosphere
    Earth 43, 72, 74, 75, 86, 87, 239
    Io 158
    Jupiter 152, 154, 166, 167, 168, 239
    Mars 20, 110, 113, 126, 130, 239
    Mercury 44, 47, 48, 55, 239
    Moon 105
    Neptune 204, 205, 206, 239
    Pluto 221
    Saturn 172, 173, 178, 193, 239
    Sun 24, 37
    Titan 184, 192
    Triton 210
    Uranus 198, 199, 239
    Venus 58, 60, 61, 62, 68, 70, 71, 239
atmospheric pressure
    Earth 82
    Jupiter 150, 152
    Mars 116
    Sun 26
    Uranus 198
    Venus 58
AU (astronomical units) 12
aurorae 29, 33, 36, 158, 167, 179
Australia 72
Australian Plate 77

B

B ring (Saturn) 176-7, 193
Baby Red Spot (Jupiter) 154
Babylonians, ancient 54, 68, 69, 104, 166
bacteria 83
Baikonur Cosmodrome (Kazakhstan) 242
Baily's beads 35
barchans 124
Barnard, Edward E. 230
Barringer Crater (Arizona) 138, 143
Bartok Crater (Mercury) 44
basalt 47, 61, 75, 101, 102, 103, 124
basins, multi-ring (Mercury) 48, 49
Bayeux Tapestry 231
Beethoven Basin (Mercury) 48
Bela Crater (Moon) 96
Belinda 201
BepiColombo missions 46, 56
Bessel, Friedrich 54
Bethlehem, Star of 230
Bianca 201
Biermann, Ludwig F. 37
Big Bang 14
Big Muley 101
biodiversity 83
bipolar outflow 14
bismuth sulfide (bismuthinate) 67
blue planets
    timeline of discoveries 212-13
    see also Neptune; Uranus
Bode, Johann 213
Boscovich, Roger 105
Bradley, James 167
Brahms Crater (Mercury) 44
Bryce Canyon (Utah) 79
Budh Planitia (Mercury) 48
Burke, Bernard 167
Burney, Venetia 212
butterfly diagram 33, 36

C

C/2001 Q4 comet 228-9
calderas
    Io 161
    Mars 120, 123
    Venus 63
calendar, prehistoric 104
Callisto 156, 157, 160-61, 163
Caloris Basin (Mercury) 44, 48, 49
Calypso 183
"canals" (Mars) 131
canyons
    Earth 84
    Mars 110, 111, 118-19, 126, 130
    Miranda 203
    Tethys 185
Cape Canaveral (Florida) 195, 242
Cape Palliser (New Zealand) 232-3
captured "centaur" planetoids 182
carbon 206, 236
carbon dioxide 43, 60, 62, 68, 75, 113, 130, 220
    ice 110, 113, 124, 126, 127, 220
carbon-14 levels 32

carbonaceous asteroids (C-type) 138
carbonaceous chondrites 15
Carina Nebula 14-15
Carme group (Jupiter) 156
Carnegie Rupes (Mercury) 52-3
Carrington, Richard 36
Cassini, Giovanni 68, 167, 193
Cassini division 174, 188, 193
Cassini spacecraft 21, 148-9, 167, 177, 178, 179, 181, 186, 187, 188-9, 191, 192, 193, 194, 195
catastrophism 87
Catholic Church 21
Cavendish, Henry 87
cells, primitive 82
Ceres 20, 138, 140, 145, 220
Cernan, Gene 109
Chao Meng-Fu Crater (Mercury) 44
Charon 221
Chasma Boreale (Mars) 126
Chelyabinsk meteorite 143
ChemCam tool 135
Chichen Itza (Mexico) 69
China
    ancient astronomers 230
    missions to Mars 132-3
    missions to Moon 105, 107
    space exploration 242
chondrule meteorites 15
Christianity 36
Christy, James 221
chromosphere 24
Chryse Planitia (Mars) 131
Churyumov-Gerasimenko, Comet 227, 231
cities 85
Clairaut, Alexis 104
Clavius Crater (Moon) 88
Clementine orbiter 105
cliffs
    Callisto 163
    Mercury 48, 52-3
    Miranda 203
climate change 33, 87
clouds
    Earth 72
    interstellar 14, 16
    Jupiter 150, 151, 154, 155, 166,167, 168
    Mars 117
    Neptune 204, 205
    Saturn 170, 172, 173, 178, 179
    Uranus 196, 197, 198, 199
    Venus 58, 59, 60, 61, 62, 68, 69, 71
clumping 177, 208
comas, comets 222, 224, 226, 227
Comet 1892 230
comets 11, 12, 37, 82, 222-9, 241
    composition of 222, 227
    discovery of 37, 222
    haloes and tails 222
    impacts/collisions 72, 87, 88, 163, 165, 166, 225, 230
    missions to 226-7
    nuclei 21, 200, 222, 225, 226, 227, 229, 230
    orbits 12, 222, 224-5
    as primordial planetary building blocks 225, 230
    as prophets of doom 230, 231
    samples from 226, 227

comets (continued)
    short-period 218, 241
    timeline of discoveries **230–31**
command module 108
compression 66
cones, volcanic 121
continental crust 74, 75
continental drift 86, 87
continents 72, 73, 76
convection cells 24
convection currents 60, 61,
    75, 78
convection cycle, Jupiter 155
convective zone, Sun 26, 27
convergent boundaries 77
Copernicus, Nicolaus 21, 36,
    37, 69
Copernicus Crater (Moon) 88, 101
coral atolls 80
Cordelia 199, 201
core
    asteroids 39
    Earth 74, 86, 87, 238
    Ganymede 162
    gas giants 149, 238
    Jupiter 152, 167, 238
    Mars 110, 112, 238
    Mercury 16, 46, 238
    metal 16
    Moon 90, 91, 92
    Neptune 206, 238
    Pluto 220
    rocky planets 43, 238
    Saturn 172, 238
    Sun 26, 36
    Uranus 198, 238
    Venus 60, 238
coronae
    Earth 75
    Sun 24, 29, 35, 37, 39, 223
    Venus 63
coronagraphs 35
coronal mass ejections (CMEs) 28,
    29, 33, 36
Cortés, Hernán 69
cosmic dust 14
cosmic rays 211
Crabtree, William 69
crater chains 164–5
craters
    asteroid belt 141
    Callisto 163
    Dione 185
    Earth 143
    Europa 159
    formation of 100
    Galilean moons 160–61, 164–5
    Ganymede 162, 164–5
    Mars 111, 130, 134, 136–7
    Mercury 43, 44, 45, 47, 48, 49,
        55, 56
    Mimas 185
    Moon 88, 89, 91, 94, 96, **100–101**,
        102, 103, 104
    Phobos and Deimos 128
    Phoebe 192
    Pluto 221
    Rhea 184
    Venus 59, 62, 63, 71
    Vesta 139
Cretaceous period 87

crust
    asteroids 39
    Callisto 163
    Earth 74, 75, 238
    Europa 21, 159
    Ganymede 162
    Mars 110, 112, 113, 118, 119, 120, 122, 238
    Mercury 47, 48, 49, 52, 238
    Miranda 203
    Moon 91, 92, 102
    Pluto 220
    Venus 59, 61, 238
cryovolcanoes 184, 185, 221
cubewanos 218
Cupid 201
Curiosity rover 131, 133, 134–5, 136
currents, oceanic 80, 198
Cuvier, Georges 87
cyanobacteria 83
cyclones 81

D

daguerreotype 36
Dali Chasma (Venus) 59, 63
Daphnis 174
Dawn spacecraft 139, 145
Deep Impact spacecraft 227, 231
Deimos **128**, 131
Delta IV Heavy rocket 242
deltas 80, 84, 116
density waves 177
Desdemona 201
deserts 85
Despina 207, 209
Diaconus, Leo 37
diamond rings 35
diamond sea/liquid 198, 206
Diana Chasma (Venus) 58
differential rotation 33
dinosaurs, extinction of 87
Dione 183, 184, **185**, 188–9
Discovery Rupes (Mercury) 45
divergent boundaries 77
DNA 82
domes, volcanic 120
Doppelmayr, Johann 105
dorsa, Mercury 48
double-ring craters 45
Dresden Codex 69
dune fields, Mars **124–5**, 126
dust 14
dust clouds, Mars 110, 113
dwarf planets 11, 20, 140, 212, 218, **220–21**
    definition of 220
    gravity 220
    orbits 220
    size and scale 18–19
Dyce, Rolf 54
dynamo effect 60
Dysnomia 213

E

Eagle Crater (Mars) 116
Earth 11, 12, 43, **72–87**
    age of 86
    asteroids and meteorites 138
    atmosphere 43, 72, 74, 75, 86, 87, 239
    axial tilt 72
    circumference 87

Earth (continued)
    data **72**, 236
    density 87
    diameter 72
    distance from Sun 12, 72
    Earth–Moon system **92–3**
    Earthrise **98–9**
    energy 82
    from above **84–5**
    future of 24
    geocentrism 20, 21, 36,
        37, 39
    gravity 72, 87, 88, 92
    hit by protoplanet 16
    influence on Moon 88,
        91, **92–3**
    life on 16, 43, 72, 73,
        82–3, 230
    magnetic field 75, 82,
        87, 238
    Moon see Moon
    near-Earth asteroids 140, **142–3**
    orbit 72, 237
    rotation 72
    search for "twin" 233
    size of 19, 24
    structure **74–5**
    surface **78–9**
    tectonic plates 62, 72, 74, **76–7**,
        78, 79
    temperatures 72, 75, 82,
        238, 239
    timeline of discoveries **86–7**
    water and ice 16, 72, **80–81**,
        82, 84
    weather/climate 33, 81, 82
    weight of 87
earthquakes 20, 74, 76, 77, 86
earthrise **98–9**
East African Rift 76, 77
eccentric orbits 218, 219, 220
eclipses
    Jupiter's moons 167
    lunar 20, 92, 104
    solar **34–5**, 36, 37
Eddington, Arthur 36
Edgeworth, Kenneth 218
Egyptians, ancient 20, 36
Einstein, Albert 36
ejecta 63, 100, 139, 164
El Caracol observatory
    (Chichen Itza) 69
electromagnetic radiation 243
elements, in solar system 236
Ellesmere Island (Canada) 78
elliptical orbits 12, 21, 105,
    141, 224
Enceladus 21, 175, 183, 184,
    185, 189, **190–91**, 192, 194, 195
Endeavour Crater (Mars) 134
Endeavour Rupes (Mercury) 48
Endurance Crater (Mars) 134, 136
Enki Catena (Ganymede) 164–5
Epimetheus 183
Epsilon ring (Uranus) 199, 201
equator, Earth 72
Eratosthenes 87
Eratosthenes Crater (Moon) 89
Erebus Crater (Mars) 134, 137
Eris 213, 218, 220
Eros 138, 140, **142**, 144, 145

erosion 78, 79, 203
ESA
    missions to comets 226–7,
        230, 231
    missions to Jupiter 168–9
    missions to Mars 132–3
    missions to Mercury 56
    missions to Moon 105
    missions to Saturn 194–5
    missions to Sun 38–9
    missions to Venus 71
    satellites 243
    see also spacecraft by name
Eta Aquarid meteor shower 229
ethane 186
Eurasia 72
Eurasian Plate 76, 79
Europa 21, 82, 157, **159**, 160–61, 168
European Southern Observatory 219
European Space Agency see ESA
evaporation 78
evolution 82, 83
exoplanets 233
exosphere 47, 75, 113
exploration, space 242–3
extinctions 87
extrasolar planets 233
extremophile bacteria 83

F

far side of the moon 88, 90, 92,
    100, 104
farmland 85
faults, geological 79, 203
festoons, Jupiter 151
filaments 28
fireballs 166
Fizeau, Hippolyte 36
flags, on Moon 109
Flamsteed, John 212
flat Earth theory 20, 86
floodplains 116
flybys 20, 21, 54, 55, 56, 71, 128,
    130, 142, 166, 167, 168, 179, 194,
    195, 208
fossils 83
Foucault, Léon 36
fracture lines 203
Franklin, Kenneth 167
frost line 16, 17
fusion, nuclear 26

G

Gagarin Crater (Moon) 101
Galatea 208, 209
galaxies 11, 14
Gale Crater (Mars) 131, 134–5, 136–7
Galilean Moons 156, 157, **160–65**, 166
    water on 168
    see also Callisto; Europa; Ganymede; Io
Galileo (Galileo Galilei), 21, 37, 54, 55, 68, 69,
    105, 156, 166, 192, 212
Galileo Regis 160
Galileo spacecraft 145, 158, 160–61, 166, 168
Galle, John 213
Galle ring (Neptune) 207, 208, 209
Ganges River 80, 81
Ganymede 156, 157, 160–61, **162**
    Enki Catena **164–5**

gas 14, 236
gas giants 11, 12, **146–215**, 233
  atmospheres 239
  formation of 16, 17
  mass of 16
  size and scale 18–19
  see also Jupiter; Neptune;
    Saturn; Uranus
Gaspra 140, 145
Gassendi, Pierre 55
Gemini Observatory 155
Genesis spacecraft 39
geocentrism 20, 21, 36, 37, 69
geology 87
geomagnetic storms 29
George III, King 213
Georgium Sidus 213
Germany, missions to Sun 38–9
geysers
  Enceladus 21, 185, 190, 191, 192
  Io 158
  Triton 210, 211
ghost craters 44
giant planets see gas giants
Gilbert, William 87
Gill, David 230
Giotto de Bondone 230
Giotto spacecraft 21, 226, 227,
  229, 230
glaciation 78, 86
global warming 87
Goethe Basin (Mercury) 44
Goldilocks zone 82
Goldstone Deep Space Communications
  Complex (California) 69, 143
GRAIL (Gravity Recovery and Interior
  Laboratory) 90
Grand Canyon 118, 119, 203
Grand Prismatic Hot Spring
  (Wyoming) 82–3
gravity
  dwarf planets 220
  Earth 72, 87, 88, 92
  Jupiter 138, 140, 141, 150, 156,
    164, 165, 169, 224
  Mars 110, 128
  Mercury 44, 47
  Moon 88, 90, 91, 92, 105
  Neptune 204, 208
  Newton's theory of 36, 105
  Saturn 172, 174, 177
  Sun 11, 12, 24, 52, 56
  Uranus 196
  Uranus's moons 201
  Venus 58
gray silicaceous asteroids
  (S-type) 138
Great Blue Hole (off Belize) 80
Great Comet 230
great comets **241**
Great Dark Spot (Neptune) 204,
  205, 206, 212
Great Red Spot (Jupiter) 149, 150,
  154, 155, 167, 168, 212
Great Rift Valley 76, 77
Great Sunspot (1947) 32
Great White Spot (Saturn) 178,
  193
Greeks, ancient 20–21, 36, 54, 69,
  104, 105, 166, 192, 208
Greenaway Crater (Venus) 58
greenhouse effect
  Earth 62
  Venus 43, 58, 62
Grigg-Skjellerup comet 21, 226
groundwater 117
Gruithuisen, Franz von 68, 69

gullies, Mars 117
gypsum 124

H
Hadley Rille (Moon) 96–7
hailstones
  diamond 198, 206
  microscopic 14
Hale-Bopp, Comet 222–3, 228–9
Hall, Asaph 128, 131
Halley, Edmond 87, 230, 231
Halley's Comet 12, 21, 222, 224,
  226, 227, 229, 230
haloes, comets 222
Harold II, King of England 231
Hartley 2, Comet 227
Hastings, Battle of 231
Haumea 220
Hawaii 67, 120, 123
Hay, Will 193
Haystack Valles (Mercury) 48
Hayabusa spacecraft 144–5
Helene 183
heliocentrism 20, 21, 36, 37, 69
Helios 36
Helios A and B spacecraft 38
heliosphere 37
helium 12, 14, 16, 25, 47, 149, 152,
  172, 198, 206, 236
  discovery of 36
  nuclei 26
Hellas Planitia (Mars) 110
hematite 116
Hephaestus Fossae (Mars) 116
Hermes 54
Hermite Crater (Moon) 88
Herschel, Caroline 231
Herschel, William 20, 131, 196,
  200, 212, 213, 231
Herschel Crater (Mimas) 185
Herschel-Rigollet/Comet 35P 231
Hesperia Planus (Mars) 121
Hesperus 69
Hess, Harry 86
Hevelius, Johannes 105
hexagonal cloud structure
  (Saturn) 179
Himalayas 76, 79, 84
Himalia group (Jupiter) 156
Hinode spacecraft 39
Hipparchus 105
hollows, Mercury 48
Home Plate (Mars) 136–7
Hooke, Robert 167
Hooker telescope
  (California) 131
Horrocks, Jeremiah 69
hot jupiters 233
hotspots, volcanic 63, 66, 76,
  120, 122
Hough, George W. 167
Hoyle, Fred 230
Hubble Space Telescope 154, 179,
  199, 200, 201, 205, 208, 209,
  212, 220, 221
Huggins, William 230
Hurricane Isabel 81
hurricanes 179
Hutton, James 87
Huygens, Christiaan 131, 193
Huygens Titan probe 192, 194, 195
Hyakutake comet 224, 228–9
hybrid solar eclipses 35
Hydra 221
Hydraotes Chaos (Mars) 111
hydrocarbons 186

hydrogen 12, 14, 16, 25, 26, 47,
  149, 152, 172, 198, 206, 236
  compounds 16
  liquid metallic 152, 153, 172, 173
  liquid molecular 172
  nuclei 26
hydrothermal vents 83
hypergiants 24
Hyperion 182, 188

I
Iapetus 182, **184**, 193, 195
ice
  Callisto 163
  Europa 159
  Ganymede 162
  geysers (Enceladus) 21, 185, 190,
    191, 192
  Mars 110, 114, 117, **126–7**, 131
  Pluto 220, 221
  in Saturn's rings 170, 174
  Triton 210, 211
Ice Age 86
ice giants see gas giants
ice sheets 80, 81
icy pits, Mars 127
Ida 141
Imbrium Basin (Moon) 89
impact craters/basins see craters
India
  ancient astronomers 55
  missions to Mars 132–3
  missions to Moon 105, 107
  missions to Sun 38–9
infrared 30
infrared radiation 62
intercrater plains 49
International Astronomical Union
  213, 220
interplanetary magnetic field 38
interstellar clouds 14, 16
Io 21, 148, 157, **158**, 160–61,
  166, 168
ionosphere, Venus 62
iron 46, 60, 61, 74, 75, 87, 90, 91,
  110, 112, 139, 162, 198, 206
irregular satellites 156
Ishango bone 104
Ishtar 69
Ishtar Terra (Venus) 58, 64, 66
ISRO (Indian Space Research
  Organisation) see India
Ithaca Chasma (Tethys) 185
Itokawa 140, 144–5
Ixion 220

J
Japan
  missions to asteroids 144–5
  missions to comets 226
  missions to Mars 132–3
  missions to Mercury 56
  missions to Moon 105, 107
  missions to Sun 38–9
JAXA (Japan Aerospace
  Exploration Agency) see Japan
jet stream
  Earth 86
  Jupiter 154
  Saturn 170
Jesus Christ 230
Jewitt, David 219
Jove 166
Juliet 201
Juno spacecraft 167, 169

Jupiter 11, 12, 149, **150–69**
  atmosphere 152, 154, 166, 167,
    168, 239
  axial tilt 150
  belts and zones 150, 151, 154,
    155, 167
  data **150**, 236
  diameter 150, 167
  energy 152
  formation of 17
  gravity 17, 138, 140, 141, 150, 1
    56, 164, 165, 169, 224
  Jupiter system **156–7**, 166
  magnetic field 152, 153, 158,
    162, 166, 167, 169, 238
  matter 149
  missions to 21, 38, 164–7, **168–9**
  moons 148, 150, **156–65**, 166,
    167, 168, 240
  orbit 150, 237
  rings 151, 157, 166, 168, 238
  rotation 150, 154, 167
  Shoemaker-Levy 9 collision with
    165, 166, 225, 230
  size of 19, 24, 150, 166
  structure **152–3**
  temperatures 150, 153, 155, 168,
    238, 239
  timeline of discoveries **166–7**
  up close **154–5**
  weather 150, 154, 155, 166, 167,
    168, 169
  year 149, 150

K
Kaguya spacecraft 98
Kant, Immanuel 231
Karakoram Range 79
Kasei Valles (Mars) 111, 116
KBOs (Kuiper belt objects)
  see Kuiper belt
Keeler gap (Saturn's rings) 174
Kepler, Johannes 21, 130
Kerberos 221
Kircher, Athanasius 87
Kleopatra 138
Kotelnikov, Vladimir 54
Kuiper, Gerard 130, 200, 218
Kuiper belt 12, 16, 17, 168, 200,
  208, 210, 212, 213, 215,
  218–19, 220
  composition of KBOs 218
  discovery of 218, 219
  distance from Sun 218, 219
  formation of 218
  size and scale of KBOs 18–19

L
Lada Terra (Venus) 58
Lagrangian point 38
lakes 116
  Titan 186–7, 193
Lakshmi Planum (Venus) 66
landing sites, on the Moon 107
landslips/landslides 118, 163
Larissa 208, 209
Lassell, William 200, 208
Lassell ring (Neptune) 208, 209
Late Heavy Bombardment 103
launch sites, rocket 242
lava flows, Mars 120, 121
lava plains
  Mars 110, 114, 120, 121
  Mercury 48, 49
lava plains (continued)

Moon 88, 90, 100
Venus 59, 62
lava tubes
Mars 121
Moon 96, 102
Le Verrier, Urbain 204, 213
Le Verrier ring (Neptune) 207,
208, 209
lead isotopes 86
lead sulfide (galena) 67
Lexell's Comet 231
life
on Earth 16, 43, 72, 74,
82–3, 230
on Europa 159
on Mars 131, 134, 135
origins of 82
on Venus 68
Ligeia Mare (Titan) 186–7
light 24
pollution 85
speed of 167
Sun bends 36
Sun rays 30–31
lightning 71, 168, 170,
172, 178
lineae, Europa 59
lithosphere 74
Little Ice Age 33
lobate scarps 45, 52
Lockhead C-141A Starlifter
transport plane 200
Lockyer, J. Norman 36
Lomonosov, Mikhail 68
Long March 2F rocket 242
long-period comets 224
Lowell, Percival 131, 213
Lowell Observatory (Arizona)
131, 213
Luna 2 spacecraft 106
Luna 3 spacecraft 20, 104
Luna 9 spacecraft 104
lunar landing module 108–9
lunar month 93
Lunar Reconnaissance Orbiter
101, 105
lunar rocks 101, 104
lunar rovers 96–7, 107
Lunokhod rovers 107
Luu, Jane 219

M

Maat Mons (Venus) 59, 63
Mab 201, 212
McNaught comet 228–9
Magellan spacecraft 62, 69,
71
magma 63, 66, 91, 96, 100, 102,
103, 198
magma chambers 120, 123
magnesium 61, 236
magnetic fields
Earth 75, 82, 238
Ganymede 162
interplanetary 38
Jupiter 152, 153, 158, 162, 166,
167, 169, 238
Mars 238
Mercury 238
Neptune 206, 238
Saturn 179, 238
Sun 14, 26, 28, 32, 33, 39
Uranus 198, 199, 238
Venus 238
magnetic shields 82
magnetosphere

Jupiter 154, 167
Mercury 55
MAHLI (Mars Hand Lens
Imager) 134
main belt *see* asteroid belt
Makemake 220
mantle
asteroids 39
Earth 74, 75, 78, 120
Mars 112, 113, 122
Mercury 16, 47, 48
Moon 91, 103
Neptune 206
Pluto 220
Uranus 198, 199
Venus 60, 61, 63
Marduk 166
Mare Crisium (Moon)
89, 101
Mare Fecunditatis (Moon)
89, 101
Mare Humboldtianum
(Moon) 101
Mare Imbrium (Moon) 88,
96, 102
Mare Moscoviense (Moon) 101
Mare Nectaris (Moon) 89, 101
Mare Serenitatis (Moon) 89, 103
Mare Smythii (Moon) 101
Mare Tranquillitatis (Moon) 89,
101, 108–9
Margaret 200
maria (seas), Moon 88, 89, 90,
92, 94, 100, 101, **102–3**, 105
Mariner 2 spacecraft 20, 69
Mariner 4 spacecraft 20, 130, 133
Mariner 9 spacecraft 118, 122,
130, 133
Mariner 10 spacecraft 54, 55, 56
Mars 11, 12, 42–3, **110–37**
age of 130
atmosphere 20, 110, 113, 126,
130, 239
axial tilt 110, 131
data **110**, 236
diameter 110
dunes **124–5**, 136
exploring **136–7**
gravity 110, 128
landings 134–5
life on 131, 134, 135
magnetic field 238
mapping **114–15**, 131
missions to 20, 42–3, 110, 112,
116, 117, 118–19, 130,
131, **132–7**
moons **128–9**, 131, 240
Olympus Mons **122–3**
orbit 130, 237
polar caps 110, 117, **126–7**, 131
rotation 110, 131
rovers **134–5**
seasonal change 124, 126, 131
similarities to Earth 110
size 19
structure **112–13**
surface of 130, 131
temperatures 112, 116, 126, 127,
131, 238, 239
timeline of discoveries **130–31**
Valles Marineris **118–19**
volcanoes 110, 112, 113, 114,
116, **120–23**
water on 43, 110, 113, **116–17**,
130, 134
weather 20
Mars (god of war) 130

Mars 3 spacecraft 132
Mars as the Abode for Life
(Lowell) 131
Mars Express 118–19, 128, 133
Mars Global Surveyor 117, 133
Mars Reconnaissance Orbiter
(NASA) 42–3, 126, 134
Marsquakes 118
mass extinctions 83, 87
Mathilde 138
Mauna Kea (Hawaii) 67,
123, 219
Mauna Loa (Hawaii) 121
Maunder, Edward 36
Maxwell, James Clerk
192, 193
Maxwell Montes (Venus) 66–7
Mayan civilization, astronomers 69
Mead Crater 63
meltwater 80, 81, 116
Mercury 11, 12, 43, **44–57**
atmosphere 44, 47, 48,
55, 239
axial tilt 44
Carnegie Rupes **52–3**
data **44**, 236
diameter 44, 55, 57
formation of 16
gravity 44, 47
magnetic field 238
mapping **50–51**, 55, 57
missions to 44, 46, 48, 54,
55, **56–7**
naming of 54
naming system (features) 48
orbit 44, 54, 55, 237
phases of 55
rotation 44, 46, 52, 54
structure **46–7**
surface **48–9**, 55, 56, 57
temperatures 44, 238, 239
timeline of discoveries **54–5**
transit of 55
Mercury Laser Altimeter
(MLA) 57
Meridiani Planum (Mars) 134
Mesopotamia, ancient 20
mesosphere 75
Messenger spacecraft 44, 46, 48,
49, 50, 54, 55, 56, 57
metallic asteroids (M-type) 138
meteor showers 223, 229, **241**
meteorite streams 231
meteorites 15, 48, 49, 62, 90, 104,
139, 143
lead isotopes 86
Martian 131
meteoroids 88, 100, 113
methane 12, 14, 16, 153, 186, 197,
198, 204, 205, 206, 220
ice 210, 211
liquid 192
Metis 157
microbes 83
microplates 76
microwave telescopes 243
Mid-Atlantic Ridge 76
Mid-Indian Ridge 77
mid-ocean ridges 76, 77, 86
Milky Way 10–11, 14, 24, 232–3
Mimas 184, **185**, 188–9
minerals
Earth 83
Mars 116
Moon 105
Venus 66, 67
Minor Planet Center

(Cambridge, MA) 143
minor planets *see* asteroids
Miranda 200, 201, 212
Miranda Rupes **202–3**
molecular cloud 14
montes, Mars 120–21
Montes Apenninus (Moon) 89,
102, 109
Montes Taurus (Moon) 103
Moon 82, **88–109**, 128, 240
age of 90, 101
Apollo Project **108–9**
atmosphere 105
craters 88, 89, **100–101**
data **88**
distance from Earth 92, 105
Earth–Moon System **92–3**
Earthrise **98–9**
eclipses 20, 92, 104
formation of 16, 20, 92, 104
gravity 88, 90, 91, 92, 105
Hadley Rille **96–7**
highlands and plains **102–3**
influence on Earth 88, 91, **92–3**
mapping **94–5**, 105
mass 104
meteorite impact 104
missions and landings 20, 89, 90,
96–7, 104, **106–9**
orbit 34, 35, 88, 92, 93,
104, 105
phases of 92–3, 104
rotation 88, 92
samples from 101, 104
solar eclipses 34–5
structure **90–91**, 92
surface 88, 90, 104, 105
temperatures 88
timeline of discoveries **104–5**
water ice on 105
moonlets, Saturn 182, 183
moonquakes 90
moons **240**
dwarf planets 220
formation of gas giants' 17
Jupiter 148, 150, **156–65**, 166,
167, 168
Mars **128–9**, 131
Neptune 204, 207, 208–11
planetary 11, 149
Pluto 221
Saturn 174, 175, 182–3, **184–91**,
192, 193, 194, 195
size and scale 18–19
Uranus 196, 199, 200–203,
212, 213
moonshine 105
Mount Palomar (California) 218
mountains
Earth 72, 73, 76, 77, 79, 84
formation of 79
Moon 89, 90, **102–3**
Venus 58, 62, 63, 64, 66–7
see also volcanoes
Mu ring (Uranus) 201
mudstones 134
Mul.Apin tablets 54
Mutch Crater (Mars) 111
Mystic Mountain 14–15

N

N1 rocket 242
Nabu 54
Naiad 209
Napoleon I, Emperor of France 68
NASA

missions to asteroids 144–5
missions to comets 226–7
missions to Jupiter 168–9
missions to Mars 132–3
missions to Mercury 56–7
missions to Neptune 114–15
missions to Saturn 194–5
missions to Sun 37, 38–9
missions to Uranus 202–3
see also spacecraft by name
natural selection 82, 83
Nautical Almanac Office 213
Near Earth Asteroid Tracking (NEAT)
  system (California) 228
NEAR Shoemaker spacecraft 138,
  144, 145
near side of the Moon 9, 89, 90
near-Earth asteroids (NEAs) 140,
  142–3, 145
neon 236
NEOWISE survey 142
Neptune 11, 12, 140,
  198, **204–15**
  atmosphere 204, 205,
    206, 239
  axial tilt 204, 206
  blue color 204, 205, 206
  data **204**, 236
  discovery of 20, 212, 213
  energy 206
  gravity 204, 208
  magnetic field 206, 238
  migration of 17
  missions to 21, 204, 205, 208,
    212, 214–15
  moons 204, 207, 208–11, 240
  Neptune system **208–9**
  orbit 220, 237
  and plutinos 218
  rings 204, 207, 208, 212, 238
  rotation 204, 206
  seasons 204
  size 19
  structure **206–7**
  temperatures 204, 238, 239
  timeline of discoveries **212–13**
  weather 149, 204, 205,
    206, 212
  year 149, 204
Nereid 208
New Horizons spacecraft 168,
  220, 221
Newton, Isaac 21, 36, 104,
  105, 231
Nicholson, Seth 131
nickel 60, 74, 75
nitrogen 74, 113, 184, 210, 220, 236
  frost 210, 211
Nix 221
Noachis Terra (Mars) 111, **124**
North America 72, 73
North American Plate 76
North Polar erg (Mars) 124–5
Northern Lights 29
Nu ring (Uranus) 201
nuclear fusion 36
nuclear reactions 14, 15, 26
nuclei, comets 21, 200, 222, 225,
  226, 227, 229, 230

O

Oberon 200, 213
observatories 243
occultations, of Mercury by Venus 55
oceanic crust 74, 75
oceans

Earth 72–3, 74, 75, 86,
  92, 104
Uranus 198
Venus 71
see also maria; seas
Oceanus Procellarum (Moon) 102
Okavango Delta (Botswana) 84
Oldham, Richard 86
Olympus Mons (Mars) 67, 85, 110,
  120, 121, **122–3**
*On the Revolution of the Heavenly
  Sphere* (Copernicus) 37
Ontarius Lacus (Titan) 193
Oort, Jan 225
Oort Cloud 12, 224, 225, 230
Ophelia 199, 201
Öpik, Ernst 230
Opportunity rover 116, 131, 133,
  134, 136, 137
optical telescopes 243
orbits 12
  asteroids 140–41, 142
  comets 12, 222, **224–5**
  dwarf planets 220
  Earth 72, 237
  eccentric 218, 219
  elliptical 12, 21, 141, 224
  Jupiter 150, 237
  Jupiter's moons 156–7
  Kuiper belt objects 218
  lunar 34, 35
  Mars 130, 237
  Mercury 44, 54, 55, 237
  Moon 34, 35, 88, 92, 93, 104
  Neptune 220, 237
  Neptune's moons 208, 209
  Phobos and Deimos 128
  Pluto 218, 220
  Pluto's moons 221
  satellites 243
  Saturn 170, 237
  Saturn's moons 182, 183
  Saturn's rings 176
  and solar gravity 24
  Uranus 213, 220, 237
  Uranus's moons 201, 213
  Venus 58, 237
Orcus 220
organic molecules 82
organisms, multicellular 82
Orion spacecraft 145
Orionid meteor shower 229
Oschin, Samuel 218
outer planets 12, 146–215
outflow channels 116
oxygen 47, 74, 75, 83, 110, 113,
  198, 206, 236
ozone 83

P

Pacific Ocean 72
Pacific Plate 77
Palus Putridinus (Moon) 97
pancake domes 63
Pandora 183
Pantheon Fossae (Mercury) 49
Paris Observatory 193
Pasiphaë group (Jupiter) 156
Paterae 120
Patterson, Clair 86
Pavonis Mons (Mars) 121
Payson Outcrop (Mars) 137
Pele volcano (Io) 160
penumbra 32, 34, 35
Perdita 201
perigee 92

periodic comets **241**
periodite 75
permafrost 110, 127
Pettengill, Gordon 54
Pettit, Edison 131
Philolaus 21
Phobos **128–9**, 131
Phoebe 182, 192
Phoebe ring (Saturn)
  175, 184
Phoenix Mars Lander 117,
  132, 134
phosphorus 69
photography, of Sun 36
photosphere 24, 26, 27, 30,
  35, 37, 39
photosynthesis 83
Piazzi, Giuseppe 20
Pillan Patera (Io) 158, 161
Pioneer 5 38
Pioneer 10 166, 168, 194
Pioneer 11 168, 193, 194
plains (planitiae)
  Mars 110, 114
  Mercury 44, 45, 48, 49
  Moon 102–3
  Venus 58, 59, 64
Planet X 213, 220
planetary motion, laws of
  21, 130
planetary nebula 24
planetesimals 16, 17, 46
planets 11
  distance from Sun 12
  extrasolar 233
  formation of **16–17**
  orbits 11, 12
  size and scale 12, 18–19
  see also planets by name
plants 72, 82
Planum Australe (Mars) 110, **127**
Planum Boreum (Mars) 110, **126**
plasma 26, 28, 29
  eruptions of 24
plate boundaries 76, 77
plate tectonics *see* tectonic plates
Plato Crater (Moon) 101
Plum Crater (Moon) 101
plutinos 218
Pluto 11, **220–21**
  atmosphere 221
  demotion of 213, 220
  discovery of 212, 220
  missions to 220
  moons 221
  orbit 218, 220
  seasonal variation 221
  size 19, 221
  structure 220
polar regions
  Earth 72, 75, 80
  Jupiter 150, 154
  Mars 110, 114, 117, 124–5,
    126–7, 131
  Mercury 44
  Moon 105
  Neptune 211
  Saturn 179
  Sun 33
  Uranus 196
  Venus 58, 68
polar vortex (Saturn) 179
Polydeuces 183
Pope, Alexander 200
Portia 201
potassium 47
potentially hazardous asteroids

(PHAs) 142, 143
Principia (Newton) 21
Proctor, Richard 68, 104
Prometheus (Io) 158, 161
prominences 28
  loop 25, 28
Proteus 208, 209
protoplanets 16, 17, 46
protostars 14–15
Ptolemaic system 21
Ptolemy, Claudius 21, 37, 192
Puck 201
Pwyll crater (Europa) 159
Pythagoras 20

Q

QB1 218, 219
Quaoar 220

R

Rachmaninoff Crater
  (Mercury) 49
radar observations 54, 69
radio telescopes 243
radio waves 30
radioactive zone, Sun 26
radiometric dating 86
rain
  Titan 187
  Venus 62
red giants 24
Red Spot Junior (Jupiter)
  154, 167
reefs, coral 80
regio, Venus 58
regular satellites 156
relativity, general theory of 36
religion 20, 21
Rhea 183, **184**
Riccioli, Giovanni Battista 69, 105
ridges, Mercury 48
rift valleys
  Earth 76, 77
  Mars 110, 112, 118–19
ring of fire 35
rings
  Jupiter 151, 157, 166,
    168, 238
  Neptune 204, 207, 208,
    212, 238
  Saturn 149, 170, 171, 173, **174–7**,
    180–81, 182, 183, 192, 193,
    194, 195, 199, 238
  Uranus 196, 199, 200–201,
    212, 238
rivers
  on Earth 80, 81, 84
  on Mars (dry) 110, 116
  on Titan 187
Roche lobe (Saturn) 177
rock cycle 78
rockets, world's largest 242
Rocky Mountains 72
rocky planets 11, 12,
  40–137, 138
  atmospheres 239
  formation of 16, 17
  size and scale 19
  see also Earth; Mars;
    Mercury; Venus
Romans, ancient 36, 54, 68,
  130, 166
Rømer, Ole 167
Rosalind 201
Rosetta spacecraft 227,

230, 231
round Earth theory 86
rovers, robotic, Mars 132,
    133, **134–5**
rupes, Mercury 45, 48,
    52–3, 57
Russia *see* Soviet Union/Russia

S

S/2004 N1 209
Sahara Desert 85
salt pans 84
sand dunes, Mars
    124–5, 136
Santa Maria Crater
    (Mars) 136–7
Sarychev Peak
    (Kuril Islands) 85
satellites 243
    camera-carrying 86
    first 20
Saturn 11, 12, 149, **170–95**
    atmosphere 172, 173, 178,
        193, 239
    axial tilt 170
    bands 170, 171
    data **170**, 238
    density 170, 172
    diameter 170
    energy 172, 178
    formation of 17
    gravity 172, 174, 177
    magnetic field 179, 238
    missions to 21, 178, 179,
        181, 188–9, **194–5**
    moons 174, 175, 182–3, **184–91**,
        192, 193, 194, 195, 240
    name system 184
    orbit 170, 237
    rings 149, 170, 171, 173,
        174–7, 180–81, 182, 183,
        192, 193, 194, 195, 199, 238
    rotation 170
    Saturn system **182–3**
    seasons 178, 179, 193
    size 18
    structure **172–3**
    temperatures 170, 172, 173,
        178, 238, 239
    timeline of discoveries **192–3**
    up close **178–9**
    weather 170, 172, 173, 178,
        179, 181, 193
Saturn V rocket 242
scarps 163
scattered disk objects (SDO) 218, 219
    size and scale 18–19
Scheiner, Christoph 37
Schiaparelli, Giovanni 54, 68,
    131, 231
Schiaparelli Dorsum (Mercury) 48
Schirra, Wally 108
Schmitt, Jack 103, 109
Schrödinger Crater (Moon) 101
Schröter, Johannes 54, 102
Schwabe, Samuel Heinrich 32,
    37, 167
science fiction 130
Scott, David 96, 108, 109
SDO *see* Solar Dynamics Observatory
seafloor 86, 87
seas
    lunar *see* Mare; maria
    Mars 116
    see also oceans
Secchi, Angelo 131

sedimentary rock 79, 134
sediments 78
Sedna 220
seismology 86
seismometers 90
service module 108
sexual reproduction 82
Shakespeare, William 200
shepherd moons 174
Shi Shen 37
shield volcanoes 63, 120
Shoemaker-Levy 9 (Comet)
    165, 166, 225, 230
shooting stars 223
short-period comets 224
Silfra 77
silicates 47, 61, 112, 158
Silk Atlas of Comets 230
silicon 236
sinkholes 80
Skinakas Astrophysical Observatory
    (Crete) 55
Skinakas Basin (Mercury) 55
slickensides 203
snow
    carbon dioxide 126, 127
    comets 222, 230
    mineral 66, 67
Snowman craters (Vesta) 139
sodium 47
SOHO 6 (comet) 223
SOHO spacecraft 37, **38**, 222
Sojourner rover 131, 133, 134
solar cycle **32–3**, 36, 37
Solar Dynamics Observatory
    24, 28, 29, 30, 37, 39
solar flares 25, 27, 28, 30,
    33, 36, 38, 39
Solar and Heliospheric Observatory
    see SOHO
solar maximum 33
solar minimum 33
solar nebula 14, 16, 17
solar radiation 82, 110, 150
    erosion by 163
solar storms 28–9
solar system **8–21**
    age of 14, 15
    around the Sun **12–13**
    birth of **14–15**, 225
    comets in inner 224
    elements in 236
    formation of the planets **16–17**
    size and scale of bodies in **18–19**
    timeline of discoveries **20–21**
    Voyager 1's portrait of 214–15
    worlds beyond 233
solar wind 37, 39, 47, 48, 62, 82,
    179, 220, 230
solstices 36
soot 14
South America 72, 73
South American Plate 76
South Pole-Aitken Basin
    (Moon) 88, 90
Southern Ocean 72, 77
Soviet Union/Russia
    missions to comets 226
    missions to Mars 132–3
    missions to Moon 20, 104, 106–7
    missions to Venus 70, 71
    satellites 20, 243
    space exploration 242–3
space exploration 242–3
space probes *see* by name
space rubble 15
space stations 243

spacecraft
    Apollo project **108–9**
    landmark missions 242
    missions to asteroids **144–5**
    missions to comets **226–7**, 231
    missions to Jupiter 21, 38,
        164–7, **168–9**
    missions to Mars 42–3, 110, 112,
        116, 117, 130, 131, **132–7**
    missions to Mercury 44, 46, 48,
        54, 55, **56–7**
    missions to Moon 20, 89, 90,
        96–7, **106–9**
    missions to Neptune 21, 204, 205,
        208, 212, 214–15
    missions to the Sun 37, **38–9**
    missions to Uranus 21, 196,
        199, 200, 212
    missions to Venus 20, 62,
        69, **70–71**
    see also by name
spectroscopy 68, 230
spherical Earth 20
spiral galaxies 11
spiral waves 177
Spirit rover 134, 136
Spitzer Space Telescope 175
Sputnik 1 20
Stardust spacecraft 226–7, 231
stars
    birth of 14
    dying 14, 24
    size and scale 18–19
STEREO (Solar Terrestrial Relations
    Observatory) 39
Stickney Crater (Phobos) 128
Stone Age 104
Stonehenge 36
storms
    Earth 81
    geomagnetic 29
    Jupiter 149, 150, 151, 154, 155,
        167, 168, 169
    Neptune 205
    Saturn 170, 172, 173, 178, 179,
        181, 193
    Sun 28–9, 36
stratosphere 75
Styx 221
Subkou Planitia (Mercury) 45
sulfur 60, 158, 236
sulfur dioxide 62
sulfuric acid 43, 58, 60, 61, 62
Sun **22–39**
    atmosphere 24, 37
    birth of **14–15**, 16
    composition of 25, 26
    data **24**
    death of 24
    differential rotation 33
    eclipses **34–5**, 36, 37
    energy from 24, 26, 27, 36
    future death of 24
    gravity 11, 12, 24, 52, 56
    heliocentric model 20, 21, 36,
        37, 69
    life expectancy 24
    magnetic field 14, 26, 28, 32, 33, 39
    matter 149
    missions to 37, **38–9**
    position in Milky Way 11
    rays **30–31**
    size of 24
    solar cycle **32–3**, 36, 37
    solar eclipses **34–5**, 36, 37
    storms on **28–9**, 36
    structure **26–7**

Sun (continued)
    temperatures 24, 26, 32
    timeline of discoveries **36–7**
    see also solar system
Sunda Trench 77
sungrazers 223
sunspots 24, 25, 27, 30, **32**,
    33, 36, 37, 39
supernovas 14
Surya Siddhanta 55
Sycorax 200

T

tacholine 26
tails, comets 37, 222, 224, 227,
    230, 231
Taurus-Littrow Valley
    (Moon) 109
tectonic plates
    Earth 62, 72, 74, **76–7**, 78, 79,
        86, 87, 118
    Ganymede 162
    Mars 110, 111, 112
    Venus 59, 62, 63
telescopes 243
    invention of 37
Telesto 183
Tempel 1, Comet 224, 227, 231
terrae, Venus 58
terrestrial planets
    see rocky planets
Tethys 183, 184, **185**
Thalassa 209
Thales 20
Tharsis Bulge (Mars) 119,
    120, 121
Tharsis Montes volcanoes
    (Mars) 119, 121
Tharsis region (Mars) 110
Tharsis Tholus (Mars) 120–21
Thebe 157
Theia 90
thermosphere 75
Tholi 120
thrust faults 79
tick volcanoes 63
tidal despinning 52
tidal forces 54, 88, 90, 91, 92,
    100, 157, 158, 159, 166, 191
tidal heating 163
tides (Earth) 104
tiger stripes 191
Titan **184**, 186–9
    Ligeia Mare **186–7**
    missions to 21, 186–7, 192,
        194, 195
Titan-IVB/Centaur rocket
    195
Titania 200, 213
Tolstoy Basin (Mercury) 48
Tombaugh, Clyde 212, 220
total eclipses (totality) 34, 35
Toutatis 140
Trans Neptunian Objects
    (TNOs) 213
    size and scale 18–19
transform boundaries 77
transit
    of Mercury 55
    of Venus 69
tree rings 32
trenches, deep-sea 76, 77
Trinculo 200
Triton 208, **210–11**
Trojan asteroids 13, 140–41
troposphere 75, 199

troughs (fossae)
  Mars 127
  Mercury 49
Tsiolkovsky Crater (Moon) 101
Tycho Crater (Moon) 89

U

ultraviolet light 30, 171
ultraviolet radiation 30, 220
ultraviolet telescopes 243
Ulysses spacecraft 38, 228
umbra 32, 34, 35
Umbriel 200
United States
  Apollo project **108–9**
  missions to asteroids 144–5
  missions to comets 226–7
  missions to Jupiter 168–9
  missions to Mars 132–3
  missions to Mercury 56–7
  missions to Moon 104, 105, 106–9
  missions to Neptune 114–15
  missions to Saturn 194–5
  missions to Sun 38–9
  missions to Uranus 202–3
  missions to Venus 70–71
  satellites 243
  space exploration 242–3
Upheaval Dome (Utah) 84
Uranus 11, 12, 149, **196–203**
  atmosphere 198, 199, 239
  axial tilt 196, 213
  data **196**, 236
  diameter 196
  discovery of 212, 213
  gravity 196
  magnetic field 198, 199, 238
  migration of 17
  missions to 21, 196, 199, 200,
    202–3, 212
  moons 196, 199, 200–203, 212,
    213, 240
  naming of 213
  orbit 196, 213, 220, 237
  rings 196, 199, 212, 238
  rotation 196
  seasons 196, 197
  size 19
  structure **198–9**
  temperatures 196, 198, 199,
    238, 239
  timeline of discoveries **212–13**
  Uranus system **200–201**
US Naval Observatory 131

V

Valhalla impact basin (Callisto) 163
Valles Marineris (Mars) 111,
  118–19, 120
valleys, glaciated 78
Vallis Schröteri (Moon) 102
Venera 3 spacecraft 69, 70
Venera 7 spacecraft 69, 70, 71
Venera 8 spacecraft 69
Venera 9 spacecraft 70
Venus 11, 12, 43, **58–71**
  air pressure 71
  atmosphere 58, 60, 61, 62, 68,
    70, 71, 239
  axial tilt 58
  data **58**, 237
  diameter 58
  gravity 58
  life on 68
  magnetic field 238

Venus (continued)
  mapping **64–5**, 68,
    69, 71
  Maxwell Montes **66–7**
  missions to 20, 62,
    69, **70–71**
  name of 68
  naming system (features) 63
  occultation of Mercury 55
  orbit 58, 237
  phases of 58, 69
  rotation 16, 58, 60, 68
  size 19
  structure **60–61**
  surface 58, **62–3**, 68,
    69, 71
  temperatures 43, 58, 69,
    238, 239
  timeline of discoveries **68–9**
  transit of 69
  weather/climate 62, 71
Venus Express spacecraft 71
Venus Tablet 68
Verona Rupes (Miranda) 202–3
Vesta 138–9, 145
Victoria Crater (Mars) 134
Vid Flumina (Titan) 187
Viking 1 20, 129, 130, 134
Viking 2 20, 130, 132, 134
volcanic domes 59, 63
volcanoes/volcanic activity
  Earth 72, 76, 77, 78,
    85, 120
  Enceladus 191
  Europa 159
  formation and types 120–21
  Io 21, 158, 160–61, 166, 168
  and land formations 78
  Mars 110, 112, 113, 114, 116,
    119, **120–23**, 130
  Mercury 44, 45
  Moon 92, 100, 102, 103, 104
  Pluto 221
  Saturn's moons 184, 185
  Triton 210
  Venus 43, 58, 59, 60, 61, 62, **63**, 71
Voyager spacecraft 179, 194
  Voyager 1 21, 37, 157, 160, 166,
    168, 192, 193, 194, 214–15
  Voyager 2 21, 160, 166, 168, 192, 193,
    194, 196, 199, 200, 201, 204, 205, 208,
    209, 210, 212, 214–15
Vredefort (South Africa) 143
Vulcan 37
VY Canis Major 24

W

Wang Meng Crater (Mercury) 45
War of the Worlds (H.G. Wells) 130
water
  corrosive effect 78
  Earth 16, 72, 75, 82, 84
  Enceladus 189
  erosion 78
  Europa 159, 160, 166
  exoplanets 233
  Galilean moons 168
  gas giants 12
  groundwater 117
  Jupiter 153, 167
  Mars 43, 110, 113, **116–17**, 130
  meltwater 80, 81, 116
  Mercury 55
  Moon 105
  Neptune 206
  Pluto 220

water (continued)
  Saturn 172
  superionic 198, 199, 206
  three states of 72
  Uranus 197, 198
  vapor 14, 47, 113, 116,
    117, 160
  water ice 16, 105, 117, 126, 127,
    160, 173, 174
waterfalls 116
waves 73, 80, 81
weather
  Earth 72, 73, 81, 82
  gas giants 149
  Jupiter 149, 150, 154, 155, 166,
    167, 168
  Mars 20
  Neptune 149, 204, 205, 206, 212
  Saturn 170, 178, 179, 181, 193
  Titan 184, 186
  Uranus 196, 197
  Venus 7, 62
weathering 78, 79
Wegener, Alfred 86
Welles, Orson 130
Wells, H.G. 130
wetlands 84
Whipple, Fred 230
Wickramasinghe, Chandra 230
Wide-Field Infrared Survey Explorer
  telescope 142
Wild 2, Comet 227
wind erosion 79
winds
  Earth 73, 81
  Jupiter 150, 154, 168
  Mars 124, 126
  Neptune 149, 204
  Saturn 170, 172, 178, 179
  Triton 211
  Venus 62
Wollaston, William 37

X

X-ray telescopes 243
X-rays 228
Xanthe Terra (Mars) 111
Xichang (China) 242

Y

Yellowknife Bay (Mars) 134
Yerkes Observatory
  (Wisconsin) 130
Yutu ("Jade Rabbit") vehicle 107

Z

Zeus 166
Zupi, Giovanni 55

ACKNOWLEDGMENTS

Smithsonian Institution: Andrew Johnston, Geographer, Center for Earth and Planetary Studies, National Air and Space Museum.

Smithsonian Enterprises: Carol LeBlanc, Senior Vice President, Consumer Products and Education; Brigid Ferraro, Vice President, Consumer Products and Education; Ellen Nanney, Licensing Manager; Kealy Wilson, Product Development Manager.

DK would like to thank the following people for their assistance in the preparation of this book:

Shaila Brown and Sam Priddy for editorial assistance; Simon Mumford and Encompass Graphics for maps; Adam Benton for additional illustrations; Steve Crozier and Phil Fitzgerald for retouching work; Tannishtha Chakraborty, Mandy Earey, Vaibhav Rastogi, Anjali Sachar, and Riti Sodhi for design assistance; Caroline Hunt for proofreading; Helen Peters for the index.

The publisher would like to thank the following for their kind permission to reproduce their photographs:

(**Key:** a-above; b-below/bottom; c-centre; f-far; l-left; r-right; t-top)

1 NASA: 61JPL / DLR (cr). **4-5 ESA:** DLR / FU Berlin (G. Neukum) (t). **6-7 NASA:** JPL / University of Arizona. **7 Maggie Aderin-Pocock:** (tc, bc). **10 Corbis:** 68 / Paul E. Tessier / Ocean. **14-15 NASA:** ESA, and M. Livio and the Hubble 20th Anniversary Team (STScI). **14 Gemini Observatory:** artwork by Lynette Cook (br). **15 ESO:** L. Calçada (bl). **Science Photo Library:** Michael Abbey (br). **16-17 Science Photo Library:** Take 27 Ltd. **20 Alamy Images:** North Wind Picture Archives (tc). **NASA:** ESA / J. Parker (Southwest Research Institute), P. Thomas (Cornell University), L. McFadden (University of Maryland, College Park), and M. Mutchler and Z. Levay (STScI) (c); (bc); JPL (br). **21 Alamy Images:** Stock Montage, Inc. (c). **Corbis:** Alfredo Dagli Orti / The Art Archive (tc); Heritage Images (tr). **ESA:** Courtesy of MPAe, Lindau (bc). **NASA:** JPL (bl); JPL / Space Science Institute (br). **Royal Society:** Ben Morgan (cl). **24 NASA:**

SDO / Goddard Space Flight Center. **28 NASA:** SDO (cl); SDO / AIA (br); SDO / GSFC (bc). **28-29 NASA:** SDO / GSFC. **29 Corbis:** Mark Bauer / Loop Images (br). **NASA:** High Altitude Observatory / Solar Maximum Mission Archives (bl). **30-31 NASA:** SDO / GSFC. 32-33 NASA: SDO / GSFC. **32 BBSO / Big Bear Solar Observatory:** NJIT (bl). **33 Corbis:** Bettmann (br). **34 Corbis:** EPA / Brian Cassey - Australia and New Zealand Out (b). **35 Corbis:** Miloslav Druckmuller / Science Faction (tr); William James Warren / Science Faction (tl); Phillip Jones / Stocktrek Images (br). **36 Corbis:** Gianni Dagli Orti (ftr). **NASA:** SOHO / ESA (cr). **Science Photo Library:** Royal Astronomical Society (fcl, br). **UCAR Communications:** HAO / NCAR (fcr). G. De Vaucouleurs, Astronomical Photography, MacMillan, 1961 (cl). 37 **Corbis:** Dennis di Cicco (fbl). **Dorling Kindersley:** (tl). **Getty Images:** Hulton Archive / Print Collector (tr). **NASA:** SDO / GSFC (br). **NOAO / AURA / NSF:** N. A. Sharp, NOAO / NSO / Kitt Peak / FTS / AURA / NSF (cr). **Wikipedia:** CNX (cr). **38 123RF.com:** rtguest (l). **NASA:** (clb, fcrb). **39 NASA:** SDO / GSFC (fcrb); STEREO (b). **42 NASA:** JPL / University of Arizona. **48 NASA:** Johns Hopkins University Applied Physics Laboratory / Carnegie Institution of Washington (br); (l). **49 NASA:** JHUAPL / CIW-DTM / GSFC / MIT / Brown Univ / Rendering by James Dickson (bl); Johns Hopkins University Applied Physics Laboratory / Carnegie Institution of Washington (t, bc); Science / AAAS (br). **50-51 NASA. 53 NASA:** The Johns Hopkins University Applied Physics Lab, the Carnegie Institution for Science (fcr, br). **54 Getty Images:** DEA / G. Nimatallah / De Agostini Picture Library (tr). **NASA:** (cr, br). **Science Photo Library:** David Parker (bc). **55 Getty Images:** Apic / Hulton Archive (tr); Guillermo Gonzalez / Visuals Unlimited (cr). **NASA:** Johns Hopkins University Applied Physics Laboratory / Carnegie Institution of Washington (br); KSC (bc). **56 123RF.com:** rtguest (l). **NASA:** JPL (bl). **57 NASA:** Johns Hopkins University Applied Physics Laboratory / Carnegie Institution of Washington (br); (bl). **62-63 NASA:** JPL (b). **63 NASA:** GSFC (cr). **64-65 NASA. 67 NASA:** JPL (cr). **68 Alamy Images:** World History Archive /

Image Asset Management Ltd. (tr). **Dreamstime.com:** Dmitry Volkov / Dymon (tc). **Getty Images:** DEA / G. Dagli Orti / De Agostini (fcl). **NASA:** NSSDC / GSFC (br). **Science Photo Library:** New York Public Library (bc); Ria Novosti (c). **69 Corbis:** Werner Forman (tr). **Getty Images:** Print Collector / Hulton Archive (cl); Adina Tovy / Lonely Planet Images (tl). **Institute e Museo di Storia della Scienza di Firenze:** (cr). **NASA:** JPL (fbl); (br); JPL / USGS (fbr). **70 123RF.com:** rtguest (l). **Wikipedia:** (br). **77 Alamy Images:** Wolfgang Pölzer (fbr). **78 Alamy Images:** Wayne Lynch / All Canada Photos (tr). **79 Alamy Images:** McPhoto / vario images GmbH & Co. KG (t). **80-81 Alamy Images:** Tom Till. **81 Corbis:** Sanford / Agliolo (tr). **Getty Images:** James Balog / Aurora (cr). **NASA:** image created by Jesse Allen, Earth Observatory, using data obtained from the University of Maryland's Global Land Cover Facility (ca); Jacques Descloitres, MODIS Rapid Response Team / NASA / GSFC (br). **82-83 Corbis:** Peter Adams / JAI. **83 Alamy Images:** AF Archive (cra). **Corbis:** Dr. Robert Calentine / Visuals Unlimited (tr). **Flickr.com:** Leon Oosthuizen / leonoos (b). **84 Alamy Images:** Guenter Fischer / imageBROKER (t). **Getty Images:** Michael Layefsky / Moment Select (cr). **NASA:** JSC (cl); USGS EROS Data Center (clb). **85 Corbis:** Tibor Bognar (clb). **NASA:** GSFC / METI / Japan Space Systems, and U.S. / Japan ASTER Science Team (tl); (tr, cr). **86 Corbis:** (bl). **Science Photo Library:** NOAA (br). **87 Alamy Images:** Interfoto (tr). **Getty Images:** National Galleries of Scotland / Hulton Fine Art Collection (cr); Universal Images Group (tl); Stock Montage / Archive Photos (cl). **NASA:** JPL (bc). **90 NASA:** (fbl). **Science Photo Library:** NASA / GSFC / DLR / ASU (bl). 92 Corbis: Alan Dyer, Inc. / Visuals Unlimited (l). **93 NASA. 94-95 NASA. 96-97 NASA. 97 NASA:** (fcr). **98-99 NASA:** JAXA. **101 Corbis:** (bl). **NASA:** (bc, br); Goddard / MIT / Brown (t). **102 NASA:** (cl, cr). **102-103 NASA:** (b). **104 Getty Images:** Jamie Cooper / SSPL (tr); Sovfoto / UIG (cl). **NASA:** (cr, br); NSSDC (bl). **105 Alamy Images:** North Wind Picture Archives (tl). **Getty Images:** SSPL (tr). **NASA:** AMES (bl); GSFC (br). **Science Photo Library:** American Institute Of Physics (cl); Detlev Van Ravenswaay (cr). **107**

NASA: (br). **108 NASA:** (tl, tr); JSC / Russell L. Schweickart (cl). **108-109 NASA. 109 NASA:** (tl, r). **114-115 NASA. 116 ESA:** DLR / FU Berlin (G. Neukum) (b). **NASA:** JPL / Arizona State University, R. Luk (tr). **Science Photo Library:** NASA (cr). **117 NASA:** JPL / ASU (tc); JPL-Caltech / University of Arizona / Texas A&M University (cla). **Science Photo Library:** ESA / DLR / FU Berlin (G. Neukum) (b); NASA (tr). **118-119 ESA:** DLR / FU Berlin (G. Neukum). **119 NASA:** (cr); JPL / USGS (ftr). **120-121 ESA:** DLR / FU Berlin (G. Neukum) (b). **121 ESA:** DLR / FU Berlin (G. Neukum) (br, fbr). **123 ESA:** DLR / FU (G. Neukum) (fbr). **NASA:** JPL (cr). **124 NASA:** JPL-Caltech / Univ. of Arizona. **125 NASA:** NASA / JPL / University of Arizona (tr); JPL / University of Arizona (tl, b). **126 NASA:** JPL / STScI (tr); (cr). **127 NASA:** JPL / University of Arizona (c, b). **128 NASA:** JPL-Caltech / Univ. of Arizona (cl, cr). **129 NASA. 130 Getty Images:** DEA / M. Carrieri / De Agostini Picture Library (tc); Popperfoto (cr). **NASA:** (cl, br); JPL (bl). **Science Photo Library:** NYPL / Science Source (tr). **131 Alamy Images:** Mary Evans Picture Library (tr); World History Archive / Image Asset Management Ltd. (cl). **Getty Images:** Kean Collection / Archive Photos (cr). **NASA:** JPL-Caltech (b). **Science Photo Library:** Royal Astronomical Society (tc). **132 123RF.com:** rtguest (l). **NASA:** JPL / GSFC (bc). **133 123RF.com:** Eknarin Maphichai (cr). **NASA:** (bl). **134-135 NASA:** JPL-Caltech / MSSS. **135 NASA:** JPL-Caltech (cr). **136-137 Corbis:** NASA / JPL-Caltech / Michael Benson / Kinetikon Pictures (t). **NASA:** JPL-Caltech / Cornell / ASU (c). **137 NASA:** JPL-Caltech / MSSS (tr); JPL-Caltech / Cornell University (cra); JPL-Caltech / USGS / Cornell University (br). **138 NASA:** JPL / JHUAP (tr). **138-139 NASA:** JPL / MPS / DLR / IDA / Daniel Macháŵek. **139 NASA:** JPL (tc); NEAR Project / NLR / JHUAPL / Goddard SVS (tl). **140 NASA:** ESA / SWRI / Cornell University / University of Maryland / STScI (br); JPL (cl); JAXA (tr). **142 NASA:** JHUAPL / NEAR (tc); JPL-Caltech (c). **142-143 Wikipedia:** Stephen Hornback. **143 Science Photo Library:** David Nunuk (cr); Ria Novosti (tl). **144 123RF.com:** rtguest (l). **NASA:** JAXA (br); JPL / JHUAPL (bc). **145 NASA:** (br); JAXA (bl). **148 Corbis:** NASA / JPL / Ciclops / UofArizona /

Michael Benson / Kinetikon Pictures. **154-155 NASA**. **154 NASA**: JPL / University of Arizona (bl); JPL / Space Science Institute (br). **155 Gemini Observatory**: (bl). **158 NASA**: NASA / JPL / University of Arizona (bc, cb, cra); JPL / University of Arizona (br, tr, bl). **159 Corbis**: NASA / JPL / PIRL / UofArizona / Michael Benson / Kinetikon Pictures (tl). **NASA**: NASA / JPL / DLR (br). **160-161 NASA**: JPL / USGS. **161 NASA**: Galileo Project / JPL (t); JPL (c, b). **162 NASA**: JPL / DLR (tr); (bl). **163 NASA**: JPL (tr); JPL / Arizona State University, Academic Research Lab (bc); JPL / ASU (br, bl). **164 NASA**: JPL / DLR (cl); JPL / Brown University (fcl). **166 Alamy Images**: Igor Golovnov (cr). **Getty Images**: Oxford Science Archive / Print Collector (c); SSPL (tr). **NASA**: (cl). **167 Getty Images**: SSPL (cr). **NASA**: JPL / University of Arizona (bl); (bc); JPL (br). **Science Photo Library**: (tl); New York Public Library (tc). **188 NASA**: JPL. **189 NASA**: (br); JPL / Space Science Institute (t, l); JPL-Caltech /

Space Science Institute (cr). **191 NASA**: JPL / Space Science Institute (cr, fcr, br). **192 Getty Images**: SSPL (tc). **NASA**: JPL / Space Science Institute (cl, bl, br); JPL (c, cr). **Science Photo Library**: Royal Astronomical Society (tr). **193 Getty Images**: Ann Ronan Pictures / Print Collector (tl); Silver Screen Collection / Moviepix (c). **NASA**: AMES (cl); JPL / Space Science Institute (tr, br); JPL / University of Arizona / DLR (bl). **Science Photo Library**: Sheila Terry (tc). **194 123RF.com**: rtguest (l). **NASA**: JPL (crb, br). **195 NASA**: (crb, fbr); JPL (bl); JPL / Space Science Institute (br). **203 NASA**: JPL (cr); JPL-Caltech (fcr). **208 NASA**: JPL (bl). **211 NASA**: JPL / USGS (cr, fcr, fbr). **212 Corbis**: Bettmann (cr). **NASA**: ESA, and M. Showalter (SETI Institute) (br); JPL (cl, bc). **Science Photo Library**: Royal Astronomical Society (tr). **213 Corbis**: Richard Cummins / Robert Harding World Imagery (fcl). **Getty Images**: Ann Ronan Pictures / Print Collector / Hulton Archive (cl); DEA Picture

Library / De Agostini (tl); Hulton Archive (tr). **NASA**: IAU / ESA / Hubble Space Telescope, H. Weaver (JHU / APL), A. Stern (SwRI), the HST Pluto Companion Search Team and M. Brown (bl). **Science Photo Library**: Sebastien Beaucourt / Look At Sciences (cr); Royal Astronomical Society (tc). **214-215 NASA**: JPL (t, b). **218 Corbis**: Roger Ressmeyer (tr). **219 ESO**: (t). **220 NASA**: (br); HST (cr). **221 ESO**: L. Calçada (b). **NASA**: ESA / HST / M. Showalter / SETI Institute (tr). **222-223 Daniel Schechter**. **223 NASA**: SOHO (bl). **Wikipedia**: (br). **225 NASA**: H. A. Weaver, T. E. Smith (Space Telescope Science Institute) / J. T. Trauger, R. W. Evans (Jet Propulsion Laboratory) (bl). **226 123RF.com**: rtguest (l). **ESA**: (bc). **227 NASA**: (bl); JPL-Caltech / UMD (bc). **228-229 NASA**. **229 ESA**: (br). **Getty Images**: Yoshinori Watabe / Amana Images (c). **NASA**: (cr); NEAT (tr). **230 Alamy Images**: World History Archive / Image Asset Management Ltd. (tr). **ESA**: (bc). **Science Photo Library**:

Emilio Segre Visual Archives / American Institute Of Physics (bl); Royal Astronomical Society (c). **231 Alamy Images**: The Art Gallery Collection / Tielemans, Martin Francois (1784-1864) (c); Walter Rawlings / Robert Harding Picture Library Ltd (tl); North Wind Picture Archives (tr). **ESA**: J. Mai (bc). **NASA**: JPL-Caltech / UMD (bl). **Science Photo Library**: Detlev Van Ravenswaay (cl). **232-233 National Maritime Museum, Greenwich, London**: Mark Gee. **238 NASA**: JPL-Caltech / SSI / Cornell (br). **239 NASA**: (tr); JPL (crb). **241 Corbis**: Tony Hallas / Science Faction (br). **242 NASA**: (cra). **1666 NASA**: HST (bl).

Jacket images: Endpapers: **ESA**: DLR / FU Berlin (G. Neukum).

All other images © Dorling Kindersley
For further information see:
www.dkimages.com

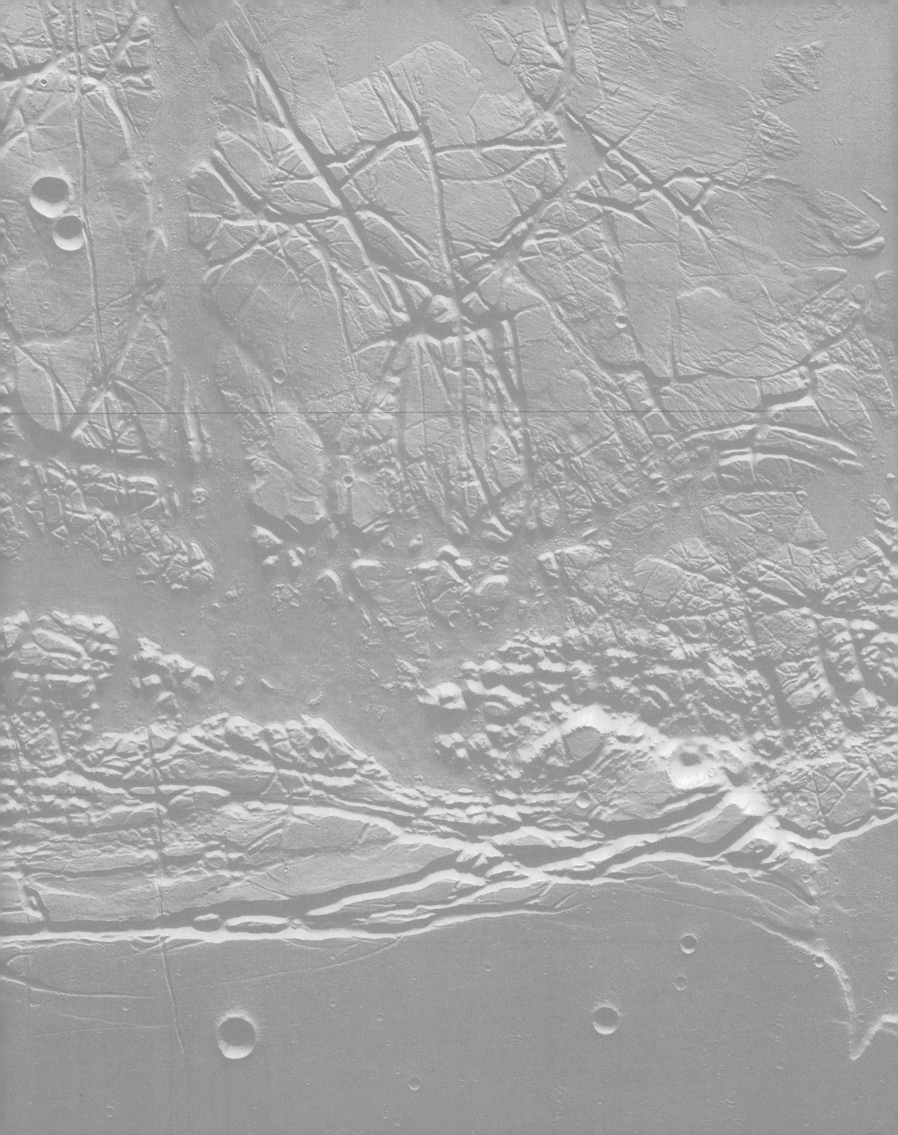